Changes in the Sheep Industry IN THE UNITED STATES

MAKING THE TRANSITION FROM TRADITION

Committee on the Economic Development and Current Status of the Sheep Industry in the United States

Board on Agriculture and Natural Resources

Division on Earth and Life Studies

NATIONAL RESEARCH COUNCIL
OF THE NATIONAL ACADEMIES

THE NATIONAL ACADEMIES PRESS
Washington, D.C.
www.nap.edu

THE NATIONAL ACADEMIES PRESS 500 Fifth Street, NW Washington, DC 20001

NOTICE: The project that is the subject of this report was approved by the Governing Board of the National Research Council, whose members are drawn from the councils of the National Academy of Sciences, the National Academy of Engineering, and the Institute of Medicine. The members of the committee responsible for the report were chosen for their special competences and with regard for appropriate balance.

This study was conducted in response to a Congressional request to the National Academies in the Agriculture Appropriations Bill (P.L. 109-97). Financial support was provided by Contract No. AG-3K06-C-06-0002 between the National Academies and the Economic Research Service of the U.S. Department of Agriculture. Any opinions, findings, conclusions, or recommendations expressed in this publication are those of the author(s) and do not necessarily reflect the views of the organizations or agencies that provided support for the project.

International Standard Book Number-13: 978-0-309-12161-3 (Book)
International Standard Book Number-10: 0-309-12161-2 (Book)
International Standard Book Number-13: 978-0-309-12162-0 (PDF)
International Standard Book Number-10: 0-309-12162-0 (PDF)

Additional copies of this report are available from the National Academies Press, 500 Fifth Street, N.W., Lockbox 285, Washington, DC 20055; (800) 624-6242 or (202) 334-3313 (in the Washington metropolitan area); Internet, http://www.nap.edu

THE NATIONAL ACADEMIES

Advisers to the Nation on Science, Engineering, and Medicine

The **National Academy of Sciences** is a private, nonprofit, self-perpetuating society of distinguished scholars engaged in scientific and engineering research, dedicated to the furtherance of science and technology and to their use for the general welfare. Upon the authority of the charter granted to it by the Congress in 1863, the Academy has a mandate that requires it to advise the federal government on scientific and technical matters. Dr. Ralph J. Cicerone is president of the National Academy of Sciences.

The **National Academy of Engineering** was established in 1964, under the charter of the National Academy of Sciences, as a parallel organization of outstanding engineers. It is autonomous in its administration and in the selection of its members, sharing with the National Academy of Sciences the responsibility for advising the federal government. The National Academy of Engineering also sponsors engineering programs aimed at meeting national needs, encourages education and research, and recognizes the superior achievements of engineers. Dr. Wm. Wulf is president of the National Academy of Engineering.

The **Institute of Medicine** was established in 1970 by the National Academy of Sciences to secure the services of eminent members of appropriate professions in the examination of policy matters pertaining to the health of the public. The Institute acts under the responsibility given to the National Academy of Sciences by its congressional charter to be an adviser to the federal government and, upon its own initiative, to identify issues of medical care, research, and education. Dr. Harvey V. Fineberg is president of the Institute of Medicine.

The **National Research Council** was organized by the National Academy of Sciences in 1916 to associate the broad community of science and technology with the Academy's purposes of furthering knowledge and advising the federal government. Functioning in accordance with general policies determined by the Academy, the Council has become the principal operating agency of both the National Academy of Sciences and the National Academy of Engineering in providing services to the government, the public, and the scientific and engineering communities. The Council is administered jointly by both Academies and the Institute of Medicine. Dr. Ralph J. Cicerone and Dr. Wm. Wulf are chair and vice chair, respectively, of the National Research Council.

www.national-academies.org

COMMITTEE ON THE ECONOMIC DEVELOPMENT AND CURRENT STATUS OF THE SHEEP INDUSTRY IN THE UNITED STATES

GARY W. WILLIAMS, *Chair*, Texas A&M University, College Station
DEEVON BAILEY, Utah State University, Logan
ORAL CAPPS, Jr., Texas A&M University, College Station
LINDA A. DETWILER, Virginia-Maryland Regional College of Veterinary Medicine, Red Bank, NJ
HUDSON A. GLIMP, University of Nevada, Reno *(Emeritus)*
TIMOTHY HAMMONDS, Food Marketing Institute, Washington, DC
DOUGLAS D. HEDLEY, Canadian Faculties of Agriculture and Veterinary Medicine, Ottawa
HELEN H. JENSEN, Iowa State University, Ames
PAUL S. KUBER, The Ohio State University, Columbus
DAVID L. THOMAS, University of Wisconsin, Madison

Staff

AUSTIN J. LEWIS, Study Director
RUTH S. ARIETI, Research Associate

Acknowledgments

This report has been reviewed in draft form by persons chosen for their diverse perspectives and technical expertise in accordance with procedures approved by the National Research Council's (NRC's) Report Review Committee. The purpose of this independent review is to provide candid and critical comments that will assist the institution in making its published report as sound as possible and to ensure that the report meets institutional standards of objectivity, evidence, and responsiveness to the study charge. The review comments and draft manuscript remain confidential to protect the integrity of the deliberative process. We wish to thank the following for their review of this report:

Walter J. Armbruster, Farm Foundation, Oak Brook, IL
Keith E. Belk, Colorado State University, Fort Collins
Neil L. First, Mississippi State University, Mississippi State (retired)
Kreg A. Leymaster, U.S. Meat Animal Research Center, Clay Center, NE
Christopher J. Lupton, Texas A&M University, San Angelo
Larry R. Miller, Cooperative State Research, Education, and Extension Service, Washington, DC (retired)
Charles F. Parker, The Ohio State University (retired)
Timothy A. Petry, North Dakota State University, Fargo
James G. Robb, Livestock Marketing Information Center, Lakewood, CO
Susan Schoenian, Western Maryland Research and Education Center, Keedysville
Dawn D. Thilmany, Colorado State University, Fort Collins
Clement E. Ward, Oklahoma State University, Stillwater
Cynthia Wolf, University of Minnesota, St. Paul

Although the reviewers listed above have provided many constructive comments and suggestions, they were not asked to endorse the conclusions or recommendations, nor did they see the final draft of the report before its release. The review of this report was overseen by Dale E. Bauman, Cornell University, coordinator, appointed by the Division of Earth and Life Studies, and Harley W. Moon, Iowa State University (*emeritus*), monitor, appointed by the NRC's Report Review Committee. The coordinator and monitor were responsible for making certain that an independent examination of this report was carried out in accordance with institutional procedures and that all review comments were carefully considered. Responsibility for the final content of this report rests entirely with the author committee and the institution.

The committee was assisted in its task by the invaluable contributions of a number of people. Fitzhugh Elder, IV, U.S. Senate Appropriations Committee; Janet Perry and Demcey Johnson, Economic Research Service, USDA; Tosha Clark, Foreign Agricultural Service, USDA; Peter Orwick, Glen Fisher, Paul Rogers, and Tom McDonnell, American Sheep Industry Association; Jim Hodges, American Meat Institute Foundation; Bo Donegan, American Lamb Board; Maurice Shelton, Texas A&M University (*emeritus*); Jim Robb and Erica Rosa, Livestock Marketing Information Center; Gary Brester, Montana State University; David Gage, Bob Harlan, and Brad Bonner, Mountain States Rosen; Tony Catelli, Catelli Brothers Veal and Lamb; Bill Brennan, Iowa Lamb Corporation; Eran Wajswol, Valley Shepherd Creamery; Jean Walsh, Mohawk Valley Suffolks; Greg Nunn, Hackettstown Livestock Auction; Susie and Omar Mady, American Halal Fresh Meats; Sam Ferrara, SMF Enterprises; Michael Anthony, Gramercy Tavern; Katherine Marshall, Ann Seitzinger, and Diane Sutton, USDA, APHIS, Veterinary Services; and William Shulaw, The Ohio State University, College of Veterinary Medicine all provided important insights into the U.S. sheep industry. In addition, representatives of wool marketing warehouses and other wool experts responded to surveys about the U.S. wool industries. Those who participated in these surveys are listed in Footnotes 1 and 4 of Chapter 5. The committee appreciates the contributions of all these individuals and organizations.

Finally, the committee expresses deep appreciation to the staff of the NRC Board on Agriculture and Natural Resources who worked closely with the committee to facilitate every step in the process of bringing this project to fruition. We owe a particular debt of gratitude to Austin J. Lewis, Program Officer, and Ruth S. Arieti, Senior Program Assistant, for their unrelenting efforts to keep the committee on schedule and moving forward. Their highly professional management of the committee and the creative process was an indispensible component of the work of the committee. They arranged for meetings, facilitated conference calls, formatted drafts, managed the committee website, assisted with committee meetings, and, in general, made the committee's work easier and more effective. The committee is also grateful for the oversight of Robin A. Schoen, Director of the NRC Board on Agriculture and Natural Resources. Her support, guidance, and gentle encouragements were an important contribution. Laurian J. Unnevehr, professor at the University of Illinois, served as the NRC Board on Agriculture and Natural Resources liaison to the committee.

Contents

APPENDIXES

TABLES AND FIGURES

Tables

Figures

Summary

The U.S. sheep industry is complex, multifaceted, and rooted in history and tradition. The fortunes of each element of the industry are inextricably linked to the others and depend particularly on the economic health and prosperity of the sheep production enterprise.

The dominant feature of sheep production in the United States, and, thus, the focus of much producer and policy concern, has been the steady decline in sheep and lamb inventories since the mid-1940s. From a record high of 56 million head in 1942, inventories on January 1, 2007, declined to 6.2 million head, the lowest level in recorded history. No one factor, event, or policy change is responsible for the contraction of the industry but rather a confluence of forces against which U.S. sheep producers have had to struggle. Although producers have little control over many of the forces for change in the industry, such as globalization and the growing competition from other meat and fiber industries in the United States and abroad, the prosperity of the industry is not entirely dependent on external forces. The various components of the sheep industry have made adjustments, invested in new technologies, and improved efficiency. These changes are transforming the industry toward a more efficient and competitive future. Signs of the transition include the recent slowing of the long-term decline in inventories in many range sheep states and the modest growth in many farm flock states. Other signs of the industry's transition are the introduction of hair sheep, the growth in direct marketing, and the emergence of the dairy sheep industry. Although often described as "an industry in decline," this report concludes that a better description of the current U.S. sheep industry is "an industry in transition."

THE COMMITTEE'S TASK

The charge to the committee (see Appendix A) was to review the development and current status of the sheep industry in the United States and to examine challenges and opportunities for the future. The study was to examine factors that have led to the current status (e.g., regulation, product pricing, and international trade) and discuss the various challenges the industry faces (e.g., disease, predation, genetic resources, land stewardship, and international trade and exchange rates). All sectors of the industry were to be examined, including meat, wool, live animals, pelts, milk, byproducts, and supporting businesses.

The study was in response to a congressional request to the National Academies in the Agricultural Appropriations Bill (P.L. 109-97) and was supported by funds received from the Economic Research Service of the U.S. Department of Agriculture. To prepare its report, the committee reviewed previous studies and scientific publications on the sheep industry, examined historical and current statistical data, conducted statistical analyses, held in-person interviews with representatives of key segments of the industry (both traditional and emerging), and conducted a survey of wool experts.

KEY FINDINGS

Although specific events, such as the end of World War II and the repeal of the National Wool Act, are often given as the cause of the decline of the industry, in fact many events and issues have contributed. Some of the more often cited factors are:

- Labor loss during World War II;
- A negative American G.I. experience with mutton during World War II;
- Changes in regulations and permits for grazing on public lands and endangered species regulations;
- Competition from other meats and other fibers;
- Changes in consumer preferences;
- Predation losses;
- Loss of the National Wool Act and the Incentive Payment programs;
- Foreign wool production subsidies;
- Competition from imports along with an appreciation of the U.S. dollar against Australian and New Zealand currencies in the 1990s; and
- Concentration in the U.S. packing and feeding industries.

Despite the continuing decline in the U.S. sheep industry, there are reasons for optimism about the future. Developments have occurred that have made the sheep industry more profitable for some. These developments include the following:

- Improvements in production efficiency;
- Leaner lamb;
- Development of further processing of lamb and new packaging techniques;
- Decline in Australian and New Zealand sheep numbers;
- Depreciation of the U.S. dollar; and
- Emergence of new and niche markets.

Development and Structure of the U.S. Sheep Industry

Most commercial sheep production in the United States consists of two types of operations. Range sheep operations are mostly in the western states and consist of relatively large flocks that graze native and unimproved pastureland. Farm flock operations are mostly found in the Midwest and East. Flocks are smaller than those in range operations and are raised on improved pastures and in feedlots. Each system accounts for approximately 50 percent of the U.S. lamb crop.

Lambs from these two systems are marketed via one of three distinct channels, which are described in Chapter 1: (1) traditional, (2) early harvest, and (3) direct marketing. Official government data capture information about the traditional channel but provide incomplete data about the other two channels. As a consequence, it is difficult to obtain reliable information on these two industry channels. Anecdotal indications are that both channels are active and growing in size.

Other products of sheep production include wool, pelts, and milk. Wool production, once the mainstay of the industry, has declined even more than lamb production. Much of the current U.S. wool production is exported to countries with expanding textile industries, such as China and India. Sheep pelts enter industrial markets and are either used for various consumer products or are exported, primarily to Asia and Europe. There is also a small but growing dairy sheep industry in the United States.

The U.S. Live Sheep Industry

Although the sheep industry has been in decline since at least World War II, an analysis in Chapter 2 of this report indicates that the decline in sheep numbers has slowed significantly and even reversed in some regions of the country, especially since 2000. While inventories in some states continue to erode, those in many other states have now halted their long-term decline and show modest growth in aggregate terms. Even in states with declining inventory, the rates of decline have been slower in recent years than a decade ago.

One contributor to this transition has been the increase in hair sheep and goats. To some extent, hair sheep and meat goats have been substituted for wool-producing sheep breeds and mohair goats. Hair sheep are easier to care for, more adaptable to humid climates, and eliminate the need for shearing.

The survival of the U.S. sheep industry depends on the potential for profitability which is affected by various economic factors, such as scale of operation, production efficiency, and labor and feed costs. The profitability and viability of the industry, however, is also dependent on the potential for continued scientific advances to improve profitability in various areas, including sheep breeding and genetics (e.g., the introduction of new breeds and mapping the sheep genome), improvements in reproductive efficiency, and improvements in nutrition. Also important is the sheep-environment interface, including sheep grazing behavior, interactions with wildlife (especially bighorn sheep), and the management of predators.

Commercial success in feeding lambs depends on numerous factors, some of which are beyond the control of the producer. A recent example is the increase in corn price because of the growth of the ethanol industry. New technologies and more efficient management practices are showing promise in terms of enhanced production and efficiency and cost reduction.

The process by which live sheep prices are negotiated and determined has been an important issue in the commercial sheep production industry in recent years. Like other livestock industries, the major issues related to pricing are the market dominance of retailers, keen competition from other livestock meats, and concentration in feeding and packing. Unique features relating to live sheep pricing are the relatively low demand at the consumer end of the market, the relative importance of ethnic markets, the loss of a large portion of consumer demand to imports, and large losses to predators. According to a recent report, most slaughter lambs are now being priced using formulas or are packer fed. Thus, rather than being driven by live markets, the lamb price discovery process now largely reflects carcass or cutout values so that price is determined by negotiation or formula related to carcass quality. As a result, a significant amount of the risk is shifted from the buyer to the seller, especially for pricing based on quality. Mandatory price reporting, fully implemented in 2001, has resulted in more price reporting for live lambs and lamb meat and made the process of price discovery more transparent to all participants.

Productivity increases have been achieved through genetic progress and improved nutrition, health, and management practices that have increased both the number of lambs weaned per ewe and the harvest weight of lambs.

The use of sheep and goats for vegetation management—from targeted or prescribed grazing to the control of invasive and nonnative weed, grass, and small shrub species and similar practices to protect ecologically sensitive areas and residential areas near the urban/wildlands interface—has also increased widely. Sheep breeding and genetic improvements have resulted in new breeds that perform at higher levels.

Key opportunities for enhanced industry efficiency and competitiveness include:

- Continued improvements in productivity through further advances in genetics (including gene biotechnology), nutrition, health, and management programs;
- Forage finishing to enhance lamb's competitive position relative to other red meats as the cost of grains and concentrate feeds continues to rise; and
- Direct marketing of high-quality, lighter-weight lambs to emerging and rapidly expanding ethnic markets.

Challenges to sustainable growth and economic competitiveness include:

- The sheep industry infrastructure suffers the consequences of decades of decline in volume;
- Increasing predation problems in wildlife populations in many states indicate the need for sheep industry alliances to work closely with wildlife agencies and interest groups and with those involved with threatened and endangered species management programs;
- Problems in pricing and price determination continue to plague the industry;
- The sheep industry continues to lag behind other livestock industries in the adoption of genetic improvement technology resulting in a competitive disadvantage of sheep with respect to other livestock species; and
- The share of public and private support for new technology development and educational activities received by the sheep industry continues to decline relative to other livestock sectors.

Sheep Health Issues

Diseases reduce sheep viability, overall growth, rate of gain, immunity, and reproductive performance, which together reduce production efficiency and income. There are few data, however, on the economic impacts or the true prevalence of most of the disease conditions affecting sheep in the

United States. This lack of information makes it difficult to make decisions about the allocation of resources and to determine research and policy priorities.

Most producers do not implement any type of quarantine, require premovement testing, or take precautions to prevent nose-to-nose contact at shows. The industry has the opportunity to improve biosecurity practices that will help to reduce the spread of endemic diseases and prevent the entry of highly contagious diseases.

According to national surveys, more producers have identified stomach/intestinal parasites (worms) as a great or moderate concern than any other disease condition. Although not usually fatal, these internal parasites cause substantial economic losses. In addition, the intestinal parasites have begun to develop resistance to the anthelmintic drugs used to treat them. In the United States, there is a shortage of approved animal drugs intended for less common animal species (minor species). The limited availability of effective anthelmintics is an example of this shortage. Other diseases that have the potential to have a significant economic impact on the U.S. sheep industry include Johnes disease, foot rot, and any foreign animal disease such as foot and mouth disease.

Scrapie is an insidious, degenerative disease that affects the central nervous system of sheep. Although scrapie is relatively rare in the United States (< 0.2 percent of animals are affected), it is always fatal. In addition, there is some concern of public health risk, although there is no current evidence as such. An outbreak of scrapie may also impede or restrict U.S. sheep exports. As a consequence, the United States has had a National Scrapie Eradication Program in place since 1952. Over the last 5 years, there has been substantial progress in the effort to eliminate scrapie from the United States.

Important accomplishments in sheep health include reducing the percentage of sheep that test positive for scrapie at slaughter (fewer new infected and source flocks and fewer animals indemnified) and a more complete animal identification system than exists for any other livestock species in the United States.

Opportunities for minimizing the economic impacts of diseases include:

- The use of the identification system to monitor animal movement as the foundation for an overall flock health program; and
- An improvement in biosecurity practices to help reduce the spread of endemic diseases and prevent the entry of highly contagious diseases.

Challenges remain however and include:

- Lack of data on the economic impact or the true prevalence of most sheep diseases;
- Unavailability of many critically needed drugs; and
- A shortage of large-animal veterinarians in many areas of the country.

The U.S. Lamb Industry

Sheep producers sell their lambs to a relatively small number of relatively large commercial feeders. As a consequence, commercial feeders have some potential market power relative to producers. Feeders sell to an even smaller number of packers. Feeders usually have few alternative buyers, giving them little bargaining power in dealing with packers. However, packers are at a distinct disadvantage in bargaining on price with large retailers and large foodservice buyers. Finally, food retailers sell to a large number of consumers. This market structure makes it difficult for value preferences to migrate from consumers to producers. Congress has attempted to address this issue in two recent pieces of legislation: (1) the Livestock Mandatory Reporting Act in 1999 and (2) the Lamb Promotion, Research, and Information Order (American Lamb Checkoff Program) in 1996. Research has indicated that price changes at the producer/feeder level are almost fully transmitted to the wholesale level but not from the wholesale level to the retail level. Although there has been concern that the decreased numbers of packers (the four largest slaughtering firms accounted for almost 70 percent of the federally inspected lamb slaughter in 2005) have affected price spreads or margins, studies to date have provided limited evidence that packer concentration exerts a significant negative effect on slaughter prices.

Lamb is a relatively minor product in most food stores, occupying only 1 to 3 percent of meat cases. However, the proportion of meat cases that contain lamb has increased and now exceeds 80 percent. Some increased penetration of lamb in retail meat cases is good news for the lamb industry, in addition to the growing custom and ethnic specialty markets not captured in surveys. Buoyed by growing lamb imports, U.S. lamb consumption has grown slowly during the last decade, leading to a relatively steady annual per capita consumption level of 0.50 to 0.55 kg since the mid-1990s.

Excess fat on lamb carcasses has been a persistent problem for the sheep industry. The current market structure and pricing system generally reward producers and lamb feeders for weight rather than value based on quality and yield grades. The evolution of a yield grade system coupled with carcass merit pricing has helped to improve this situation. Numerous modern tools to assess the yield of boneless closely trimmed retail cuts are available but have not been widely adopted by the industry because of cost and the potential to slow the rates of slaughter and processing. Although foodborne illnesses have not affected the lamb industry in the same way as they have other meat industries, lamb packers and processors have implemented multiple screening systems to minimize pathogens. Efforts such as development of controlled atmospheric packaging and products that are easy and quick to prepare have also been made to enhance the value of lamb products. Lamb has a more distinct flavor than most competing meats. Recent research studies have investigated the source of these sensory properties.

Increasing the demand for lamb is key to the growth and expansion of the industry. The traditional argument that American tastes and preferences have moved away from lamb may no longer be applicable given the steady level of consumption in recent years despite declining production. More appropriate now may be the argument, borne out by recent research, that lamb is purchased fairly consistently by a small group of consumers and not at all by most consumers. Efforts to promote lamb consumption in recent years by the American Lamb Board have been shown to be effective. Competition from imports is an important concern for the U.S. lamb industry. Between 1990 and 2005, imports increased from approximately 18 million to 82 million kilograms. By 2005, imports made up almost half of domestic consumption. Competitive advantage in global lamb markets is influenced by many factors, such as costs of production, industry infrastructure, and currency exchange rates. The U.S. sheep industry has some advantage in producing sheep in terms of a natural resource base and a strong cadre of supporting industries. Competing successfully with Australia and New Zealand may require the U.S. lamb industry to focus on differentiating American lamb from imported lamb.

The hanging carcass represents only about half the live weight of a lamb. Thus, byproduct markets are an important part of the lamb industry. Major byproducts (in addition to wool) include various edible portions of the carcass, rendered byproducts, pelts, and lanolin. Additional uses that have developed recently include pharmaceutical, research, and waste management applications. Margins in the lamb packing industry are thin so that the profit is often in the sale of the byproducts.

The functioning of lamb markets is affected by a wide range of policies and regulations designed to regulate the business practices of those who engage in the buying and selling of lamb, including meat inspection laws, the Packers and Stockyards Act, and mandatory price reporting. The USDA Livestock Risk Protection-Lamb Insurance Policy program is a new price risk management tool that provides producers and feeders of lambs with the opportunity to insure lambs they own against unexpected price declines.

The U.S. lamb industry has opportunities for growth, development, and enhanced competitiveness:

• Depreciation of the U.S. dollar has enhanced the competitiveness of domestic lamb against imports;
• The use of alternative marketing arrangements, such as forward contracts or marketing agreements, as opposed to the use of cash transactions, is also helping the industry to compete more effectively with imports;
• Promotion efforts have been shown to enhance lamb demand; and
• Along with a growing presence of lamb in the foodservice sector, new markets for lamb products are emerging.

Realizing the opportunities, however, will require the industry to tackle a number of difficult challenges:

• Continued enhancements of production efficiency and reductions in cost;
• Development of an automated system to assess accurately and uniformly carcass value across processors and over time;
• Collection of data, such as a retail lamb price series, lamb consumption by country of origin, away-from-home vs. at-home consumption of lamb, socioeconomic profiles of consumers in different market areas; and
• Research on the retail demand for specific cuts of lamb to assist retailers and foodservice purveyors in pricing and price-based promotion of lamb cuts.

The U.S. Wool Industry

At one time, wool was considered the primary product of sheep production, with lamb and mutton as byproducts. Today, the situation is reversed. Wool is sometimes even considered a "liability" in sheep production. The growth of hair sheep production is a reflection of the decline in the relative profitability of wool production vs. lamb production. Wool currently accounts for 10 to 30 percent of sheep production income in range production systems and < 5 percent in intensive farm flock production systems.

Wool producers use one of three primary market preparation methods at shearing (original bag, bellies out and the fleeces not tied individually, and table skirted and classed), and one of four primary sales mechanisms (wool warehouses, wool cooperatives, wool pools, and private treaties). Most U.S. wool is sold as original bag even though the table skirted and classed method is recommended for international and premium domestic wool markets. The value of wool is influenced by several objective measurements (e.g., fiber diameter, yield, staple length, and vegetable matter) as well as several subjective measurements (e.g., color or stains, condition of staple tips, crimp, and style or handle).

A number of natural fibers (e.g., cotton, flax, and silk) and synthetic fibers (e.g., nylon, rayon, acrylic, and polyester) compete with wool. From 1995 to 2005, wool accounted for an average of only 0.6 percent of all fibers used by U.S. mills. Furthermore, the domestic mill use of all fibers, including wool, has been in a general downward trend reflecting the growing globalization of the textile industry. International trade treaties, such as the General Agreement on Tariffs and Trade, have fostered the decline of textile manufacturing in the United States.

The world wool market is dominated by China, which purchases almost 50 percent of the world market supply of raw wool and produces 39 percent of the world's wool apparel products. More research on pricing and income elasticities for wool is critical to ensure an accurate understanding of the U.S. and world demand situation for wool. Demand analyses involving wool, cotton, and manufactured fiber are necessary both at the mill level and at the retail level. World wool market prices are primarily established by Australian supply and world demand. Australia produces approximately 50 percent of the world's wool sold at auction. Prices for U.S. wool range from 60 to 75 percent of imported Australian wool prices. An important question for wool producers is whether price differences between Australian and U.S. wool are related to differences in the extent of preparation of wool for marketing. A hedonic price model was developed by the committee to test the hypothesis that skirting and classing wool generally produced higher prices compared to original bag wool. The results provide a basis for determining premiums or discounts relative to the current practice of marketing original bag wool. Clearly, producers who do minimal preparation of their wool before sale are losing substantial premiums paid for wool that has been more prepared for sale.

Wool production and marketing have been affected by several national and international policies and regulations. The most significant was the National Wool Act (in place from 1954 to 1994), which provided support to the domestic wool industry. Other significant legislation has included the Multi-Fiber Arrangement (1974), the Agreement on Textiles and Clothing (1995), the establishment of the American Wool Trust (2000), and the Farm Security and Rural Investment Act (2002). The United States is not the only wool-producing country that has provided price support for wool producers. Price support programs in Australia, New Zealand, and South Africa set wool prices well above market levels in the 1980s and early 1990s, leading to growing wool stockpiles during that period. These stockpiles are now largely gone.

The U.S. wool industry has made some progress in responding to the pressures it has faced, including movement toward further preparation of wools by producers and public/private collaborative research to develop fabrics and garments that are machine washable, more breathable and less prickly when worn against the skin, shrink proof, and flame resistant.

A survey of key wool industry leaders identified several important opportunities currently facing the U.S. wool industry:

- Economic growth in China and other developing countries leading to potentially greater foreign demand for wool apparel goods and less import pressure on U.S. markets;
- Increased wool premiums through increased preparation of wool; and
- Growing markets for specialty wools for hand spinners, yarn for weavers and knitters, and other wools such as naturally colored wools.

Survey respondents also identified ongoing challenges to the future growth and development of the U.S. wool industry:

- Fragmented selling systems, fewer domestic wool buyers, distance-to-market challenges for producers, greater concentration of the wool processing industry;
- A shortage of qualified sheep shearers;
- A need to continue the trend toward preparing wool to international standards;
- Contamination from hair sheep breeds; and
- A lack of financial support for critically needed research.

The U.S. Dairy Sheep Industry

Although currently a relatively small segment of the U.S. sheep industry, dairy sheep production is in the early stages of becoming an economically important agricultural industry. Sheep milk is generally produced in areas that are great distances from willing milk processors. The formation of marketing cooperatives and the practice of freezing milk in plastic bags in large commercial freezers on the farm are helping to resolve the distance issue.

The United States is the world's largest importer of sheep milk cheese, accounting for about half of world sheep milk cheese exports. A lack of local commercial processing factories has led many U.S. sheep producers to make cheese on their farms in small batches for direct marketing to individuals, food stores, and restaurants. Domestic sheep milk cheeses can also be found in many specialty cheese stores and even stores of national grocery chains.

The growth of the domestic industry is the result of the production of high-quality milk, the manufacture of high-quality cheeses, and the promotion of those cheeses by both national and state organizations.

Opportunities for the dairy sheep industry include:

- The increasing consumption of sheep milk cheese by U.S. consumers;
- The growing movement among consumers to eat locally raised and produced items; and
- An opportunity for small-scale farms because relatively little land is required and sheep are small enough to be handled safely by most family members.

Future growth of the industry could be limited by:

- The lack of either regional or national genetic improvement programs for dairy sheep traits in the United States;
- Limited research and extension support for dairy sheep production, sheep milk processing, or sheep milk product marketing;
- Limited marketing options; and
- Imports of lower-priced sheep milk cheeses.

Alternative and Emerging Markets

Although the full extent of the changes taking place in the industry is difficult to determine, sheep and lamb inventories indicate growth in eastern and mid-Atlantic states and in the Midwest, where alternative and emerging markets are particularly important. Much of the lamb moving to these emerging markets is sourced directly from farms or small abattoirs, especially during religious holidays, and is not inspected in federally or state regulated facilities or included in official USDA slaughter data. These alternative or "niche" markets represent a growth area for the industry. Although the East and West coasts are the largest U.S. markets for lamb, the consumption of lamb in those areas is skewed toward religious and ethnic groups and away from groups with origins in northern Europe. For some religious groups, lamb is consumed on a regular basis and has special significance for holidays or holy days.

Because there is relatively little information on the importance of the ethnic lamb market in the United States, an analysis of the effects of ethnic markets on aggregate U.S. lamb demand was conducted as a part of this study. The primary conclusions were that Muslim holiday periods and Christian and Orthodox Easter increase slaughter levels of lamb and yearlings significantly and that the impact of these holidays appears to be increasing with time. To the extent that some of this lamb is purchased directly from farms and not recorded in official tallies of slaughter, official data could underestimate the impact of these religious periods. Other ethnic markets of growing importance in some regions of the country include Hispanic, Italian, and Greek communities.

A number of other alternative markets for the sheep and lamb industry have emerged, especially during the last 10 years. These product markets offer the opportunity for higher producer prices and profit through specialized, value-added products, such as organic lamb, specialty cheeses, fine wool, and other specialty products with perceived unique qualities. Many sheep producers are taking proactive approaches to sell their own products directly to customers and foodservice operators and bypass traditional marketing channels. Advances in communication systems, especially the Internet, have facilitated such direct marketing.

Some high-end restaurants and specialty meat retailers have begun buying whole carcasses or quality live animals for custom slaughter from producers who meet such conditions as sustainable production, "locally grown," organic, and related characteristics relevant to their customers. They price meat cuts to satisfy the local demand and in a way that sells the entire carcass.

Opportunities in the nontraditional, emerging, and alternative markets include:

- The emerging ethnic and religious markets in the major U.S. metropolitan areas;
- Organic, natural, and locally produced lamb; and
- Specialty wools.

Challenges to the expansion of these alternative and emerging markets include:

- The limited information on these markets, the value chains, and product pathways make it difficult to determine market opportunity; and
- Regulatory systems, operated either by government or as industry-established standards, will be needed to support sustained growth.

CONCLUSIONS

Although continuing declines can be expected in some areas of the U.S. sheep industry, the changes currently taking place offer grounds for optimism. The emergence of new and alternative markets for sheep products signifies that the industry may be on the brink of a transition from traditional practices and marketing channels to new markets, new technologies, new products, and a new consumer base. This offers the potential to arrest the decline experienced over the last several decades. Expanding alternative and emerging markets will create considerable challenges to the industry and to policy makers. However, all these challenges can be addressed by a concerted effort from within the industry.

1

Development and Structure of the U.S. Sheep Industry

The U.S. sheep and lamb industry is one of the most complex in animal agriculture. The industry provides lamb meat for U.S. domestic consumption (retail, and hotel, restaurant, and institutional trade) and some exports, mutton from older animals, exports of live mature animals, wool, pelts, and a variety of byproducts. A relatively new and growing industry built on milk from sheep is offering many sheep cheese varieties and yogurt, as well as some products blended with bovine milk. Hair sheep, which do not require annual shearing, have been introduced into the industry over the past few decades, thereby altering the fixed proportion of products common in the lamb-wool-pelt enterprise.

Another part of the U.S. sheep industry is made up of purebred flocks, producing high-quality breeding stock for other purebred breeders and the commercial flock. Considerable variation in breeds and breed characteristics is evident across the U.S. purebred sheep and lamb industry. Traditional European breeds have been the dominant bloodlines in North America since the introduction of sheep with the earliest settlers. In the last several decades, new and different breeds have been introduced, bringing greater variation in animal characteristics, as breeders and commercial sheep farmers have sought to adapt production methods, product characteristics, wool length and quality, milk efficiency, and the like to regional conditions. The purebred industry is a major source of high-quality breeding animals for the commercial flock, particularly rams, used for crossing with commercial ewes. In commercial operations, most producers maintain purebred lines in their ewe flock, relying on terminal sire breeds to achieve the characteristics desired in their market lambs.

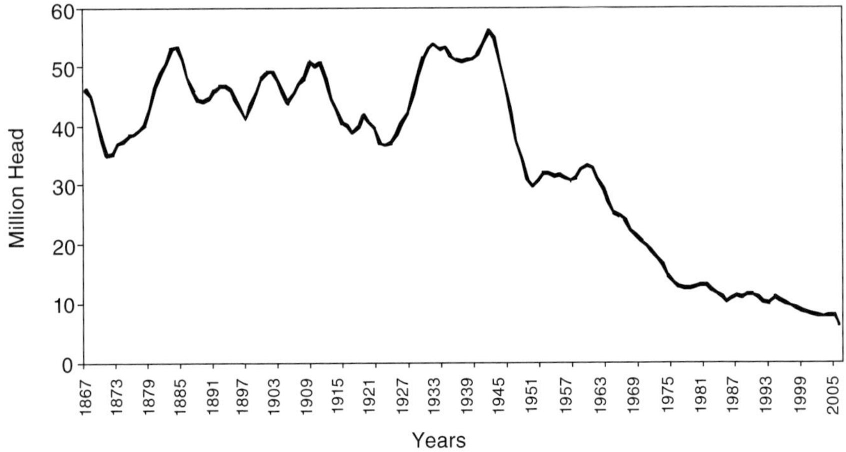

FIGURE 1-1 U.S. sheep and lamb inventories (January 1), 1867–2007.
Source: USDA (2007b).

The dominant feature of the sheep industry and the focus of much producer and policy concern over the years, however, has been the steady decline in sheep and lamb inventories since the mid-1940s (Figure 1-1). From a record high of 56 million head in 1942, inventories on January 1, 2007 reached 6.2 million, the lowest level in recorded history. In turn, the decline in sheep and lamb numbers has created difficulties in the flow of sheep and lamb products through a shrinking marketing system as producers have struggled to respond to market signals while maintaining profitability.

This chapter provides some historical background to the discussion of the current status of the various segments of the U.S. sheep industry in subsequent chapters. As well, the chapter provides an overview of the linkages and interdependencies in the industry inherent in the marketing channels or value chain through which sheep and lamb products flow from producer to end user.

HISTORICAL DEVELOPMENT OF THE U.S. SHEEP INDUSTRY

The first domesticated sheep were brought to the United States in 1493 with the second voyage of Columbus. These were largely Spanish Churro sheep, which were also later introduced to the southwestern United States by the Spanish conquistadores. They were small, hardy sheep with a poor-quality fleece by today's standards. English breeds were introduced by colonists to be used primarily for wool for home-produced textiles and, to a lesser extent, meat. Some colonies passed laws requiring young people to learn how to spin and weave.

Following the Revolutionary War, the United States began to develop a viable automated woolen manufacturing industry. With growing importation of Spanish Merino sheep in the late 1700s and early 1800s, the U.S. wool clip began to grow substantially. Woolen cloth manufacturing experienced its greatest growth during the period from 1830 to 1870, particularly during the Civil War years of the early 1860s.

Although many eastern states had large sheep populations in the 1800s, sheep production began shifting westward with improved rail transportation, feed production, and lower production costs in the West. As the industry moved west, wool production from the French Rambouillet, originally developed from Spanish Merino genetics, expanded rapidly. By 1870, about 80 percent of all U.S. sheep were of Merino origin (ASI, 2002).

The U.S. sheep and lamb inventories reached a high of 54 million head in 1884 and then declined slowly to a low of about 37 million head in 1923 (Figure 1-1). Inventories quickly turned around again in the 1920s, peaking at 54 million head in 1932 and then reaching an all-time high of 56 million head in 1942. That rapid growth spurred an equally rapid development of a marketing system (including feeding, slaughter, milling, and breaking facilities, and distribution and transport systems) to meet the rapidly growing demands for meat, wool, and other sheep and lamb products during that period. At the same time, the emphasis on sheep production began to shift toward meat rather than wool in response to the demand for protein to feed U.S. troops during World War II.

As the war drew to a close, U.S. sheep industry fortunes changed drastically as inventories plummeted over the next several years, bottoming out at 30 million head in 1950, a nearly 50 percent decline from the record high set in 1942. Wool production fell commensurate with the drop in overall sheep numbers (Figure 1-2). Following a little more than a decade of relative stability, inventories began to decline once again in the early 1960s, a trend from which the industry has yet to recover. Although specific benchmark events, such as World War II and loss of the National Wool Act, are often cited as the cause of the decline in the industry since the 1940s, in fact, many events and factors have combined to limit opportunities for growth in the industry. Some of the more often cited of these include the following:

- *Labor loss during World War II.* World War II drew a great deal of labor out of American agriculture on a permanent basis. This shift affected all of agriculture, including the subsequent availability of labor for sheep and lamb production.
- *American GI experience with mutton during World War II.* Lamb was relatively common on American dinner plates before World War II. The

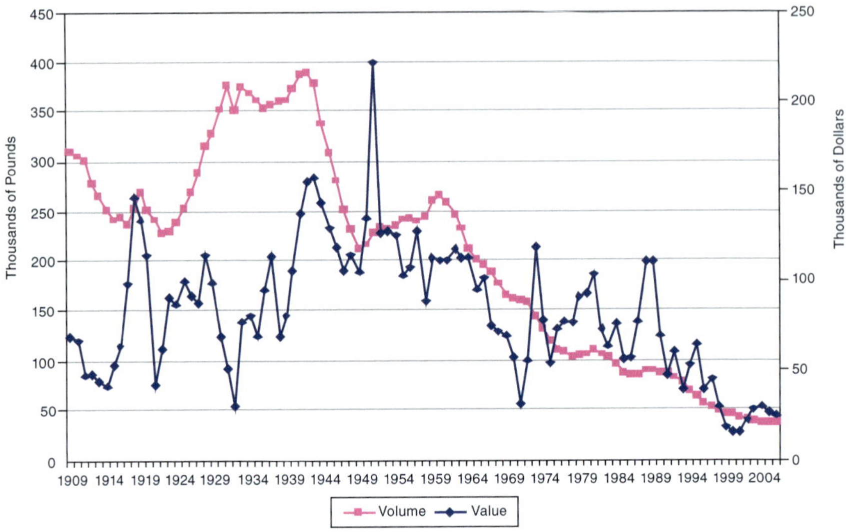

FIGURE 1-2 Volume and value of wool production, 1909–2006.
Note: 1 pound (lb) = 0.4536 kg.
Source: USDA (2007b).

experience of American GIs with poor-quality and poorly prepared mutton during the war, however, negatively affected lamb consumption in the households of returning military veterans. This trend is widely recognized anecdotally. For examples of the attitude toward mutton following the war, see Garcia (2004) and Apple (2006).

- *Grazing permits and restrictions.* The regulations and permits for grazing on public lands have changed considerably over the last several decades, with impacts on the availability of land for sheep production. This challenge to the industry is discussed in more detail in Chapter 2.
- *Competition from other livestock and meats.* The increased production of low-cost feed grains beginning in the 1950s resulted in a rapid expansion in the production of low-cost and high-quality meat products from poultry and pork. Similarly, the development of the cattle feedlot industry increased both product quality and the profitability of beef production, providing an incentive to shift from sheep to beef production in many areas.
- *Competition from other fibers.* The advent of synthetic fibers (such as polyester, rayon, and acrylic) in the 1950s and 1960s resulted in a growing substitution of the lower-cost synthetics for wool, and to a lesser extent cotton, in apparel, carpet, and industrial goods. Chapter 5 provides more discussion of this issue

- *Predation losses.* Predation losses have been a concern for many years in the sheep and lamb industry, both for range and farm flocks. Sheep, goats, lambs, and kids are particularly susceptible to predation, and the changing availability or restricted use of many methods used to control predators since the late 1970s has been cited as a cause for increased losses from predation (Hawthorne, 2004; Shelton, 2004). This issue is explored more fully in Chapter 2.
- *Loss of the National Wool Act and the Incentive Payment programs.* The National Wool Act, in place since 1955, was repealed in 1993. The act provided direct payments and other support through government programs to wool producers. Wool sales plus support payments together represented 28.1 percent of total sheep, lamb, and wool industry revenues in 1990; by 1997, wool alone accounted for only 6.6 percent of industry revenue, compared to 11.2 percent in 1990. As discussed more fully in Chapter 5, the termination of direct payments was a significant loss to wool and mohair producers, resulting in a substantial increase in the rate of contraction of the sheep and lamb industry (USDA, 1999).
- *Foreign wool production subsidies.* Support price schemes in Australia, New Zealand, and South Africa also contributed to sheep and wool market difficulties over the years. Because support prices in those countries were set well above market levels, stockpiles of wool occurred, particularly induced by the Australian Wool Council in the 1980s and early 1990s. The stockpiles of wool, however, gradually were placed on the world market over the next decade, which seriously weakened world wool prices through at least the year 2000. Prices began to rebound somewhat in 2001 when the stockpile liquidation was complete.
- *Competition from imports.* Between 1990 and 2005, imports of lamb, primarily from Australia and New Zealand, grew from 18.60 million kilograms, about 10 percent of domestic lamb supply, to 81.65 million kilograms, nearly equal to U.S. domestic production and half the total domestic supply (USDA, 2007a). In response, the U.S. imposed a tariff rate quota (TRQ) on lamb imports from Australia and New Zealand in 1999, which was struck down by a World Trade Organization ruling just 2 years later. Nonetheless, imports maintained the availability of lamb for U.S. consumers at about 0.5 kg per person throughout this period. The role of lamb imports in the U.S. sheep industry is explored in more detail in Chapter 4.
- *Appreciation of the U.S. dollar against Australian and New Zealand currencies.* Despite imposition of the TRQ in 1999, lamb imports from Australia and New Zealand did not slow down. A possible reason for this phenomenon was that the import-restricting effect of the TRQ was offset to a large extent by a growing appreciation of the U.S. dollar against Australian and New Zealand currencies during that period. For example, between January 1989 and July 2001, the U.S. dollar increased in value from 1.12

to 1.98 Australian dollars, resulting in a more than 50 percent increase in the purchasing power of the U.S. dollar in Australia (RBA, 2007). The consequence was a reduction in the cost of foreign lamb to U.S. buyers and a more profitable U.S. market for Australian and New Zealand lamb exports. See Chapter 4 for more detail.

- *Concentration in the U.S. packing and feeding industries.* The percentage of U.S. lamb slaughter accounted for by the largest four lamb packers (the four-firm concentration ratio) jumped from approximately 50 percent in the mid-1980s to 77 percent in 1988 as a result of various mergers and acquisitions in the industry (Williams et al., 1991). One consequence was growing concerns that concentration in lamb packing and feeding gave market power to packers to control lamb markets and prices. The four-firm concentration ratio in the lamb packing industry has since receded to 57 percent in 2005. Various studies (e.g., Williams et al., 1991; RTI, 2007) have failed to find overwhelming evidence of concentration as a primary driver in the U.S. sheep and lamb industry. See Chapter 4 for more detail.

Recent Developments in the Industry

Despite the continuing decline in the U.S. sheep industry, there are some reasons for optimism about the future. Developments have occurred that have made the sheep industry more profitable for those producers who have survived. Genetic improvements, breed resources, management changes, and production and technology efficiencies, along with improved postharvest marketing strategies, have improved profitability. These and other developments currently working against decline in the U.S. sheep industry are discussed in more detail in later chapters. Some of the more salient of these developments include the following:

- *Improving production efficiency.* The average number of lambs produced per ewe per year has increased from 0.88 in 1940 to 1.12 in 2006 (USDA/NASS). During the same period, the average carcass weight of lambs increased from 18 to 32 kg. Consequently, on an annual basis, sheep producers are now producing 124 percent more lamb by weight per ewe than in 1940. As noted in Chapter 2, the genetic potential exists to further increase productivity within the various production systems.
- *More stable level of national sheep inventory.* The decline in sheep numbers appears to have slowed significantly and even reversed in some regions of the country, as discussed in more detail in Chapter 2. While inventories in some states continue to erode, many states have now halted the long-term decline and show modest growth in aggregate terms. The last few years exhibit a transition period with growth in the states with farm flocks coupled with slower overall decline in the range sheep flocks, contributing

to a more stable level in the national flock than has occurred in the past 60 years.

• *New packaging technologies.* New gas flush and vacuum packaging processes, such as SecureFresh™, are extending the shelf life of fresh and chilled lamb by significantly delaying the growth of spoilage organisms. Such packaging systems reportedly can increase the storage life of case-ready meats by 12 weeks (Food and Pack, 2002). With most existing packaging, fresh chilled meat lasts only 4 to 5 days on the shelf. Most U.S. "supermarkets have to dump 20 to 30 percent of their fresh meat" because of spoilage (Food and Drink, 2002). These new packaging systems have increased the versatility of lamb marketing options. Retail and food service markets with lower lamb volume now have improved opportunities to merchandise lamb. Chapters 2 and 4 discuss this and other new technologies affecting the sheep and lamb industry.

• *Growth of the cull ewe market.* The development of the cull ewe market in Mexico over the last 20 years, and to a lesser extent the domestic market for cull ewe products, has increased the market value of cull ewes. As a result, ewe depreciation costs are lower, thus improving profitability for sheep producers. As detailed in Chapter 3, the announcements of positive tests for bovine spongiform encephalopathy (BSE or "mad cow disease") in the United States in 2003 and again in 2005 led to some reduction in U.S. live sheep exports to Mexico. Many producers retained 5- to 6-year-old ewes for an additional breeding season during this expansion period. Those exports have recently shown signs of recovery with an increase in 2006, the first since 2002.

• *Recent depreciation of the U.S. dollar against Australian and New Zealand currencies.* Following its rapid strengthening against the currencies of Australia and New Zealand in the late 1990s and early 2000s, the U.S. dollar has depreciated just as rapidly against those currencies over the last few years as discussed in Chapter 4. For example, from the high of 1.98 Australian dollars in July 2001, the value of the U.S. dollar dropped to only 1.13 Australian dollars by September 2007, a level not seen since May 1984 (RBA, 2007). When combined with increased transportation costs and increased lamb market prices in Australia, the declining purchasing power of the U.S. dollar in Australia is helping boost the competitiveness of U.S. lamb in U.S markets.

• *Decline in Australian and New Zealand sheep numbers.* The decline in U.S. sheep inventories is not a unique phenomenon among leading sheep-producing countries, as pointed out in Chapter 2. Australian sheep inventories peaked at 173.8 million head in 1990 but declined to 86.8 million head in 2007 (Figure 1-3) (ABARE, 2007; ABS, 2007). At the same time, New Zealand sheep inventories peaked at 70 million head in 1982–1983 but declined to 46 million head in 2006 (Figure 1-4) (SNZ, 2006). While

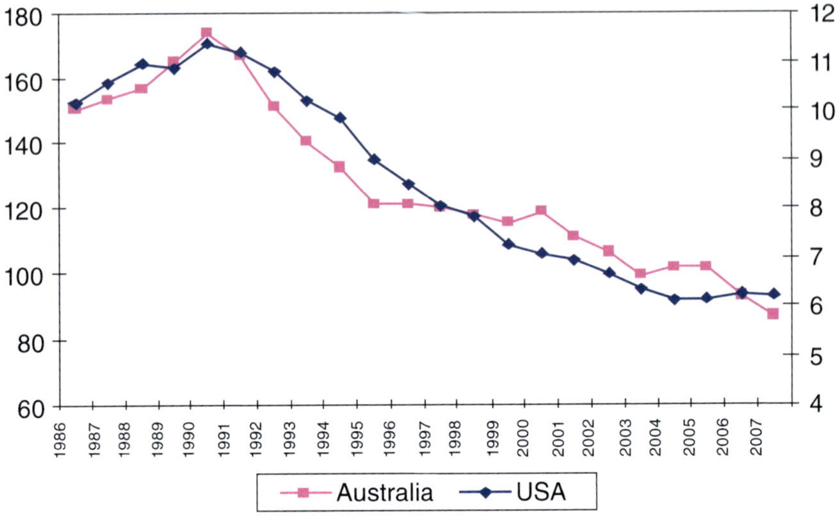

FIGURE 1-3 Australian and USA total sheep numbers, 1986–2007 (million head)
Note: Australia is left axis, United States is right axis.
Sources: USDA (2006, 2007b); ABARE (2007); ABS (2007). Copyright 2007 by ABARE (Australian Bureau of Agricultural and Resource Economics). Used with permission.

declining by 45.5 percent in an absolute sense between 1990 and 2007, U.S. sheep inventories have remained steady in relationship to those of Australia and New Zealand over that period. In 2006, the U.S. sheep flock was 6.7 percent of the size of the Australian sheep flock, the same as in 1991 (ABARE, 2007; ABS, 2007; USDA, 2007b).

• *New and emerging markets.* Perhaps the most optimistic aspect of the U.S. sheep and lamb industry is the emergence of new and niche markets for sheep and lamb products, as discussed in more detail in Chapters 5, 6, and 7. Among others, these markets include the following:

1. *A dairy sheep industry* in the early stages of becoming an economically important agricultural industry. From virtually nothing in the early 1980s, the dairy sheep industry has developed into a small but growing industry, with an estimated 6,478 ewes producing an estimated 1.15 million kilograms of milk in 2003.

2. *Purebred flocks* used in show for club lambs and raised for specialty wools. The growth of these markets is supported through active producer organizations, local marketing systems, statewide fairs, and new technologies, such as the Internet, that have reduced the costs of exchanging infor-

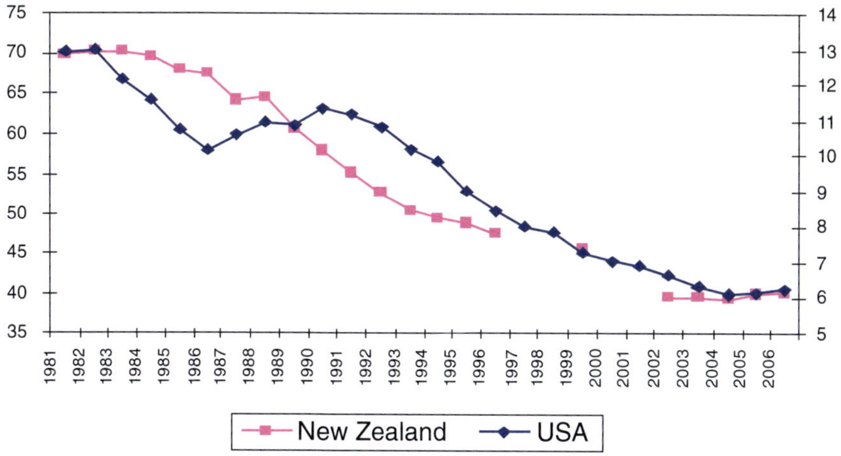

FIGURE 1-4 New Zealand and USA total sheep numbers, 1981–2006 (million head).
Note: New Zealand is left axis, United States is right axis.
Source: SNZ(2006).

mation in the market and facilitated identification of buyers and sellers of the products.

3. *Direct marketing* by producers with access to inspected slaughter plants. This market is being pushed by an expanding ethnic demand for lightweight and younger lamb carcasses. However, direct marketing customers may include local friends and neighbors or purchasers via the Internet or other direct mail-order approaches to a broader market. Individual producers have cultivated retail and restaurant outlets for their lamb, often exceeding their production capacity, and are purchasing lambs from other producers that meet their specifications. Although to a lesser extent than for lamb, ethnic markets for mutton are also increasing in the United States.

4. *Organic and natural lamb and wool products* featured by niche marketers at a substantial market premium. Although many western range lambs would likely meet the requirements for organic or natural products, no lamb slaughter plant of significant capacity currently qualifies for the production and processing of organic or natural products.

5. *Small wool mills*, often called mini-mills, that produce for specialty wool yarn, wool fabric, and finished wool product markets. Some producers sell raw wool, often naturally colored, to hand spinners. The number

of small retail outlets and hand spinners that are listed on the Internet is rapidly increasing.

THE U.S. SHEEP INDUSTRY VALUE CHAIN[1]

The marketing system or value chain through which sheep and lamb and their products flow from production to end use is complex and rooted in tradition and history. As depicted in Figures 1-5 and 1-6, the U.S. sheep industry value chain consists of six primary components: (1) farm production, (2) feedlot finishing, (3) harvesting and further processing, (4) retailing and food service, (5) trade (exporting and importing), and (6) end use (consumption and industrial use). Figure 1-5 depicts the entire sheep industry value chain, including trade. Figure 1-6 provides details of the linkages between domestic lamb production and use.

Farm Production

Sheep production in the United States is characterized by some very large flocks and many very small flocks. In 2005, 0.1 percent of all operations with sheep accounted for 14.2 percent of the national flock and 1.5 percent of operations accounted for nearly one-half (47.7 percent) of the national flock (USDA, 2007a). For the smaller operations, 92.1 percent of the operations accounted for only 31.7 percent of the total flock.

There are two primary types of commercial sheep operations in the United States: (1) range sheep operations and (2) farm flocks (Boxes 1 and 2 in Figure 1-5). Purebred sheep operations are a third type of sheep operation located throughout the sheep-producing states. Some producers maintain small purebred flocks as well as large commercial herds for the production of purebred breeding rams for sale or replacement, purebred ewes for sale to other producers, and for showing. Others specialize in purebred sheep production.

Range sheep operations (Box 2 in Figure 1-5) are found principally throughout the central and western states where flocks are maintained on native and improved pastureland. Sheep production in Texas, Wyoming, Colorado, Utah, Montana, Idaho, New Mexico, Arizona, and Nevada typify these extensive, large-scale range operations. Often the sheep enterprise on these operations is the primary source of income.

There are two general types of range operations. *Range band* operations are typically located in the 11 western states and South Dakota where there are vast areas of unfenced public grazing lands. Since the majority of

[1]The live animal marketing channel of the sheep industry value chain is discussed in more detail in Chapter 2, and the lamb meat marketing channel is discussed in more detail in Chapter 4.

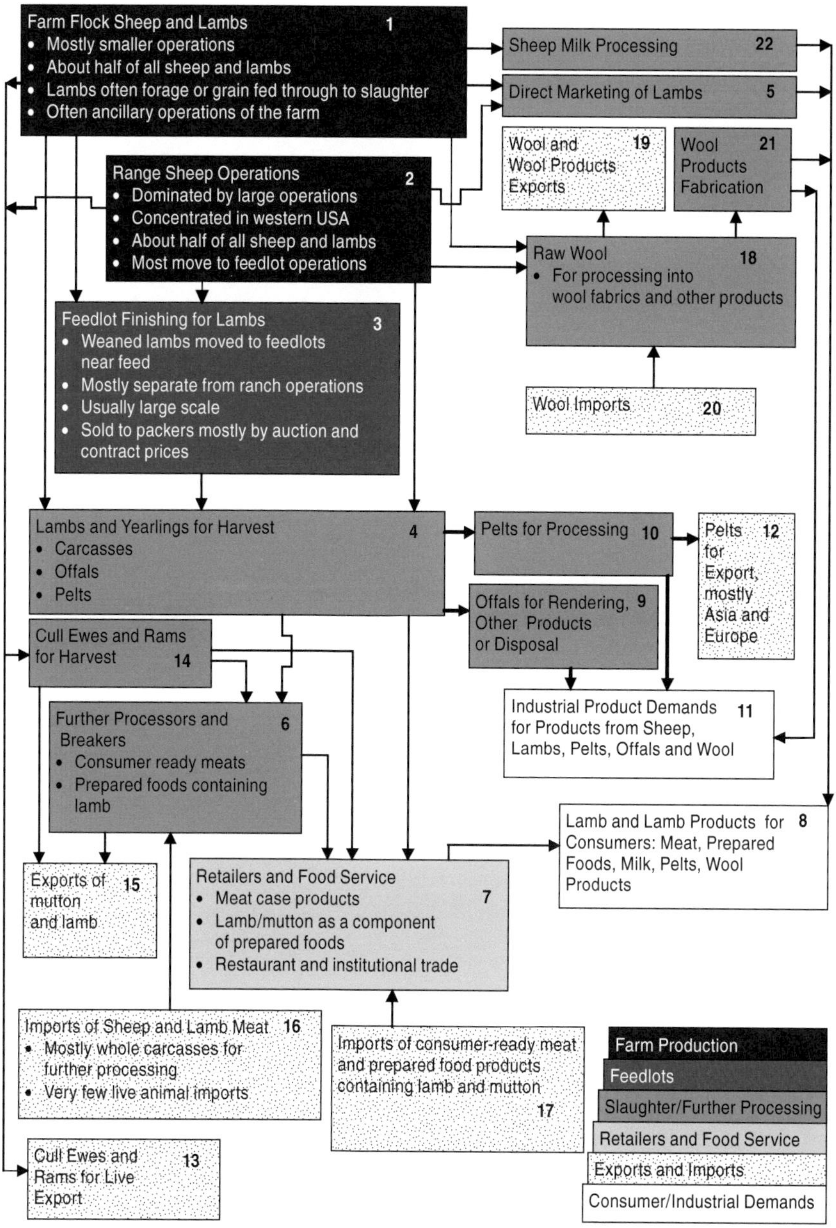

FIGURE 1-5 Sheep and lamb value chain diagram.

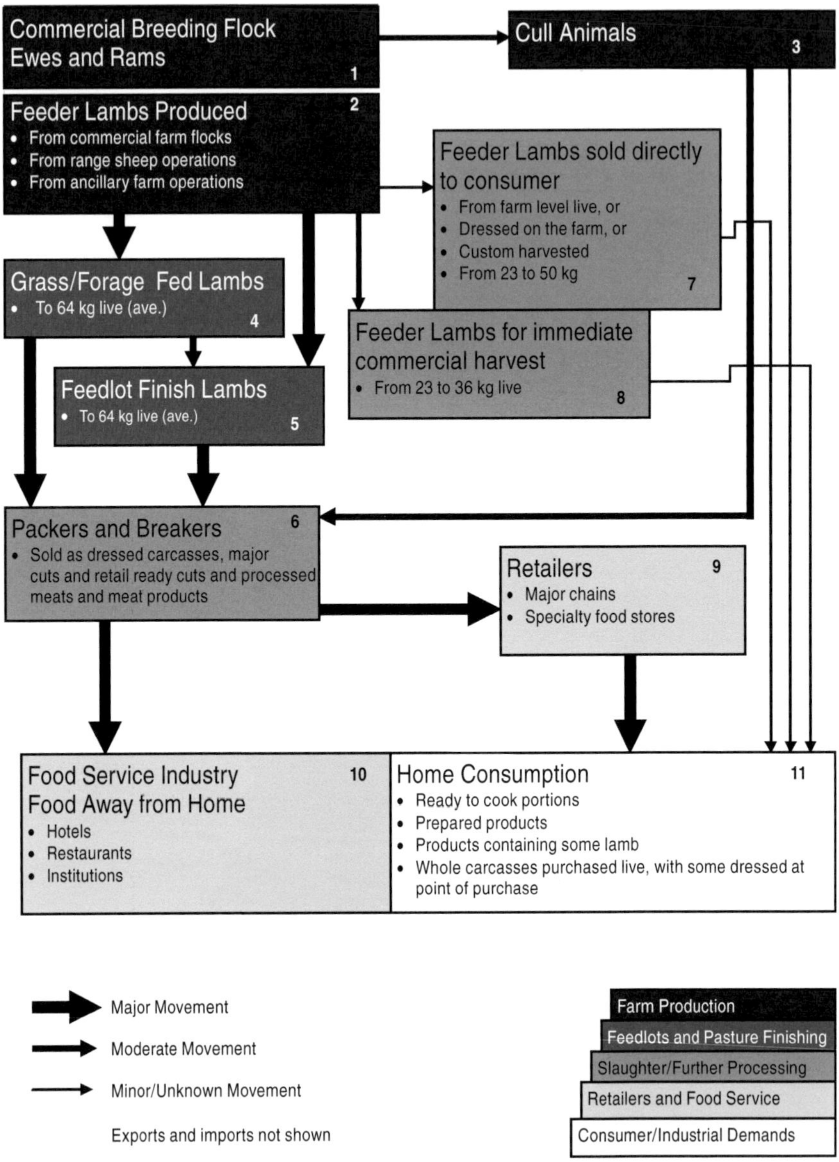

FIGURE 1-6 Farm-to-market linkages for lamb.

the land utilized by range band operations is unfenced, unimproved, native high-mountain and desert pastures, range bands often must move long distances from season to season and, thus, require on-site herders (Ensminger, 2002). The competitiveness of range band operations is affected by government controls over public grazing lands and changes in public grazing fees. *Fenced range* operations are located mainly in Texas, New Mexico, Kansas, North Dakota, Nebraska, and Oklahoma, where there is relatively less publicly owned land and the ranges are mostly fenced. Unlike range band sheep operations, fenced range producers do not normally utilize on-site herders (Ensminger, 2002).

Farm flock operations (Box 1 in Figure 1-5) are found throughout the United States but predominate in the Midwest and in the East. Farm flocks are typically found on confined, higher-quality pastures, with considerably smaller flock sizes than found in range operations. No state east of the Mississippi has more than 50 head in average flock size. In California, flock sizes are much larger than in the eastern states but are raised in both confined pasture conditions and extensive range conditions.

The range flocks are generally much larger than the more confined production systems, with the exception of California. Average flock inventory per operation as of January 1, 2005, exceeded 200 head in only a few states: Wyoming (478), Arizona (438), California (262), Nevada (250), Colorado (225), Idaho (200), and New Mexico (200) (USDA, 2007a). Range operations in the more arid rangeland states have little capacity to finish lambs for harvest on the ranches where the lambs are dropped. These lambs generally move to large-scale feedlot systems for finishing on high-quality rations or to high-quality pasture for finishing before harvest. About half of the U.S. lamb crop comes from range operations.

Lambs are normally born in the spring months and remain on pasture throughout the summer and early fall with their dams in the majority of range production systems. Fall lambing is the norm in California, Arizona, parts of Texas, and the southeastern United States. Most of the lambs are born in the fall and pastured through the winter months before moving into the light lamb trade or as feeder lambs into feedlots or high-quality pasture for finishing. Most farm flock operations lamb in the late fall and winter months, when farming activities are minimal. While it is possible to use hormones and management to control the estrus cycle in ewes to have a crop of lambs more frequently than once per year, this practice is not common in the commercial U.S. lamb flock.

Lamb Marketing Channels

There are three distinct marketing channels for the U.S. lamb crop: (1) the traditional lamb marketing channel, (2) the early harvest lamb channel, and (3) the direct marketing channel.

The Traditional Lamb Marketing Channel

In the traditional lamb marketing channel, lambs move from range and farm flock operations to higher-quality feeding systems (pasture and/or feedlot) at 23–32 kg, and are grown to about 64 kg at harvest (Box 3 of Figure 1-5 and Boxes 4 and 5 of Figure 1-6). From discussions with industry representatives during this study, about three-quarters of all lambs produced in the United States pass through this marketing channel.

Range lambs weigh about 25–40 kg in the fall as they come off pasture, at which time they are weaned and conditioned for sale through auction markets or direct sales to feedlots throughout the western United States (Box 3 in Figure 1-5 and Box 5 in Figure 1-6). These lambs are fed to an average weight of 64 kg at harvest. In some cases, the feeding operations are owned by ranchers who feed out their own and/or additional lambs, either by custom feeding or purchase of feeder lambs from other ranchers. In other cases, the larger-scale packers own and operate their own feedlots to ensure continuity of supply and quality of finished lambs. Major auctions for feeder lambs take place throughout the central United States, with the largest markets serving all aspects of the sheep and lamb trade located in San Angelo, Texas.

From the smaller confined systems, lambs may move by auction or direct sale to feedlots for finishing, although they are often finished on the same farm where they were born. Many of the smaller confined flocks represent an ancillary operation on farms along with other large enterprises or off-farm employment, filling a niche either by using available pasture not needed for the other operations or as a sideline where the primary family income is generated from off-farm employment. In the smallest of these operations, lambs are often sold as live animals directly to customers at the farm and may be dressed by the customer before leaving the farm (Box 5 in Figure 1-5 and Box 7 in Figure 1-6).

Not all lambs end up in feedlots. Many lambs are forage finished (Box 1 in Figure 1-5 and Box 4 in Figure 1-6) and sold directly off pasture for harvest (Box 4 in Figure 1-5 and Box 6 in Figure 1-6). Higher-quality feeder lambs, weighing 23–36 kg, are regularly direct marketed through brokers and buyers to the light lamb markets for commercial harvest, most often for consumers in the large urban areas of the East coast, and increasingly in the urban areas of Chicago, Detroit, Houston, Florida, and California (Box 5 in Figure 1-5 and Box 8 in Figure 1-6). Additionally, there is a market for lambs and mature animals across a wide range of weights sold directly to consumers at the farm gate. Ewe lambs are retained as replacement ewes equivalent to 20 percent of the mature ewe flock each year, both in range and farm flock operations. Finally, 4-H Club members represent another source of demand for commercial lambs for meat, as well as purebred and crossbred replacement lambs.

In the meat component of the traditional marketing channel, lambs move directly from farms and feedlots to harvesters principally through auction markets and contract arrangements with growers (Box 4 in Figure 1-5). Carcasses may be further broken and processed at the harvest facility or moved to breakers and further processors in other locations for further processing (Box 6 in Figure 1-5). The outputs from these facilities represent the meat-case ready consumer products (Box 7 in Figure 1-5), prepared food products containing lamb (Box 8 in Figure 1-5), offals (Box 9 in Figure 1-5), and pelts (Box 10 in Figure 1-5). The meat-based food products from the processing industry move directly to retailers and the hotel, restaurant, and institutional food markets and for distribution to consumers (Boxes 10 and 11 of Figure 1-6). Offals generally move to renderers for disposal or additional processing, generating industrial and some consumer-level products (Box 11 in Figure 1-5). Processed pelts are either sold domestically into several industries including auto and leather goods (Box 11 in Figure 1-5) or exported, mostly to Asia and Europe (Box 12 in Figure 1-5).

Another product coming from farm flock and range operations is cull mature animals. These animals, after four to six breeding cycles, are sold for meat (mutton). A large portion of the cull animals historically has gone to Mexico as live animals, although this trade diminished substantially from 2003 to 2005 due to trade restrictions following the identification of BSE in cattle (Box 13 in Figure 1-5) and the increased retention of older ewes for breeding. The export of live animals began to recover in 2006. Cull animals are also harvested and processed in the United States, generating meats mostly for inclusion in prepared foods, as well as generating pelts and offal for further processing (Box 14 in Figure 1-5). A small share of lamb and mutton food products is exported (Box 15 in Figure 1-5).

Breakers and further processors also rely on growing imports of lamb carcasses and cuts for their supply (Box 16 in Figure 1-5). The retail and foodservice sector also directly imports consumer-ready lamb and prepared food products containing lamb (Box 17 in Figure 1-5). About one-half of the 170.5 million kilograms (retail weight equivalent) U.S. lamb and mutton supply in 2006 came from imports, almost all from New Zealand and Australia (USDA, 2007c). While lamb imports have been growing, domestic production has been declining, leaving annual domestic per capita consumption stable at about 0.5 kg in recent years, considerably below the annual domestic per capita consumption levels of broilers (39.5 kg), beef (29.9 kg), pork (22.4 kg), and turkey meat (7.7 kg) in 2006 (USDA, 2007d).

The Early Harvest Lamb Channel

The second marketing channel for lambs involves the movement of better-quality lambs, mostly 23–36 kg in weight, coming from range and farm flock production systems and moving to commercial and noncommercial harvest for direct sale to consumers (Box 5 in Figure 1-5). These lambs represent meatier, more heavily fleshed animals, raised on higher-quality pasture and forage than their counterparts raised on lower-quality range forage. Few official data systems for lambs provide detailed information on this market. According to two active participants in this market, lambs are continuously sourced across all major markets in the United States for transport to small abattoirs near large urban centers, such as Newark and New York City (committee discussions on April 12, 2007 with Sam Ferarra, New York City, and Susie and Omar Mady, American Halal Fresh Meats, Newark, NJ). One of the participants reported handling up to 2,000 head per week throughout the year. Many early harvest lambs are sourced and harvested on behalf of consumers and, hence, fall outside official data collection requirements on lamb movements and harvest. Lamb from these operations is most often moved as whole carcass to small meat shops and farmers' markets that cater to ethnic, religious, and high-end consumers. Rarely do the products from these operations enter the large grocery retail or restaurant chains. In many cases, the relationships between the broker, harvester, meat shop, and consumer are of long standing, based on a range of attributes including family, ethnic, religious, and local customs. Based on anecdotal evidence, which is the only information currently available on this growing marketing channel, the early harvest lamb market is quite active and growing, expanding beyond the traditional East and West Coast urban areas to other major U.S. urban centers.

The Direct Lamb Marketing Channel

The third lamb marketing channel is the direct sale of live lambs at the farm gate to individual consumers. In some cases, the lamb is dressed at the farm by the buyer. There are virtually no data on these operations because most fall below the minimum requirements in farm sales for reporting and most represent ancillary operations in rural areas in relation to other farm enterprises or off-farm wage employment. Again, anecdotal evidence suggests that the buyer-seller relationship is often one of long standing, with customers spreading the knowledge of the existence of sellers and the quality of the product primarily by word of mouth. With virtually no direct evidence in official statistics on this lamb marketing channel, the importance, size, and growth of this market channel are difficult to determine. Although USDA provides estimates of on-farm harvest of lambs, the figures indicate a continuing decline in numbers while market participants report a growing market. These latter two lamb marketing channels are further explored in Chapter 7.

The Effect of Lamb Pricing on Lamb Marketing

The pricing of lambs of all weights reflects the pattern common in the cattle industry. Even with most calves born in the spring, a steady supply of beef is maintained throughout the year. By earlier entry to feedlots, or by delayed entry and feeding to heavier weights, the production industry can spread the resulting product across the entire year. However, this process is more difficult for the lamb industry. While cattle arrive at harvest from 18 to 24 months on average, lambs have only up to 5 to 14 months from birth to harvest. Fall- and winter-born lambs assist in smoothing out the supply of lambs throughout the year, providing feeder and finished lambs offset from the lamb production in the rest of the country.

As with cattle, the prices of feeder and finished lambs throughout the year are closely linked. Consider the following simple example. Suppose a feeder lamb of 30 kg is purchased with the intent of feeding it to 60 kg as a finished lamb. If the expected price for the finished lamb is $2.00/kg and the total or "all-in" cost of each kilogram of gain is $1.60/kg, then the maximum price in the market for the feeder lamb must be less than $2.40/kg if the finisher expects to make money. The lower (higher) the cost of each kilogram of gain, the higher (lower) the price for the feeder lamb will tend to move. If the cost of a kilogram of gain on a light lamb of 20 kg is less than the cost of a kilogram of gain on a 40-kg lamb, then the lighter the feeder lamb, the higher its price will tend to be in the market relative to the price of the heavier feeder lamb. This relationship is demonstrated clearly by O'dell et al. (2003).[2]

The first two lamb marketing channels interact directly in the markets and pricing for feeder lambs. The traditional demand by lamb finishers dominates this market although the early harvest lamb marketing channel competes directly for the higher-quality feeder lambs ready for immediate harvest. Because of the demand by immediate harvesters for feeder lambs, this component seems to establish the top prices in the market with the poorer-quality lambs needing more time on high-quality feeds to make a quality product, establishing the average prices in the market.

With these pricing arrangements, the recent increase in corn prices is of

[2]See Figure 17 in O'dell and related discussion in particular.

considerable interest. Corn is by far the most common feed grain for finishing lambs (as is the case for cattle and swine) in the United States. Based on the increased price of corn, the value of feeder lambs can be expected to weaken. Thus, an increasing share of feeder lambs is likely to move to high-quality pasture for finishing rather than to feedlots, so long as there is not a concomitant increase in pasture cost. Given the recent forecast for a long-term rise in grain prices in a study conducted by the Organisation for Economic Co-operation and Development and the Food and Agriculture Organization (OECD-FAO, 2007), considerable pressure for change in lamb marketing channels is likely. The demand for feeder lambs for immediate harvest can be expected to strengthen because of lower feeder lamb prices with fewer feeder lambs going to feedlots.

The Wool and Pelts Marketing Channel

For almost all breeds of sheep, wool is a product from farm and ranch sheep operations. Mature animals are sheared annually in the spring months with the wool moving into raw wool processing through auctions, cooperatives, and private brokers (Box 18 in Figure 1-5). Little wool is graded, sorted, and baled at farm level, although the percentage is increasing in the larger western range flocks. A considerable proportion of the wool produced is exported, primarily to China, India, and Germany, where it is fabricated into textile products, either as pure wool or blended with other fibers (Box 19 in Figure 1-5). These products enter clothing, carpet, and industrial product markets.

A large share of the wool products consumed in the United States is imported, both as raw and processed wool as well as finished clothing and other products containing wool (Box 20 in Figure 1-5). Over half the U.S. wool production is exported and U.S. imports of wool are roughly equal to domestic production (USDA, 2006). The proportion of U.S. wool production sold as exports is increasing, while wool imports are decreasing, indicating a declining level of wool textile and carpet production in the United States (Box 21 in Figure 1-5). For example, U.S. exports of wool in 2003 were double the export level in 2000, while imports of raw wool fell by 25 percent over the same period. Chapter 5 provides more information on the development and current status of the wool industry.

A small market exists for cottage industry spinning and weaving for locally produced and specialty wools. While tiny in comparison to the traditional U.S. wool market, this market provides some evidence of an ongoing transition of the U.S. wool industry as a whole. See Chapter 7 for more detail.

Sheep and lamb pelts are also produced from the harvest of sheep and lambs (Box 10 in Figure 1-5). Pelts are the hide with unshorn wool. These pelts enter industrial markets and some are used in consumer products (Box 11 in Figure 1-5). Other pelts are exported, primarily to Asia and Europe (Box 12 in Figure 1-5).

The Dairy Sheep Products Marketing Channel

Finally, there is a small but growing dairy sheep industry in the United States, based mostly on breeds of sheep with good milk production efficiency (Box 22 in Figure 1-5). Farm flocks for milking sheep are located primarily in Wisconsin, Minnesota, New York, and Pennsylvania, although new flocks are being established in a number of areas. The ewes begin milking as early as 48–72 hours after lambing, and the lambs are fed milk replacements. In some cases, partial milking begins soon after lamb drop and may continue for a month before full milking begins.

Sheep milk cheeses and yogurts are sold at the farm/cooperative level as well as at farm markets and small outlets in many cases. Some of the emerging cooperatives are now selling through the large retail chains. The United States is the largest importer of sheep milk cheeses in the world, largely from Europe, which are mostly sold through the large retail chains to consumers. The U.S. Department of Agriculture provides no data on this emerging industry. Chapter 6 provides a review of the dairy sheep industry in the United States.

THE U.S. SHEEP AND LAMB INDUSTRY IN TRANSITION

The long-term decline in sheep inventories is the dominant feature of the U.S. sheep industry to most observers. Although continuing declines can be expected in many areas of the industry, there are some reasons for optimism. The decline in animal numbers seems to have slowed substantially and even reversed in some regions for a number of reasons, as discussed in subsequent chapters. Furthermore, new and different products are emerging from this industry, such as dairy sheep products. Sheep and lamb marketing channels are changing as well. The traditional marketing channel may expect further declines in volume while new and different marketing channels are emerging and growing stronger. This transition in the industry is relatively new and will take several years to be fully realized.

REFERENCES

ABARE (Australian Bureau of Agriculture and Resource Economics). 2007. Australian Commodities 07.3, Table 22, *Crop areas and livestock numbers,* www.abareconomics.com/interactive/ac_sept07/excel/table _22.xls. Accessed November 12, 2007.

ABS (Australian Bureau of Statistics). 2007. Selected agricultural commodities, catalogue nos. 7112.0, 7113.0, and 7121.0. Canberra, Australia.

Apple, R. W. 2006. Much ado about mutton: But not in these parts. New York Times, March 29, 2006.

ASI (American Sheep Industry Association). 2002. Sheep Production Handbook. Englewood, CO: American Sheep Industry Association.

Ensminger, M. E. 2002. Sheep and Goat Science, 6th edition. Danville, IL: Interstate Printers and Publishers.

Food and Drink. 2002. Meat packing idea. Online at: http://www.foodanddrinkeurope.com/news/ng.asp?id=15507-meat-packing-idea. Accessed November 12, 2007.

Food and Pack. 2002. Shelf-stable meat. VATIS, Food processing: packaging. Online at: http://www.techmonitor.net/techmon/03may_jun/fpr/fpr_package.htm. Accessed November 12, 2007.

Garcia, K. L. 2004. World War II in microcosm: One soldier's experience. MA Thesis, Angelo State University, p. 48.

Hawthorne, D. W. 2004. The history of federal and cooperative animal damage control. Sheep Goat Res. J. 19:13–15.

O'dell, D., D. J. Marshall, D. Singh, G. F. Plaugher, P. E. Lewis, E. K. Inskeep, and D. Smith. 2003. Final report: West Virginia lamb marketing information project. Prepared for the West Virginia Department of Agriculture, Marketing and Development Division, Charleston, WV.

OECD-FAO (Organisation for Economic Co-operation and Development and the Food and Agriculture Organization of the United Nations). 2007. OECD-FAO agricultural outlook, 2007–2016. Online at: http://www.oecd.org/dataoecd/6/10/38893266.pdf. Accessed November 10, 2007.

RBA (Reserve Bank of Australia). 2007. Exchange rates. Online at: http://www.rba.gov.au/Statistics/Bulletin/F11hist.xls. Accessed November 10, 2007.

RTI (Research Triangle Institute International). 2007. Livestock and Meat Marketing Study. Volume 5: Lamb and Lamb Meat Industries. RTI Project Number 0209230. Final report prepared for the U.S. Department of Agriculture Grain Inspection, Packers and Stockyards Administration, Washington, DC, February.

Shelton, M. 2004. Predation and livestock production: Perspective and overview. Sheep Goat Res. J. 19:2–5.

SNZ (Statistics New Zealand). 2006. Agricultural production statistics. Various Issues. Assembled from databases online at: http://www2.stats.govt.nz/domino/external/pasfull/pasfull.nsf/0/4c2567ef00247c6acc256e83007303b5/$FILE/Part%202_01%20-%202_36.xls; http://www.stats.govt.nz/NR/rdonlyres/FD1CE3D6-10AB-44DD-9267-6822AA3D787E/0/agriculturalproductionyejun06hotp.pdf; and http://www2.stats.govt.nz/domino/external/pasfull/pasfull.nsf/0/4c2567ef00247c6acc256fe2001333dd/$FILE/alltabs.xls. Accessed November 12, 2007.

USDA (U.S. Department of Agriculture). 1999. Economic Impact of the Elimination of the Wool Act. Economic Research Service. Report prepared for the House Committee on Appropriations, as requested in House Report Number 105–588, p. 16 (1998). Online at: http://www.ers.usda.gov/Briefing/Sheep/WoolActStudy.pdf. Accessed November 10, 2007.

USDA. 2006. Cotton and Wool Yearbook 2006. Economic Research Service. Online at: http://usda.mannlib.cornell.edu/MannUsda/viewDocumentInfo.do?documentID=1282. Accessed November 10, 2007.

USDA. 2007a. Food availability spreadsheets: Beef, pork, lamb and mutton, and total red meats. Online at: http://www.ers.usda.gov/Data/FoodConsumption/spreadsheets/mtredsu.xls#lamb!a1. Accessed November 10, 2007.

USDA. 2007b. Quick Stats. National Agricultural Statistics Service. Online at: http://www.nass.usda.gov/QuickStats/index2.jsp. Accessed November 10, 2007.

USDA. 2007c. FASonLine: U.S. trade, FATUS commodity aggregations. Online at: http://www.fas.usda.gov/ustrdscripts/USReport.exe. Accessed November 10, 2007.

USDA. 2007d. Livestock, dairy and poultry outlook, October 22, 2007. Online at: http://usda.mannlib.cornell.edu/usda/ers/LDP-M//2000s/2007/LDP-M-10-22-2007.pdf. Accessed November 10, 2007.

USDA/NASS. Several years. Chapter VII: Statistics of cattle, hogs, and sheep. National Agricultural Statistics Service. For 1998 as an example, see: http://www.usda.gov/nass/pubs/agr98/acro98.htm. Accessed November 26, 2006.

Williams, G. W., O. Capps, Jr., R. Dietrich, J. W. Edwards, R. A. Field, H. L. Goodwin, Jr., D. B. Griffin, K. Litzenberg, J. P. Nichols, J. W. Savell, G. Smith, D. Tatum, and J. B. Ward. 1991. Assessment of marketing strategies to enhance returns to lamb producers. Commodity Research Report No. CM-1–91. Texas Agribusiness Market Research Center. Texas A&M University, College Station, TX.

2

The U.S. Live Sheep Industry

An evaluation of the current economic status of the overall U.S. sheep industry must begin with an examination of the live sheep component that anchors the U.S. sheep industry value chain as outlined in the previous chapter. The live sheep component of the supply chain encompasses all functions and processes required to raise and feed sheep and lambs and deliver them to packers for slaughter. The process begins with breeding, which generally occurs only during specific times of the year. The typical biological cycle results in the majority of lambs being born in the spring. After weaning, some lambs intended for specific markets for young, lighter-weight lambs are sent directly to slaughter. Other lambs are put on forages to increase frame size and body weight, before being sent to feedlots and placed on grain-based rations. The remaining lambs are finished on high-quality forages. Finished lambs, also known as slaughter or fat lambs, are sent to packers where they are slaughtered, and the pelts and offal are separated from the edible products.

Although affected by a wide variety of forces relating to demand, policy, trade, price, and much more, the economic condition of the downstream components of the sheep value chain is most critically dependent on the economic fortunes of the live sheep component of the chain. Changes in sheep production technology and pricing, or health issues, for example, ripple all the way downstream from live sheep markets through the various components of the value chain. Consequently, this chapter evaluates the current status of the live sheep industry with a focus on the primary factors driving the industry with the exception of factors relating to sheep health, which are examined in detail in Chapter 3. Following an examination of live sheep production, feeding, trade, pricing, and policies in the United States, the chapter concludes with a summary discussion of the major accomplishments and future opportunities as well as the key challenges facing the live sheep component of the U.S. sheep industry.

U.S. SHEEP PRODUCTION

In the United States, sheep and lambs are raised primarily in small farm flocks in the Midwest and the East and on larger ranching operations in the West. Market lambs are the primary source of income for sheep producers, with some additional income being generated from the sale of wool and cull ewes and rams.

Sheep Inventory and Operations

The five largest states in terms of sheep inventories are Texas, California, Wyoming, Colorado, and South Dakota (inventories by region are shown in Table 2-1). Over one-third of all sheep (34.1 percent) are found in the mountain-range states of Colorado, Montana, South Dakota, Utah, and Wyoming. Together, these states, along with Texas and New Mexico and the western states of Arizona, California, Idaho, Nevada, Oregon, and Washington, account for nearly 70 percent of all U.S. sheep and lambs, but only 37.1 percent of U.S. sheep operations, indicating clearly that the U.S. sheep industry is still primarily dominated by range sheep production systems. An estimated 25 percent of the national sheep inventory spends a significant portion of the year grazing on western public land permits managed by the Forest Service of the U.S. Department of Agriculture (USDA) and the Bureau of Land Management (BLM) of the U.S. Department of the Interior (USDI) (ASI, 2002). In contrast, midwestern and eastern states account for roughly 30 percent of all sheep and lambs but nearly two-thirds (62.9 percent) of all sheep operations, indicating that more intensive smaller farm flock production systems are the norm in these two regions (Table 2-1).

Productivity (lambs produced per 100 ewes) is much higher in the confined, intensive systems of the Midwest and East (Table 2-1). The lower level of productivity in the range states is primarily due to their extensive, low-input production systems and higher predator losses. The Mountain states have a higher level of productivity than the other range states primarily because they still utilize shed lambing systems.

Approximately half of all U.S. sheep are found on farms with fewer than 500 head of sheep and the other half on farms with more than 500 head (Table 2-2). Even though sheep numbers are evenly divided between large and small operations, the latter account for most of the sheep operations in the United States (98.4 percent). Consequently, only about 1.6 percent of all

TABLE 2-1 U.S. Sheep Inventories, Productivity, and Operations by Region, January 1, 2006

	All Sheep and Lambs		Breeding	Market	Lambs per	Sheep Operations	
	Number	Share	Sheep	Sheep	100 Ewes	Number	Share
Region[a]	(1,000 head)	(%)	(1,000 head)	(1,000 head)	(head)	(1,000)	(%)
1	1,280	16.5	800	480	104	9,060	13.5
2	1,470	18.9	980	490	83	8,000	11.8
3	2,650	34.1	1,460	1,190	121	8,000	11.8
4	1,200	15.4	665	535	135	14,270	21.1
5	1,170	15.1	730	440	125	28,250	41.8
Total U.S.	7,770	100.0	4,635	3,135	112	68,280	100.0

[a]States included in regions as follows: 1 = Arizona, California, Idaho, Nevada, Oregon, and Washington; 2 = New Mexico and Texas; 3 = Colorado, Montana, South Dakota, Utah, and Wyoming; 4 = Iowa, Kansas, Minnesota, Missouri, Nebraska, North Dakota, and Oklahoma; 5 = All other states.
Source: USDA (2007g).

sheep operations account for half of all U.S. sheep inventories. These larger operations are almost exclusively range sheep operations in the western part of the United States (Figure 2-1).

Industry sources suggest that sheep and lamb numbers may be as much as 10 percent higher than reported by USDA.[1] Although not well documented, this view is widely held in the industry and may be due to the following:

- National Agricultural Statistics Service surveys include operations with farm income of at least $1,000. An increasing number of small-scale sheep flocks exist on farms with little or no earnings from their sheep operations or do not report the small income earned from sheep for various reasons.
- Animal slaughter is reported for inspected slaughter plants where animals and carcasses are individually inspected. Sheep and lambs may be slaughtered at home or in local noninspected plants. Custom slaughter for specialty markets such as halal and kosher slaughter in inspected plants may not be reported if the carcasses are not individually inspected, which is not required for custom slaughter.

[1]Committee members interviewed several sheep producers, livestock market representatives, lamb order buyers, specialty lamb market processors, wool handlers and processors, sheep product vendors at various levels, and others on an informal basis as part of this project to obtain industry views on the status of the U.S. sheep industry.

TABLE 2-2 U.S. Sheep Inventories and Operations by Flock Size, January 1, 2006

	Flock Size (head)			
Item	1–99	100–499	500–4,999	5,000+
Inventories (%)	28.7	22.0	33.8	13.5
Operations (%)	90.8	7.6	1.5	0.1

Source: USDA (2007g).

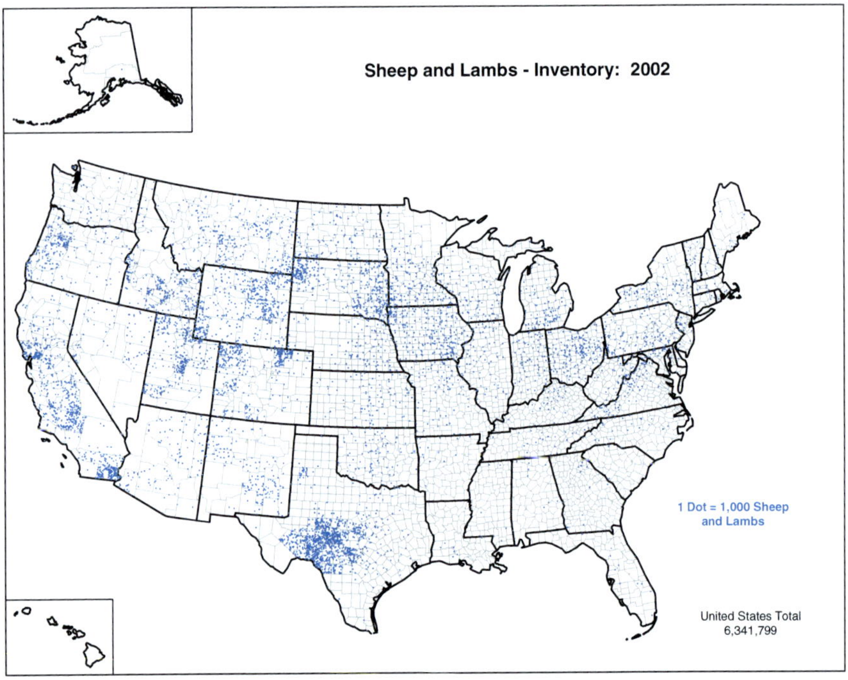

FIGURE 2-1 Geographic distribution of U.S. sheep and lamb inventories, 2002. Source: USDA (2002).

- The number of suburban small farms of 1–5 acres is increasing. Many of these farms may have 5–10 sheep as "lawn mowers" or "weed eaters." Also, they may have sheep to maintain their agricultural property tax exemption. Lambs produced from these flocks may be primarily used for home consumption or bartered for other goods or services.

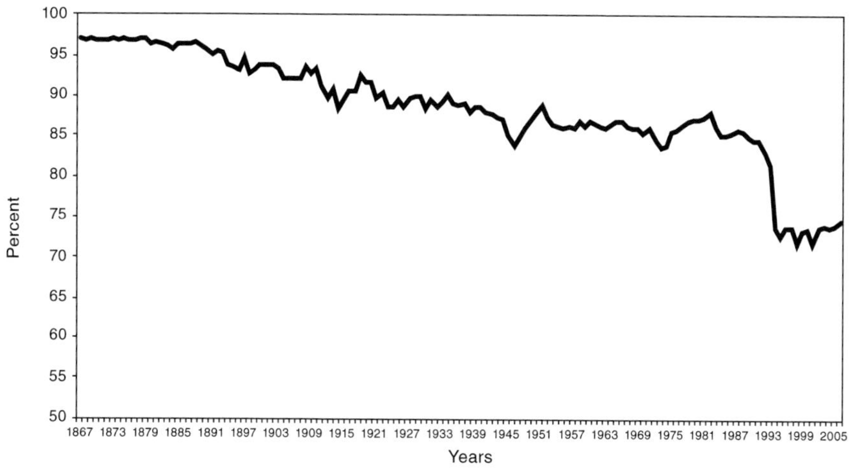

FIGURE 2-2 Breeding stock as a percent of total sheep and lamb inventories, 1867–2007.
Source: Calculated from data in USDA (2007g).

- Youth 4-H or Future Farmers of America sheep projects may be viewed by farm owners as youth income, rather than farm income, and may or may not be reported.

Sheep and lamb inventories include breeding sheep (ewes, rams, and replacement lambs) and nonbreeding sheep and lambs. Breeding stock as a percentage of total sheep and lambs has been declining over the years from about 97 percent in the late 1800s to a low of 72 percent in 2001, reflecting the increasing productivity of U.S. sheep breeds and a general trend toward disinvestment in sheep production (Figure 2-2). The phase-out of wool price supports under the Wool Act between 1992 and 1994 led to a major sell-off of breeding stock as the profitability of wool production declined sharply. As a consequence, the share of total inventories accounted for by breeding stock tumbled from around 85 percent in the early 1990s to 73 percent in 1995 (Figure 2-2). The breeding stock share of inventories has stayed at about that same level since that time.

Transition in the Regional Distribution of Sheep Inventories

A dominant characteristic of the sheep industry since at least World War II has been the steady decline in sheep and lamb inventories (see Chapter 1).

Focusing on the change in aggregate numbers over time, however, conceals an important recent phenomenon in the industry that is changing the traditional national pattern of sheep production. A simple trend analysis of the sheep inventory data indicates that the decline in sheep numbers has slowed significantly and even reversed in some regions of the country.[2] Analyzing the two periods of 1989–2000 and 2001–2007, the compound annual rate of change in sheep and lamb inventories by state was calculated and compared in the two time periods and across all states for which there were continuous data series. The remaining states were aggregated as a single region. For the 1989–2000 period, the estimated coefficients of the time trend variable for all states and the total U.S. are negative, and strongly so in many cases, indicating a continued negative rate of change in inventories during that period.

For the more recent years of 2001–2007, however, the results differ substantially, indicating that many states are no longer in decline in sheep and lamb inventories. Figure 2-3 shows a comparison of the rates of change in inventories by state over the two periods. In four states (California, Utah, North Dakota, and Nevada), the rates of decline are larger for the more recent period of 2001–2007 than for the 1989–2000 period. As well, New Mexico shows little change in the rate of decline between the two periods. These five states represented 19.6 percent of total U.S. sheep and lamb inventories on January 1, 2007. For most states with larger shares of the national flock, the rates of decline have abated substantially. The United States as a whole appears to be still in decline for the period, although the rate of decline in inventories has slowed from over 4 percent to less than 2 percent annually. In addition, over the last 4 years (2004–2007), total U.S. inventories have been relatively stable.

Several states showed positive rates of inventory change for the 2001–2007 period, including Pennsylvania, Wisconsin, Michigan, Oklahoma, Missouri, New York, Virginia, and the "Other States" category in Figure 2-3. This group of states represents over 16 percent of U.S. sheep and lamb inventories. For all other states, the annual rates of change in 2001–2007 were less than half the rates during the 1989–2000 period. In the case of Texas, for example, the rate of decline in inventories during 2001–2007 was –1.7 percent compared to –4.6 percent during the earlier period. Although the smallest in terms of sheep and lamb inventories, the Other States category, which includes those states for which continuous data were not

[2]The analysis used a simple time trend, semi-log regression to calculate the rates of change in U.S. sheep inventories by state for the two periods of 1989 to 2000 and 2001 to 2007. The regression model was: $\log Y_{ij} = \alpha_i + \beta_i X_{ij} + \varepsilon_{ij}$ where Y_{ij} = January 1 inventory of all sheep and lambs for state i in year j and X_{ij} = time trend (1 for first year, 2 for second year, . . .) for state i in year j.

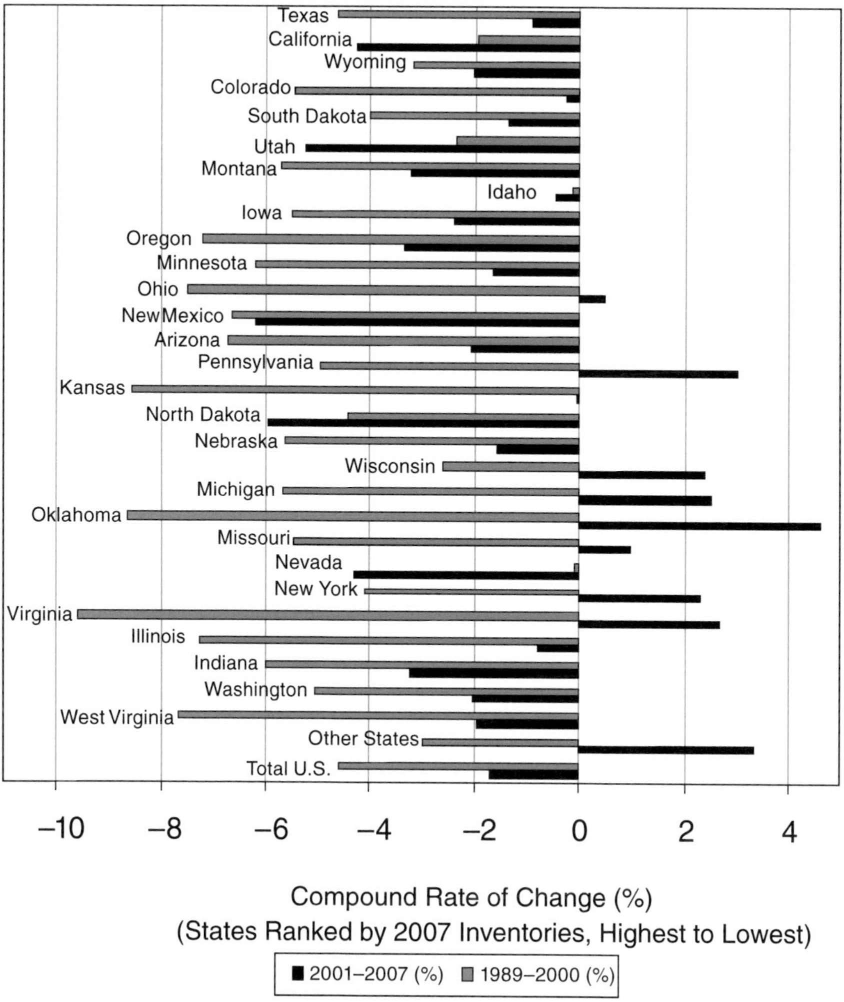

FIGURE 2-3 Rates of change in sheep and lamb inventories by state, 1989–2000 and 2001–2007.

available for the full period of 1989 to 2007, exhibited the second fastest inventory growth during the more recent years.

Clearly, considerable change is taking place in growth rate and location of U.S. sheep and lamb inventories. While inventories in some states continue to erode, those in many other states have now halted their long-term decline and show modest growth in aggregate terms. The last few years may be the beginning of a transition period in which growth in farm flocks is coupled with a slower overall decline in range sheep flocks, contributing to a more stable level in the national flock than has occurred in the past 60 years.

A major contributor to this transition is the increase in hair sheep production in Texas and the farm flock states. These states have also experienced increases in meat goat production. To some extent, these transitions have been due to substitutions of hair sheep and meat goats for wool-producing sheep breeds and mohair goats, but some farms have added hair sheep and/or goat enterprises. Benefits of hair sheep and goats include their adaptability to humid climates, elimination of the need for shearing (shearers are sometimes difficult to find), increased ethnic demand for lighter-weight slaughter lambs and goats, and their use in weed and brush control programs. Hair sheep and meat goat breed shows and exhibitions have also been significant contributors to this increase. However, there is concern that selecting these animals on the basis of subjective evaluation may be detrimental to selection for their desired "easy care" attributes.

The Profitability of Sheep Production

Decisions made by lamb producers determine the supply of feeder lambs each year and, ultimately, the supply of lamb meat. Each year, sheep producers assess the prices and net returns received from feeder lambs and decide whether and how many ewes to slaughter or retain as capital stock for breeding purposes. The decision to retain ewes for breeding indicates optimism on the part of producers about market conditions. By the same token, the decision to sell ewes for slaughter indicates that producers may not anticipate high enough prices to hold back a significant number of ewes for future production. Other factors, such as low rainfall and available forages, may also affect decisions on ewe retention.

Examples of production costs and returns are presented in Table 2-3 for a 3,000-ewe public land range sheep production operation in Nevada and in Table 2-4 for a more intensively managed 50-ewe farm flock enterprise in Wisconsin. Although there are numerous variations of both range and farm flock sheep production systems in terms of size and management, and profitability may differ with each variation and from year to year, these two examples suggest that sheep production has been profitable in recent years, on the average for operations of this size using the depicted management systems. The major difference in these examples is that labor and management costs are the largest expense for public range operations at 39 percent of all expenses, while family labor provides the labor for farm flock enterprises and is not considered an "out of pocket" expense. However, family labor has an opportunity cost for alternative economic activities that is not

TABLE 2-3 Enterprise Budget for Public Land Range Sheep Operation: 2006 Expense and Income Analysis for a 3,000-ewe Range Production System

Item	Per Ewe ($)	Total $
EXPENSES		
Labor (herder, plus support at lambing)	20.00	60,000
Management (owner oversight or employed manager)	10.00	30,000
Public land lease (10 months/year)	3.25	9,750
Public land maintenance and improvements	1.00	3,000
Facilities and equipment depreciation, operation, maintenance	9.00	27,000
Supplemental feed, minerals, private pasture management	8.00	24,000
Ram costs	3.00	9,000
Vaccination, medication, internal and external parasite control	5.00	15,000
Water hauling (varies from 1 to 4 months of year)	2.00	6,000
Shearing	3.00	9,000
Marketing lambs, wool, cull ewes	4.00	12,000
Guardian and herding dog expenses	1.00	3,000
Operating capital interest	2.00	6,000
Predator control (direct expenses, contributions to state programs)	2.00	6,000
Miscellaneous, including dues and subscriptions	2.00	6,000
Total Expenses	$75.25	$225,750
INCOME		
0.95 lamb sold/ewe × 40.72 kg/lamb × $2.21/kg	85.50	256,000
4.53 kg wool/ewe × $3.09/kg	14.00	42,000
0.15 cull ewe × 59.09 kg × $0.66/kg	5.85	17,550
Wool Loan Deficiency Payment @ $0.33/kg	1.50	4,500
Total Income	$106.85	$320,550
Return on investment	$30.60	$91,800
Investment costs (animals, equipment, facilities, etc.)	$180.00	$540,000
Estimated return, investment, and risk (%)		17.0

Break-even cost per kilogram of lamb sold (wool and cull ewe income constant) = $1.5465

Explanation of Budget:

- Of the 5 producers interviewed for this analysis (17,000 ewes total), lamb sales varied from 0.9 to 1.1 lambs per ewe at average weights from 38.6 to 47.6 kg per lamb at prices from $2.09/kg to $2.38/kg. Includes an average of 0.2 ewe lambs per ewe (20%) retained as replacements.
- Lambs sold and sale weights may vary ± 5% due to climatic conditions, predation, and seasonal forage quantity and quality. Lamb sale prices have been relatively stable for these producers, and wool prices have almost doubled during the last 3 years.

Source: Update of the Standardized Performance Analysis (SPA) project initiated by Oltjen and Glimp (1992).

TABLE 2-4 Enterprise Budget for a 50-ewe Farm Flock Sheep Enterprise: 2006/2007 Operating Costs and Income

Item	Per Ewe ($)	Total $
EXPENSES		
Feed		
Hay and grain	39.31	1,966
Salt and minerals	2.25	112
Supplemental feed for finishing lambs	20.16	1,008
Pasture maintenance	8.00	400
Health program		
Internal and external parasite control	8.96	448
Vaccinations	4.24	212
Other veterinary medications, services	5.20	260
Shearing	3.64	182
Ram replacement	6.00	300
Bedding straw	6.24	312
Marketing and transportation	7.95	398
Supplies	5.20	260
Manure disposal	7.00	350
Building maintenance	6.00	300
Interest on operating expenses	3.90	195
Total operating expenses	134.05	6,703
INCOME		
1.4 Lambs/Ewe @ 59.09 kg/lamb × $2.09/kg	172.90	8,645
Cull ewes and rams	16.53	826
Wool	2.50	125
Wool loan deficiency payment	1.25	62
Unshorn lamb pelt payment	1.44	72
Total operating income	$194.60	$9,730
Return on land, labor, and investment	$60.56	$3,027
Investment costs (animals, facilities, equipment, land improvements)	$338.00	$16,900
Estimated return to labor, investment, and risk (%)		17.9
Break-even cost per kilogram of lamb sold (wool and other income constant) = $1.6237		

Source: Thomas (2007). Copyright 2007 by David L. Thomas. Used with permission.

included in these calculations. If family labor costs are included, many small farm flocks may be only marginally profitable; however, a sheep enterprise may be a more economically competitive use of farm family labor than other possible enterprises.

Because most range operations up to 5,000 ewes are managed by the owner, the return on investment and management for those operations would be higher than that shown in Table 2-3 by $10 per ewe, or 18.9 percent, with a lower break-even price per kilogram of lamb sold of $1.4606/kg ($0.6625/lb). The major advantage to farm flock enterprises is their ability to sell heavier harvest-ready lambs and more lambs per ewe because of their higher feed inputs (52 percent of expenses) and low family labor costs. The public land range operation has the advantage of lower feed (15.5 percent of expenses) and facilities costs, and economies of scale such as lower veterinary and marketing costs per ewe. Increasing ewe productivity can be a major factor in enterprise profitability, but increased income from production increases must exceed cost inputs.

Variations in range sheep production systems include higher inputs of labor and facilities with shed lambing and supplemental feeding in most of the Intermountain states with higher-quality forage resources on summer rangelands that increase both number and weight of lambs weaned per ewe, to the extensive low-input private rangelands range sheep operations in western Texas and New Mexico that result in lower productivity. Although not documented in this analysis, profitability of these variations may be similar to the range enterprise presented in Table 2-3. Farm flock enterprises are the primary sheep enterprise in the eastern United States and are also quite common in the western states in irrigated valleys and in higher rainfall regions such as northern California, Oregon, and Washington. Farm flock sheep operations may vary from 5 to > 500 ewes with the norm being from 5 to 50 ewes. The major difference is that range flocks of > 1,000 ewes are considered an important business enterprise, while farm flocks of < 100 ewes may be considered a secondary enterprise on the farm.

Commodity cash receipts for sheep, lamb, and wool products at the farm level in 2006 were estimated at about $497.7 million, with wool accounting for about 5.1 percent of this amount (USDA, 2007a). In relation to all commodity cash receipts, sheep, lambs, and wool account for only 0.21 percent of commodity receipts for farms, and only about 0.42 percent of all animal/livestock receipts. In a few states, the share of total cash receipts for sheep, lambs, and wool is 1 percent or more, including Colorado (2.1 percent), Utah (1.5 percent), and Montana (1.0 percent).

Many factors impact the profitability of sheep production in the United States. Although U.S. wool prices are 20–30 percent below world market prices for comparable wool grades, prices have doubled since 2000 (see Chapter 5). Government programs, including the wool loan deficiency payment (LDP) program and payments for retained replacement ewes in 2004 and 2005, have supported profitability and encouraged industry expansion. A number of other external and internal factors contribute to industry profitability, among which the most salient are the following:

- *Scale of operation.* There are numerous advantages to sheep enterprises larger than 1,000 head. Input costs such as vaccines, drugs, transportation, and labor may be half the cost per animal for a 1,000-head flock compared to a 100-head flock. Perhaps as important is that producers with large flocks view sheep as a primary business enterprise and, consequently, strive for increased efficiency in their operations. Producers with small flocks, on the other hand, generally consider sheep as a secondary enterprise with lower returns to investments in productivity increases.

- *Distance and access to markets.* Transportation costs, feedlot and slaughter plant locations, and distance from major production areas to major areas of consumption can affect profitability. As inventories have declined and production has become more geographically dispersed over time, the numbers of auctions, feedlots, and slaughter plants have declined as well, while the average distance that animals must be hauled to market has increased. Small flock producers are particularly disadvantaged in their access to major auctions and traditional buyers.
- *Labor cost and availability.* Sheep producers are required to have herders with their sheep while their sheep are grazing on public lands. Labor is the largest expense of sheep production for these producers. Since the domestic availability of herders is low, U.S. Department of Labor H2A work visa permits have been used to bring herders in from other countries. With homeland security concerns, the time required to obtain a work visas has increased from 1–2 months to 3–6 months (WRA, 2007). A similar problem exists with obtaining temporary work visas for highly qualified sheep shearers who have historically come to the United States from Australia and New Zealand during their southern hemisphere winter off-season. Lack of available sheep shearers in states with lower sheep numbers has resulted in producers either quitting sheep production or shifting to hair sheep or meat goat breeds (see Chapter 5).
- *Feed costs.* Feed costs may account for over 50 percent of all expenses in intensive sheep production systems (Thomas, 2007). Increasing use of feed grains for ethanol production may impact lamb feedlot costs of gain. An increasing cost for feed could shift emphasis in finishing lambs for slaughter from feedlots to high-quality pasture and alfalfa and other crop aftermath grazing. Increased feed grain prices could also discourage lamb feeders from overfeeding lambs to excessive fat levels because of cheap gain costs. Higher feed grain costs could have a larger impact on the cost of poultry and pork production than on sheep production, which would improve the competitive retail margin of lamb relative to those of other meats products.
- *Wool.* Wool is now of minor importance to more intensive sheep production systems and may even be considered a liability within some breeds and locations. However, in range sheep production systems where limited feed resources can limit the weight of lamb weaned per ewe, wool is more important to the sheep enterprise. The sheep breeds best adapted

to range production systems with limited feed resources also tend to be the best wool-producing breeds (see Chapter 5).

Sheep Genetics and Breeds

Sheep are thought to have been domesticated over 10,000 years ago from captured animals from the wild breeds of sheep in central Asia. Since domestication, sheep have evolved through selection to meet humans' needs for meat, wool, and milk for cheese production. They have been selected to be adapted to specific environments as well as for specific product needs. The breeds adapted to range production systems, including the Rambouillet, Merino, Targhee, and Columbia breeds, have strong flocking instincts, meaning that they will generally be together, at least within sight distance of others in the flock when grazing and will bed down together at night. Breeds such as the Suffolk, Hampshire, Dorset, and Polypay, on the other hand, are adapted to more intensive pasture and crop production systems and generally do not have strong flocking behavior.

Genetics

Early selection was based primarily on desired phenotypic traits, depending on the needs and desires of the breeder and their communities and clientele. The 20th century was the major transition period to performance-based selection for desired traits. Advances in technology to improve selection for traits to improve growth and reproduction, for example, have resulted in the weight of lamb produced per ewe to more than double in the last 60 years. Lamb meat is now also leaner and more nutritious. Milk output by dairy sheep breeds for cheese production has experienced similar productivity increases. Although the wool yield per animal has not increased to the same extent as meat yield in the United States, wool productivity has advanced in other countries like Australia. All of the economically important wool traits are highly inheritable and can be improved more rapidly than most traits.

Production economics have, to a large extent, dictated selection emphasis within specific countries and environments. Adaptability to the environment in which the animal is expected to produce is still probably the most important economic trait. Multiple advances in environmental adaptability have been made, such as improvements in growth and reproductive traits in range breeds that evolved initially for wool production. Specialized sire breeds have been developed that permit crossbreeding all or a portion of the range flock ewes to increase the weight of lamb produced per ewe while maintaining the hardiness and the wool production of the foundation ewe flock. Hair breeds have been imported for use in the more humid and subtropical environments and in regions with inadequate number of sheep shearers, where wool may be a liability. Dairy sheep breeds, developed in the more traditional sheep milking regions of Europe, have supported the development of this new industry in the United States.

The sheep genome is being mapped and, when completed, will provide the foundation for future genetic improvement in the sheep industry. Gene markers for a number of important economic traits are being intensively investigated. A few have already been identified, such as the gene marker for superfine wool by New Zealand scientists and a group of markers that can be used to select for internal parasite resistance. An increasing number of gene markers will likely be identified in the future that will improve the industry's ability to accelerate genetic progress. The continued development of gene markers will require close working relationships between scientists, producers, economists, and investors to identify research priorities and to validate the potential markers identified. The application of this new knowledge will be essential to future global competitiveness of the U.S. sheep industry.

Sheep Breeds

The most recent edition of the *Sheep Production Handbook* lists 49 breeds of sheep in the United States (ASI, 2002). There are some recent imports, however, such as the Lacaune, Charollais, and Ile de France, that are not listed. Consequently, there are currently over 50 breeds of sheep in the United States, among which there is a large amount of variation in body size, appearance, and the amount and quality of meat, wool, and milk produced. The wide variety of breeds allows producers to choose the breeds that best meet the requirements of a specific production environment or market situation. However, the wide variety in breeds utilized in this country has resulted in wide variation in the type and quality of lamb and wool marketed.

Sheep breeds in the United States can be grouped according to principal function or use:

- *General purpose breeds.* These are breeds with a good balance between meat and wool production and adaptability to a reasonable range of environmental conditions. Such breeds may be the best choice for small flocks where the logistics of a crossbreeding program are not feasible. Some breeds in this group are Dorset, Montadale, and Polypay.
- *Maternal range breeds.* These are the predominant U.S. breeds in terms of total population numbers. In general, they are characterized by good adaptability to more difficult environments, above average fleece weight and quality, and longevity. Ewes of these breeds are found primarily in the range areas of the western United States, including the Merino, Ram-

bouillet, Targhee, and Columbia breeds, and crosses among these breeds. Maternal range breeds such as the Romney are also found in the coastal hill areas of the Pacific Northwest.

- *Prolific maternal breeds.* The Finnsheep and Romanov breeds set themselves apart from other breeds because of the very large numbers of lambs born to a ewe at each lambing. Many Finnsheep and Romanov flocks will average three or more lambs per ewe. In addition, these two breeds are noted for a very young age at puberty and for excellent newborn lamb vigor. These breeds are often crossed with general purpose and range breeds in order to produce crossbred ewes with greater lamb production for farm flock production systems.
- *Sire breeds.* The most important characteristics of these breeds, used primarily to sire crossbred market lambs, are large mature size, rapid growth rate, superior muscling, and lower carcass fat. The Suffolk and Hampshire are the dominant breeds in this group. Less common sire breeds include the Oxford, Shropshire, and Texel.
- *Dairy breeds.* These are breeds that are milked with machines as are dairy cows and dairy goats. The milk is processed into specialty cheeses and yogurts. Dairy sheep production is a relatively new industry in the United States (see Chapter 6). There were no specialized dairy breeds in this country until the 1990s. Two dairy breeds that excel for milk production have been imported from Europe: the East Friesian and the Lacaune.
- *Hair breeds.* These breeds do not produce wool but, instead, have a hair coat like that of cattle. They were developed in the humid tropics where a wool coat would be a disadvantage. They are adapted to a hot, humid climate, and tend to be more tolerant of internal parasites than many wool breeds. They are popular in the southeastern United States where wool sheep are not well adapted and among a growing number of flock owners throughout the United States who have responded to the low value for wool and the high cost of shearing. Hair breeds in the United States are Dorper, White Dorper, Katahdin, St. Croix, Barbados Blackbelly, and Royal White.
- *Specialized wool breeds.* These breeds are raised primarily to produce colored fleeces or fleeces with other unique characteristics desired by fiber artists, hand spinners, or weavers. Breeds in this group tend to be the long-wooled breeds (breeds that grow wool with a long fiber length). Longer-fibered wool is somewhat easier to spin than shorter-fibered wool. Some breeds in this group include Shetland, Icelandic, Lincoln, and Border Leicester.
- *Hobby or rare breeds.* The majority of the 50-plus breeds of sheep in the United States are not used in commercial production to any great extent and, therefore, contribute little to the national supply of meat, wool, or milk. The hobby breeds are found almost exclusively in small purebred

flocks and are generally raised for competitive exhibition. Owners of the elite flocks of these breeds sell a large portion of their sheep as breeding animals to other purebred breeders. The less successful breeders sell the majority of their animals as market animals. Although some hobby breeds are found in relatively large numbers relative to other breeds (e.g., Southdown and Cheviot), their primary reason for existing is still exhibition and not the production of food and fiber. Some breeds of sheep are recognized by the American Livestock Breeds Conservancy as rare (e.g., Romeldale, American Jacob, Cotswold, and Navajo-Churro) and in need of preservation (ALBC, 2007). While the hobby and rare breeds contribute very little to current supplies of sheep products, they may possess traits of importance to production in the future. Germplasm from several breeds of sheep, including both minor and major breeds, has been collected and preserved for possible future use through the National Animal Germplasm Program of USDA (USDA, 2008).

Crossbreeding is a common practice in the production of market lambs. In the western range flocks, only enough of the ewes of the maternal range breeds (Merino, Rambouillet, Columbia, or Targhee) or crosses among these breeds are usually mated to rams of these breeds to produce the number of replacement females needed to maintain flock numbers—about 40 percent of the ewe flock. The remaining ewes are mated to rams of the sire breeds (Suffolk or Hampshire) to produce more desirable market lambs. A similar crossbreeding system is often used in larger farm flocks but with ewes that will produce more lambs than the maternal range breeds. The ewe flock may be composed of general purpose breeds or crossbred ewes produced by mating the prolific maternal breeds of Finnsheep or Romanov with the maternal range breeds or general purpose breeds. In these farm flocks, 25 to 30 percent of the ewes must be bred to produce the replacement ewes and 70 to 75 percent of the ewes can be mated to Suffolk or Hampshire rams for the production of market lambs. One challenge with these crossbreeding systems is the poor fitness and lack of longevity among rams of the traditional sire breeds, which increases costs of producing commercial market lambs.

An increasing number of range producers are breeding all of their ewes to maternal range breed sires and becoming specialized producers of replacement ewes. These replacement ewes are sold to other range producers who prefer to purchase all their replacements ewes and breed all ewes to terminal sire breed rams for the production of market lambs.

Most breeds of sheep are represented by a national breed association. The primary purpose of the breed association is to maintain pedigree records on animals of the breed and to provide breeders and owners of these purebred animals with registration papers as proof of breed purity. These associations also promote their respective breed to the commercial sheep industry and potential new purebred breeders, set standards for acceptable breed type, and sponsor shows and sales for their breed.

An estimate of the number of registered purebred breeding sheep can be obtained by using the number of new purebred animals registered each year (an average of 81,474 for 2005 and 2006) (Table 2-5) and making some reasonable assumptions about the lamb production of registered ewes (1.20 lambs raised per ewe mated), the proportion of male lambs registered (20 percent), the proportion of female lambs registered (40 percent), and the ewe-to-ram ratio among purebred-registered breeding sheep (15:1). These assumptions yield an estimated U.S. population of registered purebred breeding sheep in 2005–2006 of 241,000, or about 5.3 percent of the total U.S. breeding sheep inventory. This relatively small population of registered purebred sheep has a large influence on the genetic potential of the U.S. flock since most of the ewes and lambs in commercial flocks are only one or two generations removed from a purebred registered ram.

A recreational purebred sheep industry has developed among the registered purebred flocks in which animals are valued more for their ability to win shows and less for their ability to efficiently produce meat, wool, or milk. Although especially the case for "hobby" breeds of sheep, many of the flocks raising registered purebred animals of the breeds commonly used in commercial sheep production also are involved in this recreational purebred sheep industry. While the use of purebred sheep primarily as show animals provides an interesting recreational activity for many adults and youth and a lucrative source of income for some breeders, the recreational use of purebred sheep provides little of benefit to the commercial sheep industry. Recreational purebred sheep are selected primarily on conformation traits that are not highly correlated with the efficient production of meat, wool, or milk. The genetics from many of these sheep eventually reaches commercial sheep flocks.

Table 2-5 presents the top 10 breeds among the number of purebred registrations in 20-year intervals from 1965–1966 to 1985–1986 to 2005–2006. In some cases, the numbers in Table 2-5 are not indicative of the relative number of animals of a particular breed in the U.S. sheep population. For example, there are more sheep of the maternal range breeds (Rambouillet in particular) in the United States than any other breed, but most are nonregistered purebreds, individuals of a high percentage of one of the maternal range breeds, or crosses among the maternal range breeds in large range flocks. The two maternal range breeds listed in Table 2-5 (Rambouillet and Columbia) never rank first in number of registrations but always follow the sire breeds of Suffolk and Hampshire, the primary breeds of rams used to sire commercial market lambs in the United States. Part of the reason for the high registration numbers for Suffolk and Hampshire

TABLE 2-5 Top 10 Sheep Breeds with Annual Number of Purebred Registrations at 20-year Intervals, 1965–1966, 1985–1986, and 2005–2006

		Average of 1965 and 1966[a]	
Rank	Breed[c]	No. of Registrations	% of Total Registrations
1	**Suffolk**	32,314	27.9
2	**Hampshire**	26,168	22.6
3	Corriedale	12,448	10.7
4	**Southdown**	8,986	7.7
5	**Columbia**	6,502	5.6
6	**Shropshire**	6,352	5.5
7	**Dorset**	6,070	5.2
8	**Rambouillet**	5,742	4.9
9	Cheviot	3,274	2.8
10	**Montadale**	2,718	2.3
Top 10 breeds		110,574	95.3
All breeds		116,011	100.0

		Average of 1985 and 1986[b]		
Rank	Breed[c]	No. of Registrations	% of Total Registrations	% Change from 1965–1966
1	**Suffolk**	65,133	44.7	101.6
2	**Hampshire**	16,325	11.2	–37.6
3	**Dorset**	13,672	9.4	125.2
4	**Rambouillet**	12,189	8.4	112.3
5	**Columbia**	7,756	5.3	19.3
6	**Southdown**	4,820	3.3	–46.4
7	Corriedale	4,477	3.1	–64.0
8	**Shropshire**	3,435	2.4	–45.9
9	Polypay	3,408	2.3	NA[d]
10	**Montadale**	2,764	1.9	1.7
Top 10 breeds		133,979	92.0	21.2
All breeds		145,638	100.0	25.5

relative to the maternal range breeds is their popularity as recreational show sheep, especially in market lamb or club lamb shows. However, changes over time in the number of registrations and rank in registrations of breeds are generally indicative of the popularity of that particular breed in commercial sheep production.

Total purebred registrations decreased by 34,537 (–29.8 percent) animals over the 40-year period from 1965–1966 to 2005–2006 (Table 2-5).

TABLE 2-5 Continued

Rank	Breed[c]	Average of 2005 and 2006[b] No. of Registrations	% of Total Registrations	% Change from 1965–1966
1	**Suffolk**	13,428	16.5	–58.4
2	**Hampshire**	8,784	10.8	–66.4
3	**Dorset**	8,729	10.7	43.8
4	Dorper	7,020	8.6	NA[d]
5	**Southdown**	5,793	7.1	–35.5
6	**Katahdin**	5,316	6.5	NA[d]
7	**Rambouillet**	3,472	4.3	–39.5
8	Columbia	2,950	3.6	–54.6
9	**Shropshire**	2,607[e]	3.2	–59.0
10	**Montadale**	2,064	2.5	–24.1
Top 10 breeds		60,163	73.8	–45.6
All breeds		81,474	100.0	–29.8

[a]Mead (1967).
[b]Deakin (2007).
[c]Breeds in bold type were among the top 10 breeds in registration numbers at all three time periods.
[d]Breed associations for the Polypay, Katahdin, and Dorper breeds were not present in 1965–1966.
[e]Number of registrations for the Shropshire breed was not available for 2006.
Sources: Mead (1967), ASI (2002), and Deakin (2007).

However, there was not a constant decrease over this period because the number of registrations halfway through this period in 1985–1986 was actually higher than in either 1965–1966 or 2005–2006. The percentage decrease in total sheep inventory during this 40-year period was over twice as great as the percentage decrease in purebred registrations. This suggests that registered sheep numbers were not reacting to the same factors or in the same manner as the total sheep inventory. Many purebred sheep are in small flocks with little potential for generating significant amounts of income or in flocks that are maintained for competitive exhibition and recreation. Such flocks are less affected by external economic factors than larger commercial flocks so it is not surprising that trends in purebred registration numbers and total sheep inventories differ.

Eight breeds (Suffolk, Hampshire, Dorset, Southdown, Rambouillet, Columbia, Shropshire, and Montadale) were among the top 10 breeds for registration numbers in all three time periods (Table 2-5). Between 1965–1966 and 1985–1986, Polypay replaced Cheviot, and between 1985–1986 and 2005–2006, the two hair breeds of Dorper and Katahdin replaced Corriedale and Polypay in the top-10 group. Both Dorper and Katahdin have had increases in number of registrations in most years from the time their registration numbers were first reported in 1996 and 1995, respectively (Deakin, 2007). The Dorset was the only breed present in 1965–1966 that posted an increase in registration numbers over the past 40 years (+43.8 percent) and is evidence of the breed's popularity as a general purpose breed for commercial production (Table 2-5).

In 1965–1966 and 1985–1986, the top 10 breeds represented over 90 percent of total registrations among all breeds. In 2005–2006, however, the top 10 breeds accounted for only 74 percent of total registrations, indicating a growing interest in recent years in the breeding of minor breeds or new breeds among hobbyists, breed preservationists, and hand spinners. There are more breed choices today than there were 40 years ago. Registration numbers were reported for 17 breeds in 1965–1966 (Mead, 1967), rising to 33 breeds in 2005–2006 (Deakin, 2007). The most significant change in breed composition of the national flock in the past 10 years has been the increase in number of hair sheep and hair sheep crosses during a time when the national sheep inventory and purebred sheep registrations have declined. Population numbers of hair sheep and their crosses are not collected by USDA; however, an indication of their popularity can be determined by examination of purebred registrations (Deakin, 2007). From 1997 to 2006, the total number of registrations of animals of all purebred breeds decreased by 19 percent (from 95,523 to 77,340 head), and the top four wool breeds (Suffolk, Hampshire, Dorset, and Southdown) decreased by 32 percent (from 52,686 to 36,035 head). Hair sheep breed registrations, on the other hand, increased by 246 percent from 1997 to 2006 (from 4,032 to 13,944 head) and accounted for 18 percent of all purebred sheep registrations in 2006. The popularity of hair sheep has been due to their "easy care" nature. They do not require shearing, are more tolerant of internal parasites (Wildeus, 1997), and are comparable or superior in ewe productivity (weight of lamb weaned per ewe mated) to many wool breeds, especially in hot, humid environments (Bunge et al., 1995). A negative effect of the growth in hair sheep numbers has been increased hair contamination of the national wool clip with fleeces from hair-wool sheep crosses and with fleeces from wool sheep in mixed flocks of wool sheep and hair sheep (Talley, 2008).

The National Sheep Improvement Program (NSIP) is available to breeders of purebred sheep to calculate estimates of genetic merit of individual sheep across flocks for the economically important traits of fleece weight, staple length, fleece grade, direct weaning weight, indirect weaning weight (milk), postweaning weight, and number of lambs born. In addition, some breeds have developed specific estimates of genetic merit for traits with special importance to their breed such as ewe productivity (weight of lamb weaned per ewe lambing) and fecal egg count (indicator of internal parasite resistance). Access to NSIP is through a sheep breed association or a group of breeders of a particular breed who will take responsibility for collection of the performance data, put it in a form needed by NSIP, and submit it to NSIP for analysis.

Flocks enrolled in NSIP have shown positive genetic changes over time for economically important traits. As an example, Figure 2-4 presents the expected progeny difference (EPD), the estimate of improved genetic merit identified by NSIP, for 60-day and 120-day weights from 1986 to 2005 of Polypay sheep in flocks enrolled in NSIP. A doubling of the EPD is an estimate of genetic merit of an individual animal. Polypay lambs born in 2005 were expected to be about 0.91 kg heavier at 60 days and 1.81 kg heavier at 120 days than Polypay lambs born in 1986 because of genetic improvement. Most of this gain came in the past 8 years, probably because the breeders had more confidence in the use of EPDs when making selection decisions in later years after they had some experience in their use and value.

The earliest users of NSIP were the Targhee, Suffolk, and Polypay breeds. More recently, the Dorset, Hampshire, Katahdin, Rambouillet, Columbia, and Romney breeds have joined NSIP. In 2005–2006, records from approximately 110 flocks, including 7,000 ewes and 11,000 lambs, were processed by NSIP across these nine breeds. The Katahdin breed was represented with the most flocks (34). Four of the breeds (Hampshire, Rambouillet, Columbia, and Romney) had five or fewer flocks represented (Notter, personal communication, 2007). While estimates of genetic merit from NSIP will aid in the genetic improvement of the U.S. sheep population, the participation numbers indicate that fewer than 20 percent of U.S. sheep breeds and a minority of breeders within these breeds are using the genetic

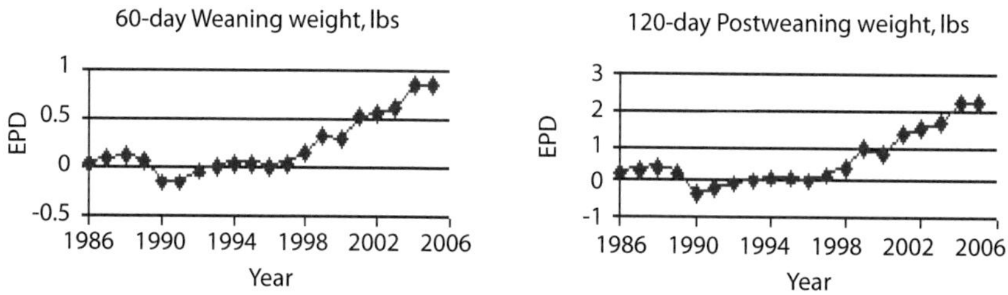

FIGURE 2-4 Expected progeny differences (EPD) for 60-day weight and 120-day weight of Polypay lambs in flocks enrolled in the National Sheep Improvement Program and born between 1986 and 2005 (1 pound [lb] = 0.4536 kg).
Source: NSIP (2007). Copyright 2007 by NSIP (National Sheep Improvement Program). Used with permission.

technology available through NSIP. For the most part, selection decisions in U.S. purebred sheep flocks are based on a visual, subjective evaluation of the appearance of an animal or within-flock performance records (e.g., fleece weight, fleece grade, lamb body weights, and/or ewe litter size) that yield less accurate estimates of genetic merit than those available from NSIP. As a result, genetic progress in the U.S. sheep population is less than could be the case if existing genetic technology were more widely utilized.

Reproductive Efficiency and the Sheep Production Cycle

Reproductive efficiency is the most important economic trait in sheep production. It is a complex trait that includes seasonality of mating, ovulation rate, fertilization of ova by sperm, embryo implantation and survival to parturition, and the birth of live, healthy lambs. The gestation length in sheep ranges from 140 to 155 days, depending on breed, fetal number, and other management factors. Sheep are generally considered seasonal breeders. As the day length decreases (increases), estrus and ovulation rates increase (decrease).

Although there are exceptions, breeds developed and managed closer to the equator tend to have less seasonal breeding restrictions than breeds developed and managed in regions with greater seasonal variation in day length. Breed variations in reproductive efficiency range from the Finnsheep breed that averages over 3 lambs per ewe to breeds developed in more arid, harsher environments that may average less than 1.5 lambs per ewe. Because the gestation length averages 150 days, and lambs may be early weaned at 60 days or less, the potential for at least 3 lamb crops in 2 years exists. However, most production systems are planned for lambing to occur when the highest-quality forage resources will be available to meet the increased nutrient needs for lactation and lamb growth.

The sheep production cycle may vary due to production system, available nutrient resources, climatic constraints, available labor, and other enterprise activities on the farm or ranch. Reproduction is the first stage of the cycle and includes several critical points (NRC, 2007):

- *Premating "flushing."* Producers may set aside or provide higher-quality feed resources for 3–5 weeks before the breeding season to improve ovulation rate and mating success.
- *Pregnancy.* The breeding season is normally 35–40 days or two ewe estrus cycles. Embryo implantation occurs approximately 21–25 days after the ewe becomes pregnant. Higher-quality feed resources and minimum stress are critical during this period.
- *Gestation.* Gestation length may vary from 140 to 155 days. After implantation, ewes are managed to maintain or slightly increase body

weight for the next 75–80 days. Most of the increase in fetal weight is during the last trimester of pregnancy. Good nutrition is more critical during this period.

- *Parturition.* Lambing systems vary from "lock the gate and do not disturb" in much of New Mexico and Texas, to managed range lambing systems in portions of the western states, to intensive shed lambing systems in many of the Mountain states and the Midwest. The lambing season normally lasts 40–45 days. Managing nutrient needs, minimizing climatic stress, and controlling predators are critical during this period.
- *Lactation.* The peak nutrient demands in all sheep production systems are from the last 2–4 weeks of gestation through the first 6–8 weeks of lactation to optimize fetal and lamb growth and ewe milk production. Most production systems plan lambing to be synchronized with peak production of high-quality forages. This may result in April and May lambing in the western and northern states, and November and December lambing in the South, Southwest, and Mediterranean climate regions where higher-quality pasture and forage crops are available in the fall and winter.
- *Weaning.* Weaning age may be as young as 5–30 days in sheep dairies and may range from 60–100 days in more intensive production systems with more expensive feed costs. Most range production systems wean lambs at an average age of 120–150 days.

Although many lambs may be ready for slaughter at weaning, most will require additional growing and/or feedlot finishing before slaughter (Figure 2-5). Lambs may reach slaughter weight and condition postweaning by grazing on high-quality forages such as improved grass-legume pastures, wheat and other small-grain pastures, and crop residues such as alfalfa. Heavier feeder lambs at weaning and lighter-weight lambs after postweaning grazing will be finished in feedlots where they will receive concentrate-based diets. After weaning their lambs, ewes are usually culled for age and other production problems such as physical defects or poor lamb performance. The ewe flock will then be managed to maintain body weight until the premating flushing period.

Environmental Impact and Management

Livestock species strive to survive, thrive, reproduce, and be productive in the environment within which they are placed. Appropriate grazing or improper overgrazing may occur depending on management practices followed by the producer. Historically, sheep overgrazing occurred on both private and public rangelands. On private lands, producer experience, education programs, and state and federal incentives for range and pasture improvements, such as water capture and distribution, fencing, and

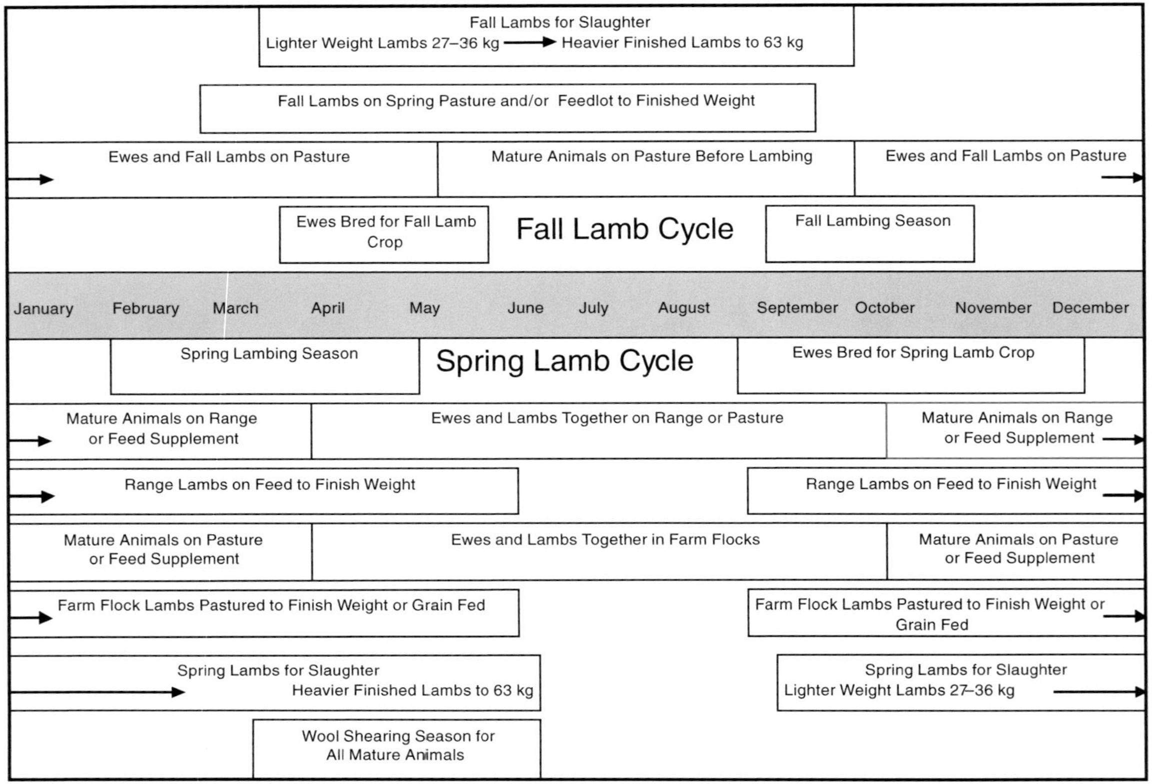

FIGURE 2-5 The sheep-lamb annual production cycle.

other restoration practices, have generally improved both land and animal productivity. Public lands were inappropriately grazed in many areas prior to the Taylor Grazing Act of 1934, which resulted in issuance of grazing permits to producers that control season of use, stocking rates, flock management, and herding requirements. Federal public land grazing permits are administered by the USDI BLM and by the U.S. Forest Service (USFS). State-owned lands may also issue livestock grazing leases on certain lands. This section reviews both the potentially positive and negative impacts of sheep grazing management practices and the impacts of public policies and their implementation.

Sheep Diet and Grazing Behavior[3]

Because of diet preferences and their resulting selective grazing patterns, individual animal species may cause shifts in plant communities (Rector, 1983). For example, sheep may prefer forbs and grasses in the spring and early summer and shift to shrubs and mature grasses in the late fall and winter, while cattle may consume some forbs in the spring and some shrubs in the fall but primarily prefer grass communities. Grazing only sheep may result in a shift to grass plant communities, whereas grazing only cattle may result in a shift to shrub and forb plant communities. Goats will consume some grasses and forbs, but generally consume a larger amount and a wider range of shrub species than do sheep.

Dramatic increases in invasive plant species and their impacts on the integrity and health of natural resources in the United States have been documented (Westbrooks, 1998; Pimentel et al., 2000; Gaskin and Schaal, 2002). Others have shown that prescribed grazing using sheep and/or goats can control invasive plant species and restore healthy plant communities (e.g., Glimp and Haug, 2004). A recently published handbook on grazing was designed as an education manual for producers interested in using prescribed grazing practices and for land managers who may want to utilize sheep and goat grazing for invasive species control, grazing firebreaks, and other rangeland restoration practices (Launchbaugh, 2006).

The dietary overlap between sheep and other herbivores including cattle, goats, deer, and elk is generally considered moderate to low, depending on the available plant community (Cook, 1985). Optimum plant species diversity and animal performance tend to occur when properly managed multispecies grazing is utilized (Glimp, 1988; Walker, 1994). In states with

[3]The Sheep & Goat Research Journal published a special issue in 1994 titled "The Role of Sheep Grazing in Natural Resource Management" which stimulated both research and education programs on appropriate sheep management practices to enhance natural resource management.

mostly private lands, where fee hunting is economically important, foraging relationships between domestic and wildlife species are important to the land manager. In western states with large public land holdings, wildlife are generally managed by the state, yet federal land managers are responsible for land management. Conflicts often occur when wildlife populations are not controlled, thus requiring land managers to reduce livestock grazing to avoid overgrazing and damage to plant communities and rangeland health. Most public land grazing permits are for single species grazing, yet most public lands would benefit from multispecies grazing management (Walker, 1994).

Wildlife Interactions

Other than wild predators, perhaps the most contentious domestic sheep and wildlife interaction issue is with bighorn sheep. The primary concern is the perception that domestic sheep will transmit diseases to bighorn sheep. T. McDonnell (personal communication with the committee in 2007) estimates that, because of this perception, grazing permits for > 200,000 sheep have been revoked in the last 10 years and permits for > 50,000 sheep are currently at risk. The principal disease issue is pneumonia and other respiratory diseases from *Pasturella* spp. infections. The basis for physical barriers between domestic and bighorn sheep is research by Foreyt et al. (1994), in which deaths of bighorns occurred following inoculation of captive bighorns with large doses of *Pasturella hemolytica* from healthy domestic sheep.

Knowles and Rink (2006) emphasized a number of facts that they claim are often overlooked or ignored in considering wildlife interactions with sheep:

- Bighorns and other wildlife species populations are often infected with various strains of *Pasturella* spp. and may transmit the organism to other members of the population. Martin et al. (1996) and Ward et al. (1997) isolated *Pasturella* from bighorns in Alaska and Idaho that had never been in contact with domestic sheep.
- *Pasturella* spp. normally will not trigger pneumonia episodes unless infected animals are stressed. Stressors in bighorn herds may include undernutrition, predator attacks, trapping and relocation, hunting and other human disturbances, climate, and other disturbances related to inappropriate habitat.
- No die-offs of bighorns in their natural habitat have ever been proven to be due to contacts with domestic sheep.
- *Pasturella* spp. do not form spores and thus, are infectious for a very

short time period, almost requiring nose-to-nose contact among animals for transmission.

- Wildlife biologists have been successful in requiring barriers of 5–8 km between bighorn herds and domestic sheep flocks. Some biologists are advocating increasing the barrier to 12–13 km. When bighorn herds recruit new habitat closer to domestic sheep grazing permits, producers are often forced to vacate the permits.
- Groups supporting wildlife interests are often in conflict with domestic sheep producers about the validity of scientific evidence.
- Knowles and Rink (2006) emphasized the need for research that addresses the genetic basis of enhanced susceptibility of bighorn sheep to respiratory disease; the percentage of bighorn sheep that carry and transmit pathogens associated with respiratory disease; and what, if any, are the conditions that trigger transmission of pathogens and development of respiratory disease in bighorn sheep in their natural habitat.

Predation Management

Historically, predators have been the largest cause of sheep and lamb deaths for many years (USDA, 2005). In 2004, sheep producers lost 224,200 sheep and lambs to animal predators (Table 2-6). This represented 37.3 percent of the total losses from all causes and resulted in an economic loss of $18.3 million to farmers and ranchers. Coyotes accounted for over 60 percent of all confirmed predator losses with domestic and feral dogs second at over 13 percent. The primary predators in the "All Other Predators" category include wolves, vultures, and feral hogs. Feral hogs are becoming an

TABLE 2-6 Sheep and Lamb Death Losses Due to Predation, 2004

Predator	Sheep Killed		Total Value of Loss ($1,000)
	Number (head)	%	
Coyotes	135,600	60.5	10,707
Dogs	29,800	13.3	2,807
Mountain lions, cougars, or pumas	12,700	5.7	1,101
Bears	8,500	3.8	769
Foxes	4,200	1.9	285
Eagles	6,300	2.8	438
Bobcats	11,100	4.9	814
All others	16,000	7.1	1,376
U.S. total	224,000	100.0	18,297

Source: USDA (2005).

increasing problem in Texas and other southern states. An excellent review of predation issues was published in a special edition of the *Sheep & Goat Research Journal* titled "Predation" (SGRJ, 2004). The issue contained research and invited review papers from leading U.S. and international authorities that were presented at a symposium sponsored by the American Sheep Industry Association (ASI).

Shelton (2004) pointed out that predation costs may be much larger than actual losses due to producer expenses for protection from predators and hidden costs due to predator injuries to animals, abortions, stress effects on performance from predator attacks, and even some preventive measures such as night penning that may affect animal performance.

The most in-depth analysis of the economics of predation management and the importance of predator control programs for livestock is provided in a literature review by Bodenchuk et al. (2002). In five studies of losses where predator control programs were absent, annual losses averaged 5.7 percent for adult sheep and 17.5 percent for lambs. In eight studies where predator control programs were present, average annual losses were 1.6 percent for adult sheep and 6.0 percent for lambs. Based on comparative losses in the absence of predation management versus losses with management activities across all livestock, Bodenchuk et al. (2002) estimated direct economic benefits to livestock producers of $62,606,770. They suggested that this estimate is conservative because of ancillary benefits to other livestock and wildlife species present where management programs are in place. They then estimated that investments in livestock predation management programs are approximately 40 percent federal and 60 percent from cooperative state programs. On the basis of estimated market values of livestock saved versus predation management program costs, they estimated the benefit-cost ratio (BCR) from predator management to be 3.1 to 1. Considering only federal investments in livestock predation management, they estimated a BCR of 6.75 to 1. Using the Jahnke et al. (1987) nonagricultural multiplier effect of 3.0, they estimated the total economic BCR from all predation management expenditures to be 12.2 to 1. Considering just federal expenditures, the total economic BCR was 27 to 1, suggesting that the tax revenues generated from the increased economic activity may be significantly greater than the federal expenditures for the predation management.

Because of increased predation by wildlife species, significant investments are being made in USDA predation management programs to protect various wildlife species (Bodenchuk et al., 2002). Mule deer, pronghorn antelope, bighorn sheep, wild turkey, and other upland game birds are examples where specific populations were being decimated by predation, and management activities have benefited their recovery. Predation management programs have also been used to protect threatened or endangered species populations such as the black-footed ferret, San Joaquin kit fox,

Utah prairie dog, and Mississippi sandhill crane. A BCR analysis of wildlife protection programs is difficult because of the intrinsic benefits of wildlife to the public. As stated earlier, wildlife populations have also benefited from USDA livestock predation management programs. In some situations, livestock producers may also benefit from predation management programs for wildlife.

Sheep producers have also made direct investments in nonlethal methods for protection of their flocks from predators. Principal strategies used include:

- *Fencing.* Permanent and portable electric fencing, as well as permanent predator-proof traditional net fencing around perimeters, are used. Predator-proof fencing may also be used for night lots, holding pens, and corrals.
- *Guardian animals.* Guardian dogs are the most widely used guardian animal. An estimated 60 percent of all producers with over 100 sheep use guardian dogs. Other guardian animals include llamas, burros, and mules. Guardian dogs have been reported as 65–85 percent effective in protection against predators in various surveys, depending on the type of predator (e.g., coyotes, dogs, lions, or bears) and management conditions (intensive or extensive pasture management, herding).
- *Shed lambing.* Controlled lambing conditions protect lambs at their most vulnerable stage. Increasing costs of labor, facilities, operating expenses, feed, and other expenses make shed lambing increasingly expensive in relation to the value of the lambs saved.
- *Herding.* Sheep producers with public land grazing permits are required to have a shepherd with each flock. Although expensive, predation and other factors would require herders on grazing permits even if they were not required by the land management agencies.

Surveys of sheep and goat producers clearly indicate that a significant portion of them would be forced to abandon sheep and goat production without USDA predation management programs (USDA, 2005). The belief that predator losses were a major factor in the decision of many former producers to abandon sheep and goat production is widely held in the industry.

SHEEP AND LAMB FEEDING

Sheep are ruminants that have a complex digestive system with a stomach containing four compartments (reticulum, rumen, omasum, and abomasum). The rumen, or primary fermentation vat in the ruminant digestive system, contains many species of bacteria and protozoa that enable the digestion and utilization of cellulose and other complex fibrous components of grasses, weeds, and certain shrubs. Once the more complex components have been digested in the rumen, they may be absorbed through the rumen wall for utilization or passed to the other stomach compartments and intestines in less complex forms and as bacterial components for further digestion and absorption. Sheep are considered opportunistic grazers in that they select the more nutritious plant species available during the various seasons of the year. Their ability to utilize a diversity of plant species means that they can be managed in a wide range of ecological environments. The digestive anatomy and physiology of sheep and their nutrient requirements were reviewed recently by the Committee on the Nutrient Requirements of Small Ruminants of the National Research Council (NRC, 2007).

The USDA has not reported the number of lambs on-feed since 1994 (Figure 2-6). This makes any statement relating to the number of lambs being placed on feed speculative. The precise effect of the loss of this information on price discovery and marketing agreements and contracts is unknown. But, the lack of this type of information may reduce transparency in the market. For example, feedlot operators, breakers, and packers appear either unable or unwilling to reveal sequencing (numbers of lambs

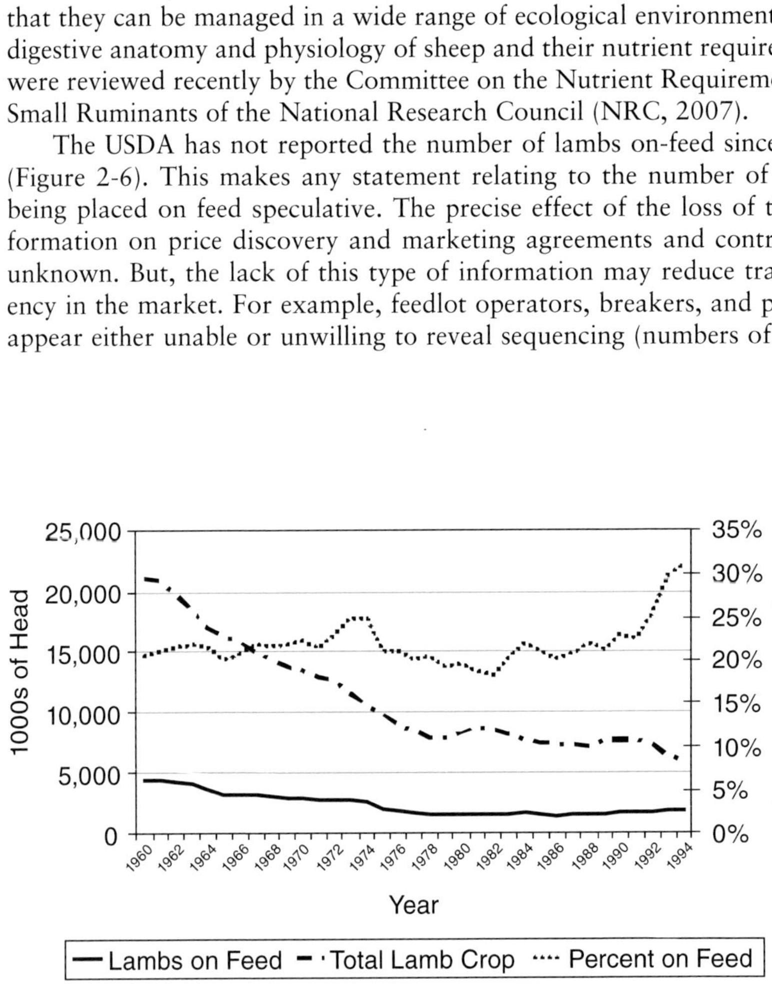

FIGURE 2-6 Lamb crop, lambs on feed, and percent on feed, 1960–1994.
Source: Data from USDA (National Agricultural Statistics Service) and compiled by Livestock Marketing Information Cener (LMIC: http://www.lmic.info/). Used with permission.

TABLE 2-7 Major Lamb Processing Plants and Feedlots in the United States

Firm/Plant Name	Location	Capacity[a] (head)
Packers:		
Swift & Company	Greeley, CO	4,000/day
Superior Packing	Denver, CO	2,000–2,500/day
Superior Packing	Dixon, CA	4,000–5,000/day
Iowa Lamb Corporation	Hawarden, IA	1,500–2,000/day
Wolverine Packing	Detroit, MI	1,500–2,000/day
Den-Franco Corporation	Chicago, IL	1,500–2,000/day
Feedlots:		
Harper Livestock	Eaton, CO	80,000–100,000
Cactus Hill Feeders	Windsor, CO	60,000
Double J Feedlot	Ault, CO	45,000
Rule Feedlots, Inc.	Brighton, CO	40,000–50,000
Mountain View Lamb Feeders	Eaton, CO	40,000
Richard Drake	Eaton, CO	25,000

[a]Capacity per unit of time for packers and one-time carrying capacity for feedlots.
Source: Boland et al. (2007).

coming out of feedlots per time period) because this might influence relative bargaining positions among the market segments.

Lambs typically are born in the spring, weaned in the fall, and then fattened for slaughter either in feedlots or on grass (see Figure 2-5). Light lambs that were recently weaned are referred to as "feeder" lambs because they are typically placed in feedlots or on grass to fatten before slaughter. Feeder lambs typically weigh between 27 and 41 kg when they are placed on feed and 50–64 kg at slaughter. Lambs that are ready for immediate slaughter are referred to as "slaughter" lambs. Carcasses are typically about 52 percent of the lamb's live weight, suggesting that most lamb carcasses weigh 26–33 kg. While lamb slaughter takes place throughout the year, there is a strong seasonal upswing each spring because demand increases during the early spring. Table 2-7 lists locations of the largest sheep feedlots and packers in the United States.

Wether lambs are fed in feedlots or on forage depends on the type of feed available (concentrate or forage) and the availability of land. Forage-based operations have been traditionally used where forage/crop aftermath (after the harvest of grain or forage) and land resources are available. Financial inputs may vary depending on the type of forage system used in either a backgrounding or finishing operation. These operations vary in size and scope and have been the focus of increasing interest with the growing popularity of alternative markets (natural and organic). Drylot feeding operations rely on grains and harvested forages to feed lambs, much like cattle feeding operations.

In both types of feeding operations (feedlot and forage), the genetic type of the sheep fed varies because of size, operation location of the feeding system, and availability of lambs in close proximity to the operation. Each of these operations differs in its goals based on the type of lamb marketing options that are available. The following discussion details the two types of feeding operations, the challenges that they face, and the direction in which they are headed.

Commercial Feeding Versus Field Finishing Operations

According to USDA (1994, 2003), of the total 1.52 million sheep and lambs on-feed in 1994, 1.02 million were in feedlot operations and 447,000 were in forage-based operations. Of the total lambs on-feed, 38 percent weighed 38.6 kg and under, 40 percent weighed 39.0 to 47.6 kg, and 22 percent were over 47.6 kg. Colorado (385,000), California (215,000), and Wyoming (194,000) were the top three states for lambs on-feed across feeding operation type (drylot or forage-based operations). Forage-fed lambs under 38.6 kg numbered 294,500, whereas lambs in the same weight category in a drylot environment numbered 268,000 head. Furthermore, as the weight of lambs increased, the number of lambs in a drylot environment increased as well when compared to forage-based systems. Across the 16 states surveyed by USDA in 1994, 752,000 head (in two categories: 38.6 to 47.6 kg, combined with 47.6 kg and above) were in drylot feeding operations, compared to 152,000 combined across the two weight categories in forage-based systems. Although these data were collected over 10 years ago, a similar ratio between drylot and forage-based operations is believed to reflect current feeding operations.

In the Midwest and Mountain states, large populations of lambs are typically fed in a drylot environment and placed on a concentrate diet (high-energy diets referred to as concentrate/grain-based diets). Although drylot feeding operations can be found on the East and West coasts, the typical lamb feeding operation in those states is forage-based. These feeding operations rely on improved pasture, natural grasses on owned land, and/or grazing rights on public lands, as well as cool-season grasses and crops (Brassicas), depending on location of the operation and resource allocation. The terrain in the forage feeding areas is typically suited to grazing rather than concentrated feeding operations. In some areas, a popular practice among sheep and lamb producers is planting and harvesting a second crop during a single growing season that grows well into the first frost (brassicas). The large feedlot operations in the central Midwest and the Mountain states have the largest number of lambs on-feed at any one given time.

Furthermore, feedlot operations are typically located near grain-growing regions, reducing the cost of transporting harvested feeds. According to USDA (2007b), packers are located near consumer outlets and feeders are typically located near the packers. The largest packing operation is in the Denver area and the largest concentrated lamb feeding operations are also located in that same region.

The types of feeder lambs used in drylot operations are typically from a western white-face ewe (Rambouillet, Targhee, Columbia, or another maternal breed that are used in a crossbreeding system to maximize maternal traits) and then bred to a meat-type sire (Suffolk, Hampshire, or a cross of the two breeds). The end result of this terminal cross is a lamb that is known for performance, carcass characteristics, and a high-quality pelt. The large forage-feeding operations use a similar type of lamb, phenotypically and genotypically. Only a few of the many breeds discussed earlier are actually used in commercial operations and make up the genetic base of the commercial U.S. sheep and lamb industry. Although there is a changing dynamic in the types of sheep various producer groups use in their operations (those breeds that are considered heritage), dairy and hair sheep are sparking interest in small farm flock communities.

Once lambs have been acclimated to a grain diet, gains on concentrates result in elevated average daily gains (Murphy et al., 2003). This enhanced performance is observed in both lambs fed grain diets from weaning direct to drylot as well as those lambs that were fed a forage-based backgrounding diet and then finished in a drylot environment prior to harvest (Murphy et al., 2003). Many studies have reported that animals that are fed an ad libitum (free-feeding) diet of concentrate generally have higher average daily gains than animals that are fed or grazed on legumes (Tatum et al., 1988; McClure et al., 1994; Murphy et al., 1994). In addition to enhanced growth rate, lambs fed on a high-concentrate diet result in more rapid intermuscular, intramuscular, subcutaneous, and internal (kidney, pelvic, and heart) fat deposition, when compared to forage-fed lambs (Crouse et al., 1978; Murphy et al., 2003). The deposition of fat is only an issue if lambs are fed for an extended period of time when efficiencies of lean tissue growth have declined and the lamb's metabolism produces fat at a more rapid rate. Although lamb fat in moderation helps in providing a distinct flavor profile, excess fat is the least desirable in lamb when compared to pork and beef. The current market structure in the lamb industry does not provide producers and feeders an incentive to produce lean lambs. Rather, the market continues to offers an incentive to maximize weight regardless of the share of the carcass weight composed of fat.

Notter et al. (1991) and Murphy et al. (2003) reported that lambs fed to slaughter weight on a high-concentrate diet when compared to lambs fed a high-concentrate/forage mixed diet or an all-forage diet resulted in higher mechanical shear force values resulting in a tougher consumer product. Crouse et al. (1978) concluded that lambs on high-energy (concentrate) diets are physiologically more mature when compared to lambs on low-energy or low-protein diets, which could explain differences in shear force values. In beef, physiological maturity has an impact on tenderness values as a result of increasing amounts of connective tissue and collagen crosslinks (Goll et al., 1963). Bouton et al. (1978) reported that in sheep, progressive increases in toughness could be seen in sheep meat from animals 2 months to 8 years of age. Subtle increases in fat deposition as a result of high-concentrate diets may help offset the effect of physiological maturity and elicit a more acceptable mouth feel and sustained juiciness, resulting in a more positive eating experience (Weir, 1960). The advantages of feeding a high-concentrate diet include less land use, more rapid rate of gain, and higher feed efficiency, resulting in an accelerated market readiness. Lambs are often fed high-concentrate diets past their optimal endpoint or slaughter weight, leading to carcasses with excess fat cover.

The typical concentrate diet includes corn as the energy component. Although the energy component may vary from region to region, the largest traditional drylot feeding operations are located near corn-growing regions. Corn prices have been on the rise in recent months as corn finds a foothold in the biofuel industry. According to Mosier (2006), U.S. fuel ethanol production is expected to exceed 7.5 billion gallons before 2012, a doubling of ethanol production over 2004 when only 10 percent of the U.S. corn crop was utilized for fuel production. The rising cost of corn has many in the livestock industry considering feeding alternatives, including substitute energy sources for animal feeds or shifting to forage feeding.

Enhancing the efficient use of solar energy, recycling nutrients to the soil, using noncompetitive renewable resources (high cellulosic), contributing to soil and water conservation, investing lower amounts of capital, and adding enterprise flexibility are some of the advantages of forage feeding systems (Ely, 1994). Forage feeding on a large scale occurs in the San Joaquin Valley and the Imperial Valley in California, the Willamette Valley in north central Oregon, and the Columbia River Basin in Washington State. These areas of the West are rich in harvested forage production (legume and grass hay), turf and grass seed production (such as rye and Kentucky bluegrass seed), legume seed production (such as alfalfa seed), and sod production for commercial and residential landscaping. Grazing lambs on these forage types during the winter months reduces the environmental impact of grazing in extensively managed rotational grazing systems. The result is a healthier forage stand, increased seed production, and other advantages for the subsequent growing season. This method of managing forage and grass production systems has replaced the use of field burning, which is now illegal in some states because of pollution and potential smoke-related respiratory complications.

The largest forage-feeding operations are in the West, but forage feeding is practiced nationwide, although the practice varies in scale and forage type. Forage-feeding operations begin stocking up on lambs during the winter and early spring, reaching maximum capacity just prior to Easter. Lambs on-feed in forage-based systems are typically either fall born or are from the previous spring crop. Fall-born lambs typically reach 5 to 7 months of age by the time of harvest with an average finish weight of 54.4 kg. Spring-crop lambs are typically 12 to 14 months at time of harvest with an average finish weight of 65.8 kg. The USDA (2007b) reports that the average weight of finished market lambs in the United States is 61.2 kg. In addition, the USDA (2001) reports an average live weight at harvest of 63.0 kg with a relatively wide range around the average, regardless of size of operation.

The main benefit of forage feeding over drylot feeding has traditionally been understood to be a reduction in total body fat. Allowing sheep to graze a feedstuff, rather than harvesting the feedstuff mechanically for feeding, also results in reduced production cost and less environmental impact. However, ASI (2002) reported (Table 2-8) a decrease in performance for forage-fed (daily gain 0.15 kg) compared to drylot-fed lambs (daily gain 0.26 kg). Even though pasture-fed lambs may not grow as fast as lambs fed concentrate diets, forage feeding is more economical with less potential health risks (Schoenian, 2007).

Research studies have concluded that to prolong skeletal and muscle growth in sheep, energy intake must be reduced (Yambayamba and Price, 1991; Berger, 1991; Shanks et al., 2000). The result is a decrease in fat deposition potentially from an increase in muscle mass. The low-input forage-feeding system requires producers to ship animals long distances to market and sometimes across state lines. The same is often required even

TABLE 2-8 Performance of Pasture-fed and Drylot-fed Lambs[a]

Feeding System	Daily Gain (kg)	Feed Consumed (kg/day)	Feed:Gain Ratio	Carcass Fat (%)
Pasture	0.154	—	—	23.9
Pasture + supplemental feed[b]	0.263	1.00	3.79	27.3
Drylot, 13% CP	0.268	2.06	7.71	33.2

[a]Initial weight = 31.75; Slaughter weight = 50 kg.

[b]13% CP (crude protein) supplement; same as drylot, 13% CP.

Source: ASI (2002). Copyright 2002 by ASI (American Sheep Industry Association). Used with permission.

with drylot feeding, however, because of the distribution of lamb production across the United States.

In some cases, fall-born lambs may stay with their mothers until they have reached an acceptable harvest weight and fatness (49.9 kg live weight). Lambs are selected based on perceived carcass fatness (market readiness based on USDA yield grades) and sorted into two groups—those ready for slaughter and those not meeting the buyer's criteria, which are sent to a feedlot. Ideally, lambs shipped to packers are between a yield grade 1.9 to 2.9 (measuring 3.81 to 6.35 mm of backfat at the 12th/13th rib interface) (USDA, 1992).

Although forage operations are used to finish lambs, they can also be used to prepare lambs to maximize frame growth (backgrounding) before finishing lambs in a drylot system. Lambs that are forage fed for backgrounding purposes will enter the drylot with less internal and external fat cover but maximum skeletal frame growth. Because of the seasonality of sheep and lamb production, the lamb feeding season must be spread out over a 12-month period, resulting in lambs of various slaughter weights and ages, which complicates efforts to optimize carcass production. Arnold and Meyer (1988) concluded that lambs grazing irrigated pastures before finishing in drylot had less fat over the longissimus muscle and a lower percentage of kidney and pelvic fat than lambs that were weaned and placed directly in a drylot. Shanks et al. (2000) reported that body weight at the end of a backgrounding period on stubble barley was greater when compared to lambs fed a concentrate ration in a drylot. They also concluded that lambs in drylot situations may experience greater stress, which affects their performance, but that, in most cases, concentrate diets increase gain and reduce time to market. Shanks et al. (2000) also found that mechanical shear force was increased for lambs on concentrate diets from time of weaning versus those that were backgrounded initially and then fed concentrate before harvest. Crouse et al. (1978) found that physiological maturation rate was potentially elevated in lambs fed a concentrate diet from weaning compared to those that are backgrounded on forage and then fed concentrate or those that are forage-fed until harvest.

New forage varieties have been investigated that yield potentially favorable results in performance when compared to more traditional grazing opportunities. Brassica species, including turnips, rape, and tyfon (a hybrid of turnip and Chinese cabbage), seem to be promising new forages (Koch et al., 2002). Brassicas are fast growing and tolerant to frost and cold, and many producers are using them as second crops after harvest of primary crops in rotation. They can be planted from mid- to late summer after the harvest of the primary crop and then grazed through the fall and winter (depending on geographical location). Tyfon and turnips are near maximum production in about 60 days and rape in 75 days. Koch et al. (2002)

reported that lambs grazing turnips for 39 days gained 0.18 kg/day and drylot lambs gained 0.20 kg/day, which initially suggests that there is little performance difference in the feeding phase. When this time was extended past 40 days, lamb performance when grazing turnips was lower compared to the performance of concentrate-fed lambs. Except for their tolerance to cold, Brassicas perform similarly to other forage-based systems.

Using forage and concentrate in combination can help maximize growth and reduce carcass fat deposition if managed using proper evaluation methods and guidelines as to when a lamb is market ready. Proper feed management is often not done because of lack of incentive since few industry groups use any type of grid marketing option to encourage feeders to send lambs to packers at the optimal point in their growth to meet specific quality and yield specifications. Forage-based systems also have limitations, including problems with availability of forage during the feeding season, since lambs often are not the primary reason that the forage crops are grown. Also, not all forages perform the same. Some breeds of lambs are more suited for the more intensive management system of commercial drylot feeding operations. Some large commercial producers that utilize public land grazing allotments in combination with private land forage resources, however, are managing sheep and lambs with little supplementation year round.

Falxa et al. (2002) reported that while there may be opportunities for diversification of land use in the coexistence of sheep and cattle, these two species have been shown to exhibit complementary, supplementary, and competitive relationships depending upon the stocking rates of the two species and the type of land resources available. After reviewing 200 studies, Van Dyne et al. (1980) concluded that sheep consume 50 percent grass, 30 percent forbs, and browse for the other 20 percent, while cattle consume 70 percent grass, 15 percent forbs, and 15 percent from browsing. Falxa et al. (2002) also concluded that improved balance in utilization of forage types may result in increased animal performance and stocking rates, as well as improved cash flow, when cattle and sheep are grazed together.

Most range sheep production systems are managed on lands that do not produce adequate forage resources to produce slaughter-ready lambs. Although backgrounding weaned lambs on high-quality forages is desirable where forage resources are available, these resources may either be unavailable to many producers or inadequate to produce gains and finish lambs for harvest. High-quality forages also tend to be seasonal in their availability, which may result in seasonal restrictions in lambs available for slaughter. Lamb feedlots provide high-quality concentrate finishing programs that augment gains provided by forage-based systems. They can also provide lambs for harvest on a year-round basis and provide an important assembly function for packers seeking adequate numbers for plant operations. However, lamb feedlots, because of their environmental and management effects on range-raised lambs and high animal concentrations per unit of land, generate issues related to animal digestive health and diseases, and environmental impacts that must be closely monitored. Typically, feeder lambs (31.8–49.9 kg) are purchased for feeding operations and placed on an 80-percent concentrate (grain-based) and 20-percent forage diet, which tends to optimize performance (see Table 2-8).

Specialty sheep operations influence commercial lamb production through the sale of breeding stock to commercial producers and the distribution of surplus lambs into the commercial system. These impacts may include, for example, purebred breeding flocks that produce rams for sale to commercial flocks, and large commercial flocks that specialize in the production of replacement females for those producers that prefer to purchase replacement ewes rather than raising their own. Other specialty operations producing lambs may include highly specialized wool producers that consider lamb as a byproduct and dairy sheep production systems whose primary objective is to optimize milk production.

Lamb Feeding Arrangements

According to the USDA (2001), 15.1 percent of all operations that reported selling lambs to a feedlot, including herded open range, fenced range, farm flock, and feedlots, retained complete ownership (contract feeding), 5.6 percent retained partial ownership (producer/feeder), and 79.3 percent retained no ownership (i.e., the drylot feeder purchased the lambs). Bastian and Whipple (1998) concluded that the options available to lamb producers for sale of their lambs included sale directly to the feedlot, producer-retained ownership through the feeding phase, and producers selling fed market lambs direct to the packer.

Bastian and Whipple (1998) also reported that packers owned about 28 percent of all lambs fed in the United States in the late 1990s. Williams and Davis (1998) reported many large range operations sell feeder lambs directly to feedlots under contract or feed their own lambs and sell to the packer under contract. Over time, the sheep industry has gradually shifted from marketing lambs through public markets to direct marketing through various types of contractual agreements. In the latter case, producers who retain possession of their animals through the feeding phase typically raise their sheep on a forage-based feeding operation. Smaller range producers concentrate on marketing lambs through public auction and through intermediaries, although some also sell finished lambs directly to packers. Farm flock producers tend to produce and feed lambs and then sell them directly to packers.

Packers generally prefer to purchase lambs by the truckload (roughly 400 head depending on the size of the lambs and the transport vehicle) to reduce transportation costs. Purchase methods may include direct purchase in truckload lots from feedlots and individual producers with larger operations, or buying lambs at auction markets that exist in several states, the largest of which is Producers Livestock Auction in San Angelo, Texas. Smaller lots of lambs from producers are usually grouped together into larger loads. Various approaches are used for assembly, including Internet auction, telephone auction, cooperatively negotiated sales for specified assembly days and locations, delivery to a designated buyer representative, and "buy days." "Buy days," increasingly popular in eastern and midwestern farm flock states, are established dates and locations agreed to by buyers and sellers. Producers deliver their lambs, which are weighed and graded by the buyer, and price is negotiated. Once the buyer has enough lambs for a truckload, the lambs are then delivered to the harvest facility.

Feedlot Lamb Nutrition and Health

Lamb feeders use a variety of methods to finish lambs. Feeding operations vary on when and how they wean their lambs and when the weaned lambs are moved into the feeding operation. In concentrate and forage-feeding systems, lambs are normally exposed to feeds or forage prior to weaning (ASI, 2002). Lambs sent to feedlots are normally introduced to concentrates through a creep-feeding system (pen arrangement so only lambs and not their dams can enter). Young lambs can handle both forage and concentrate diets because of the early development of their ruminant stomach system (ASI, 2002). In some cases, supplemental concentrates that parallel the diets found in most drylot feeding operations are fed to young lambs in conjunction with forage-based diets in order to elevate lamb performance and efficiency of feed conversion.

The success of lambs entering a feeding system is dependent on the first few weeks of the feeding period after arrival. When they arrive, lambs are usually hungry, thirsty, and stressed from travel since many must be hauled long distances to reach their final feeding destination. Typically, they are fed harvested forages to help them acclimate to their new environment with a salt supplement to aid in their off-truck water intake. Additionally, they are usually dewormed for internal parasites, topically treated for external parasites, and vaccinated for enterotoxemia Type D. Other vaccinations and preventative measures may be offered depending on the location and type of feeding operation as needed for satisfactory lamb performance. As reported by the USDA (2001), feedlot operators receive limited information related to prearrival health status of lambs being shipped. There are a number of reasons that the transfer of animal health information may be hindered. However, information related to preshipment vaccinations and other medications could enhance feedlot efficiency and reduce animal disturbances.

Prearrival records provide critical information for quality assurance, food safety, and consumer awareness of the history of the food animal products that they consume. Some pharmaceutical products used as preventative or disease treatments have required withdrawal periods to ensure that there are no drug residues in the meat that may have an effect on human health (Roeber et al., undated).

The packing industry and the ASI developed the National Sheep Safety and Quality Assurance (SSQA) program as an attempt to address concerns related to quality assurance and food animal product production. The SSQA program was designed to educate all sheep producers (commercial, farm flock, hobby farms, purebred, and club lamb) on the proper handling and practices of live animal production and helps assure consumers that the products they consume are safe and wholesome. Many states have taken leadership in developing curricula to instruct members of agricultural youth programs (such as 4-H and Future Farmers of America [FFA]) about proper handling and practices related to their livestock project experiences (Kuber, 2005). Improvements in animal welfare, food safety, and quality should continue to build consumer confidence in lamb as a safe and wholesome food product. Nevertheless, 37.5 percent of the feedlot operators responding to the USDA's National Animal Health Monitoring System (NAHMS) survey indicated that they had never heard of the ASI SSQA program, and another 40.6 percent indicated that they had heard of it but were not familiar with it (USDA, 2001). Only 6.3 percent of the responding feedlot operators indicated that they were very familiar with the program. In contrast, the National Pork Producers Council (NPPC) has successfully operated a similar program since 1989 with the backing of the packing industry, since packers require all producers from whom they purchase hogs to be certified (NPB, 2007). The NPPC program has reportedly reduced carcass blemishes and drug residues and helped assure consumers that pork products have been monitored and verified throughout the production process (NPB, 2007).

Many feedlot owners apparently have little interest in the prearrival health history of the feeder lambs entering their feedlots. The NAHMS survey results showed that 41 percent of responding feedlot owners considered the prearrival vaccination history of lambs entering their lots to be unimportant (USDA, 2001). Only 21 percent considered the vaccination history to be very important. Also, 52 percent considered prearrival deworming history information to be unimportant and only 14 percent very important. Yet a majority of those same feedlot operators responded that prearrival procedures (such as treatment for parasites and clostridial vaccinations) could reduce sickness and death in feedlot lambs. Even though the responding feedlot operators recognized the effectiveness of prearrival procedures in reducing sickness and death among feedlot lambs, few recognized the importance of obtaining prearrival information on lambs in determining the type of lambs they will accept into their feeding operations and the preventative measures needed upon arrival to prevent illness and death.

The NAHMS survey results also indicated that about 75 percent of responding feedlot operators processed new arrivals as a group within 72 hours of arrival (USDA, 2001). Nearly all feedlot owners, regardless of feedlot size, indicated that they treat new arrivals for internal parasites and vaccinate against clostridial disease (Types C and D). Only the large operations (greater than 5,000 head), however, indicated that they treat for external parasites. Internal parasites and clostridial diseases have the most impact on feedlot lamb performance and efficiency of production and are the diseases most often treated.

The NAHMS survey further reported that the leading causes of death among feedlot lambs across all feedlot sizes include respiratory disorders (29.1 percent), enterotoxemia (28.7 percent), shipping fever/pneumonia (12.8 percent), other digestive orders (6.1 percent), and crowding as well as parasites (5.8 percent). For smaller feedlot operations (less than 500 head), the NAHMS survey reported that the leading cause of death was shipping fever pneumonia (34.1 percent). Among the larger feedlots (more than 500 head), respiratory disorders and enterotoxemia were responsible for the majority of the deaths (31.1 percent and 30.1 percent, respectively).

Health concerns specific to sheep and lambs from weaning to maturity in feeding operations are reviewed in the *Sheep Production Handbook* (ASI, 2002). The health section of the handbook reviews the clinical signs, diagnosis, treatment, and prevention of the common diseases of concern in the sheep and lamb feeding industry, including rumen lactic acidosis and enterotoxemia (overeating), urolithiasis (urinary calculi), polioencephalomalacia (polio), salmonellosis (*Salmonella* dysentery), pneumonia (pasteurellosis, shipping fever), rectal prolapse, copper toxicity, copper deficiency (enzootic ataxia or swayback), selenium deficiency, contagious foot-rot, and parasites (internal and/or external). In many cases, these diseases affect production efficiencies such as growth rate, weight gain, and fat deposition. In more severe cases, they may result in irreversible damage and even death. In any case, disease affects the feeder's profit margin from loss of production, increased costs of labor and medical supplies in treating sick animals, and pharmaceutical costs for preventative measures.

Sheep Feedlot Environmental Impact and Management

The U.S. Environmental Protection Agency (EPA) defines an animal feeding operation (AFO) as a location that confines animals in an area for > 45 days in a year (EPA, 2002). A large concentrated animal feeding operation (CAFO) for sheep includes operations with 10,000 or more sheep confined to an area for more than 45 days in a year. A sheep AFO may be defined as a medium CAFO if it has at least 3,000 sheep and either a man-made ditch or pipe carrying manure or wastewater from the feeding operation or surface water with which sheep come in contact running through the area where sheep are confined. All CAFOs are required to have a permit, for which the minimum EPA requirements are as follows, although there may be additional state requirements:

- Implement a nutrient management plan;
- Submit annual reports to the permitting authority;
- Keep permit current until the operation is closed and all manure is removed; and
- Keep records of all nutrient management practices for at least 5 years.

The nutrient management plans for sheep CAFOs must include provisions to:

- Ensure adequate manure storage capacity;
- Properly handle dead animals and chemicals;
- Divert clean water from the production area;
- Keep animals out of surface water;
- Use site-specific conservation practices;
- Develop and implement ways to test manure and soil;
- Ensure appropriate use of nutrients when manure is spread on land other than the CAFO area; and
- Keep accurate records of nutrient management practices.

Although not a part of EPA standard CAFO requirements, sheep feedlots may come under pressure from communities if dust control or odors are of concern. States may also have additional requirements or standards that must be met by lamb feedlots.

New Feedlot Technologies and Management Practices

New technologies and more efficient management practices are showing promise in terms of enhanced production efficiency and cost reduction in feedlot operations. Feedlot managers/owners are making strides in improving nutrition, health, growth performance and feed efficiency, and potential market value. They are utilizing innovative approaches to select and segregate lambs upon arrival to the feedlot; utilizing feedback information from the packing plants to determine which lambs, feed rations, and management practices performed the best; utilizing live animal imaging to identify lambs that will perform the best in the feedlot; and, finally, using real-time ultrasound (RTU) to determine when lambs are properly finished for harvest. The following is a brief discussion of these feedlot applications to improve performance and product quality.

- *Feedback to origin flocks.* According to the NAHMS survey, 50 percent of responding feedlot operators never or almost never reported occurrences of disease, performance, or carcass quality back to the producers of origin. Another 25 percent of the respondents only provided such information sometimes. At least one quarter of feedlot owners, however, did provide feedback information to producers to assist them in determining needed genetic and management improvements (health and efficiency of production). Feedback information can be used for genetic improvement, health management, and other quality improvements by the producer. With this information producers can prioritize genetic and management needs that will optimize performance and profitability at both the producer and feeder level.
- *Segregating lambs upon arrival to the feedlot.* Improvements in efficiency can be attained through segregation of lambs based on sex and frame size prior to feeding in concentrate feeding operations. Commonly accepted in the industry is that lambs with more potential for frame growth will finish at heavier weights. Many feedlot operators feed lambs for the same amount of time and to the same endpoint regardless of age, sex, frame size differences, and/or fat cover. Because lambs vary in their growth potential based on their genetic makeup and metabolism, sorting feeder lambs and managing their feeding regimen according to characteristics affecting growth potential would optimize the rate of gain and profitability. The additional revenues earned from sorting, however, must be sufficiently above the additional labor and facility costs associated with sorting or there will be no incentive for feedlots to sort their lambs. If the industry were to develop scientifically based, objective criteria for sorting and evaluating lambs based on their growth potential, producers and feedlot operators would have a valuable tool for evaluating (and pricing) lambs based on growth potential. Based on such criteria, lambs evaluated as exhibiting relatively less growth potential could be marketed earlier and at lighter weights compared to those that might be judged as capable of greater gain and/or likely to produce lean carcasses. Segregation of lambs at the point of feeding could increase feed efficiency by up to 70 percent (U.S. International Trade Commission [USITC], 1999).
- *Live animal imaging.* This emerging technology provides an objective evaluation of frame score, composition (muscle, fat, and bone), and growth potential. Studies in pork (Suster et al., 2003) and beef (Nada et al., 2005) have reported a high level of accuracy for evaluation of total body composition (bone, muscle, and fat) using this process. There are a number

of ways imaging could be used, both by producers and feedlot operators to determine muscle growth potential and the type of feeding regime best suited to their needs. Sorting lambs using an objective system such as a live animal imaging unit would help in selecting lambs that are market-ready with more accuracy.

- *Ultrasound technology.* Miller (1998) reported that the use of real-time ultrasound has had considerable impact in reducing excess fat in swine herds as major breeding companies have employed that technology in their selection process. Houghton and Turlington (1992) report that the ultrasound technology is a cost-effective, noninvasive approach to estimating carcass composition and quality of live animals. Brethour (1994) reported similar results with ultrasound technology in estimating muscle quality. There are currently certification programs in the beef and the pork industries for technicians that utilize RTU to estimate live animal and carcass yield and quality grades, but no such program exists in the sheep and lamb industry. However, some technicians certified to use RTU on beef and pork also use this technology on sheep.

LIVE SHEEP PRICING

Price determination and price discovery have become increasingly important issues for the U.S. sheep industry. *Price determination* relates to the broad forces of supply and demand that lead to market-clearing prices. Changes in the market environment generated by increased lamb imports and consumer preferences and market shocks, such as the border closures that resulted from the first discovered case of bovine spongiform encephalopathy (BSE or mad cow disease) in December 2003, have all influenced the location and slope of the demand and supply schedules for lamb and mutton. *Price discovery* refers to the efficiency and accuracy with which buyers and sellers are able to gather and interpret market information, which is then incorporated into individual negotiations for trading sheep. Concerns about price discovery are often related to a lack of adequate price reporting or reporting from markets with few buyers and/or low volumes of sales, also referred to as "thin" markets.

Price Determination

Basic economic theory defines the quantity of a product demanded by consumers at the retail level as being a function of its own price, the prices of substitute and complement products, consumer income, consumer tastes and preferences, and quality (typically measured by yield grade in the case of lamb and mutton). Seasonality may also play a role in the demand for agricultural products, especially for a product such as lamb whose consumption is heavily influenced by annual holidays.

Economic theory also suggests that the quantity of lamb supplied at the retail level is a function of its own price and costs as determined by the underlying production function for firms in the industry. Seasonality and cycles also play a significant role in meat production because of gestation periods and other issues related to biological lags.

Imports influence retail supply because almost all imports are in carcass form (very few live sheep are imported into the United States) and may also influence retail demand if imported and domestic products are differentiated. Domestic and imported lamb may be differentiated to a degree because imported lamb tends to be leaner and have a lighter carcass weight than domestic lamb. Because the supply of agricultural products is typically fixed in the short run because of biological lags, demand models for food and agricultural products are often specified as being price dependent. This is also referred to as "inverse" demand and simply indicates that prices adjust to existing quantities rather than the reverse.

The demand for live sheep is an indirect or a "derived" demand that is reflected from the retail demand/supply of lamb and mutton meat back through the supply chain to the farm level. Consequently, the demand (price) for sheep ready for slaughter is a function of the quantity of sheep offered for sale; retail/wholesale meat prices; processing, packaging, and other marketing costs; feeding costs beyond the farm gate; the price of pelts (a jointly produced product); and seasonality. In addition, a recent study suggests live prices are also a function of the pricing method used to procure ewes and/or lambs (RTI, 2007). Farm-level supply for sheep is expected to be a function of the farm-level price, feeding costs (corn and/or hay), and seasonality. The price (demand) for feeder lambs is expected to be a function of the price offered for slaughter lambs and the costs (availability) of feed resources, while feeder lamb supply is determined by prices and costs of production at the farm level.

The lamb cutout value has increased in both nominal and real terms since 2001, suggesting a relatively strong market for domestic lamb (Figure 2-7). However, live lamb prices have been flat to slightly lower, on average, during the past 5 years (Figure 2-8), indicating that an increasing margin has been required during the past 5 to 6 years to pay for production and marketing costs beyond the farm gate.

Lamb is more expensive than some competing meats, specifically beef and pork. However, in relative terms, lamb prices have remained fairly constant with beef during the past 5 years, as suggested by indexes of retail prices (Figure 2-9). Beef is usually considered the closest meat substitute for lamb, and lamb appears to have maintained a relatively stable competitive position with beef since 2001. However, lamb's competitive position with pork has been eroding since 2003 based on relative indexes of retail prices as depicted in Figure 2-9.

FIGURE 2-7 Weekly average USDA lamb cutout value, October 2001–March 2007 (1 hundredweight (cwt) = 45.36 kg).
Source: Data from USDA (Agricultural Marketing Service) and compiled/adapted by LMIC. Used with permission.

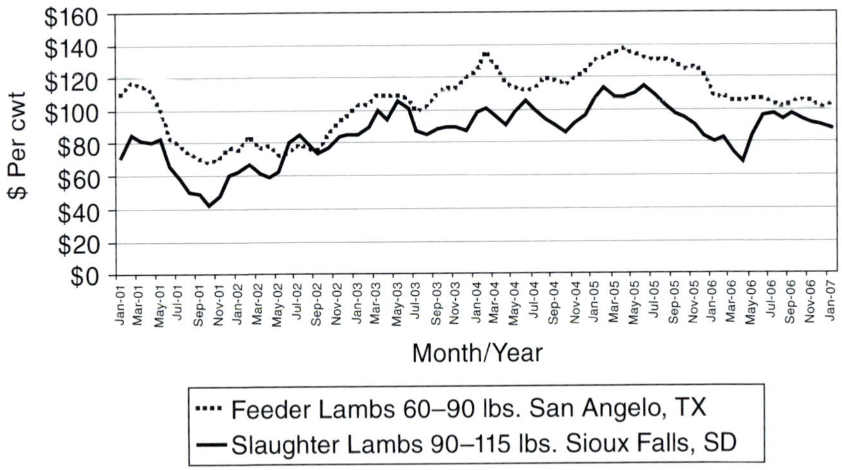

FIGURE 2-8 Feeder and slaughter lamb monthly live prices, January 2001–February 2007 (1 hundredweight (cwt) = 45.36 kg).
Source: Data from USDA (Agricultural Marketing Service) and compiled by LMIC. Used with permission.

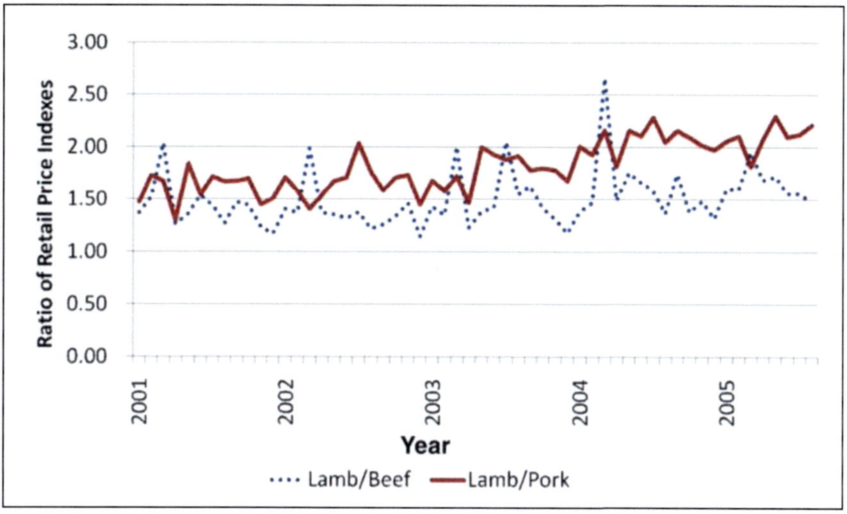

FIGURE 2-9 Monthly relative indexes for lamb, beef, and pork retail prices, January 2001–August 2005.
Source: USDA (2007c).

The USDA yield grades signal a measure of the amount of external fat over the ribeye area. As the amount of external fat increases, the percentage of meat available for retail sale decreases. Yield grades range from 1 to 5 with yield grade 1 indicating the least amount of fat (USDA, 1992). Quality measured by yield grade has been declining on the average for domestic lamb and mutton since 2000 (Figure 2-10). Packers often develop "grids" whereby they reward lamb sellers based on quality. Table 2-9 presents an example of a grid that is used by the Mountain States Lamb Cooperative.

Few publicly available data are available for assessing relative prices between domestic lamb carcasses and imported lamb carcasses. Some data are available for selected periods from the USDA weekly reports (USDA, 2007c). These data suggest that imported carcasses are less expensive, on the average, than domestic carcasses (Figure 2-11). However, the difference in price between domestic and imported lamb fluctuates considerably and time periods apparently exist when the two prices are essentially equal (Figure 2-11).

The average price for sheep pelts has dropped considerably since 2004 (Figure 2-12), while feeding costs, as measured by corn prices, have increased dramatically in the last year. The price of hay has also trended upward over the last few years (Figure 2-13). Increasing feed costs do not necessarily disadvantage domestic lamb relative to other meats, however, because feeding costs affect all segments of the domestic livestock industry.

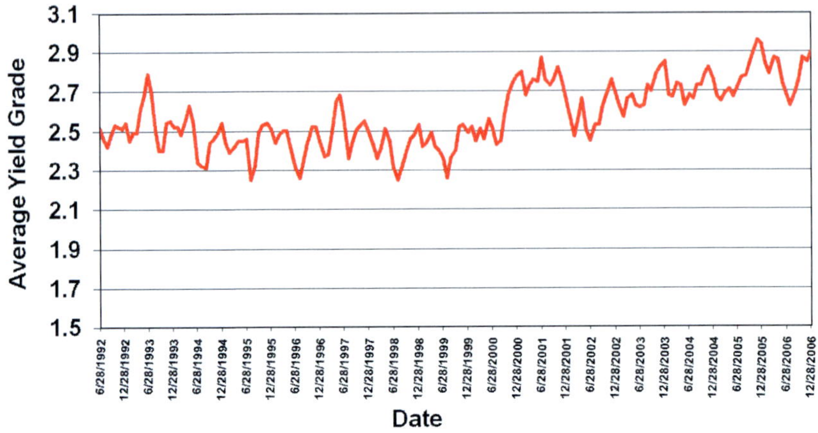

FIGURE 2-10 Average yield grade for mutton and lamb, 1992–2006.
Source: USDA (2007c).

TABLE 2-9 Mountain States Lamb Cooperative Quality Grid

Carcass Yield	Premium ($/kg)	Discount ($/kg)
1	—	—
2	0.176 (0.08/lb)	—
3	0.110 (0.05/lb)	—
4	—	0.176 (0.08/lb)
5	—	0.661 (0.30/lb)

Source: Boland et al. (2007).

Increasing corn prices, however, may place domestic lamb, which is mostly fed on concentrates in feedlots, at a disadvantage with grass-fed imports as corn prices rise.

The price for mature sheep, specifically cull ewes, is expected to be affected by retail meat prices, the number of ewes being slaughtered, processing costs, and on-farm production costs. Mature sheep slaughter follows a seasonal pattern with low points occurring in the early spring at lambing time. Both mature sheep and total sheep slaughter (mature, yearlings, and lambs) have trended slightly downward as domestic sheep numbers have declined (Figure 2-14). Recent droughts have affected on-farm feeding and

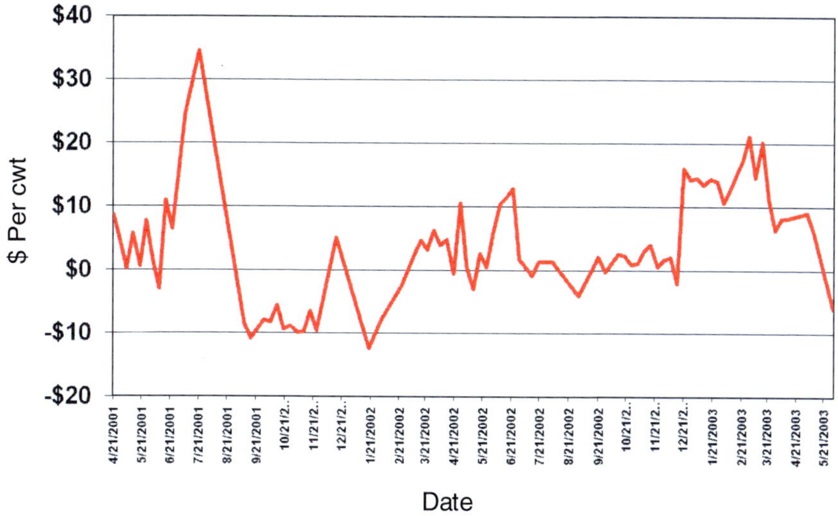

FIGURE 2-11 Western direct carcass price minus imported carcass price, April 2001–May 2003 (1 hundredweight (cwt) = 45.36 kg).
Source: USDA (2007c).

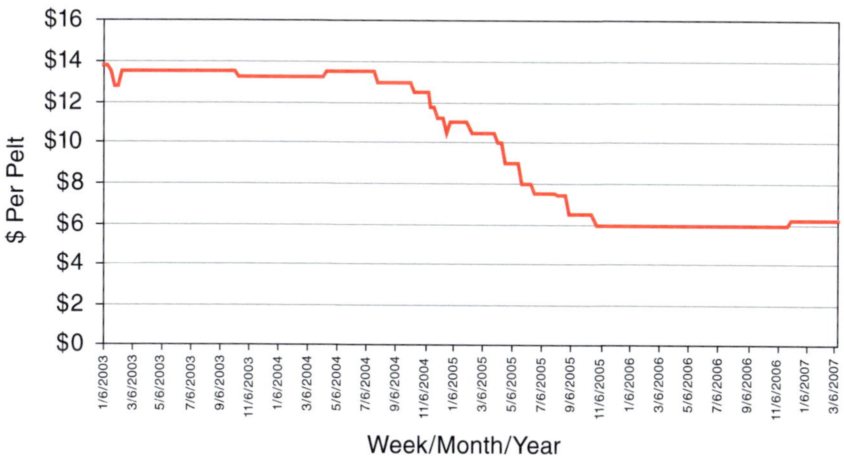

FIGURE 2-12 Average pelt price for fall clips, January 2003–March 2007.
Source: Data from USDA (Agricultural Marketing Service) and compiled by LMIC. Used with permission.

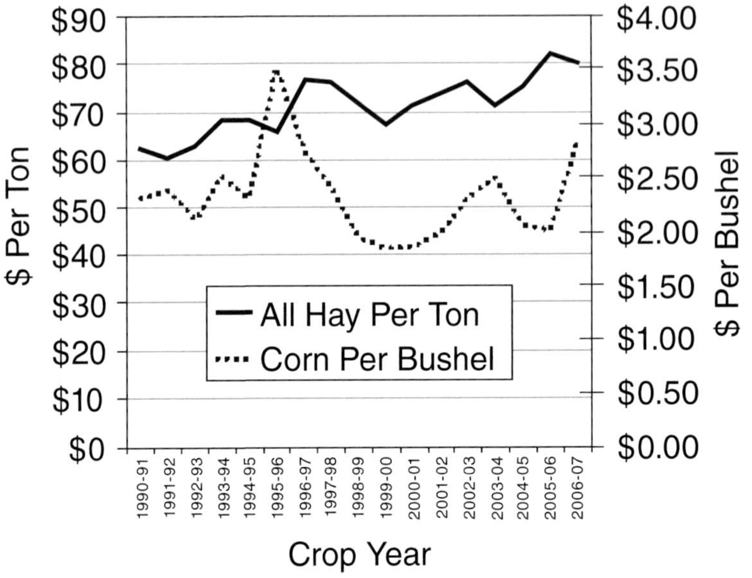

FIGURE 2-13 U.S. average feed costs, corn and hay, 1990/1991–2006/2007 (1 ton = 907.18 kg).
Source: Data from USDA (National Agricultural Statistics Service) and compiled by LMIC. Used with permission.

maintenance costs, and predators also inflict a significant cost on sheep producers as discussed earlier.

Price determination in the U.S. sheep industry appears to be affected by most of the same market forces as other domestic meat products. All meat segments in the United States are characterized by dominance by the domestic retail market, keen competition from other meats, and high levels of concentration in feeding and packing. Unique features relating to the sheep industry, compared to other meats such as beef and pork, are its low consumption level compared to other meats, relative importance of ethnic markets, loss of a large portion of market share to imports, and large losses to predators.

Price Discovery

Feeder lambs are typically sold either under contract or some other negotiated pricing method directly to feedlots or through auctions at sale barns (cash sale). Cash sales are by far the most common method for pricing feeder lambs, with only approximately 8 percent of feeder lambs being

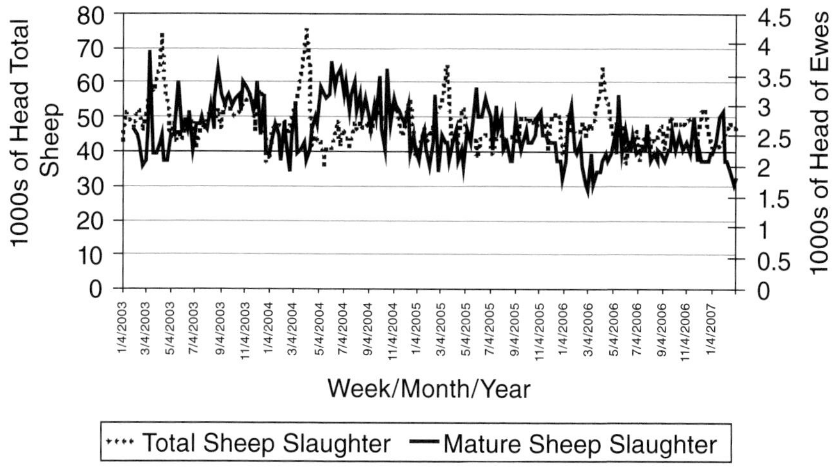

FIGURE 2-14 U.S. weekly sheep and lamb slaughter, January 2003–March 2007. Source: Data from USDA (National Agricultural Statistics Service) and compiled by LMIC. Used with permission.

purchased under contract (RTI, 2007). A recent study by RTI (2007) found that while there is a slight trend toward more direct sales for feeder lambs, auction markets remain by far the most important pricing mechanism for feeder lambs. The USDA reports live lamb prices for markets in 18 different states on a monthly basis (USDA, 2007c). San Angelo, Texas, is the largest auction market for feeder lambs in the nation. Other important live lamb auction markets include Centennial, Colorado, Newell and Sioux Falls, South Dakota, and New Holland, Pennsylvania. The San Angelo market has often been used in the past as the primary market for western live lamb price quotes, and the New Holland market is often the reference market in the eastern United States. Buyers at lamb auctions are primarily feedlot operators or buyers purchasing lambs that will be slaughtered at light weights for ethnic (primarily halal) markets. These lighter lambs, while referred to as feeder lambs, will actually be slaughtered at 27–41 kg liveweight. There are some movements toward further integration in the lamb market. For example, the Mountain States Lamb Cooperative is attempting to inject more cooperation in the lamb market by its recent joint venture with B. Rosen and Sons, a large lamb fabricator, processor, and distributor based in the New York City area.

Most slaughter lambs are now being priced using formulas or are packer fed (RTI, 2007). Thus, rather than being driven by live markets, the lamb price discovery process now largely reflects carcass or cut-out values so that price is determined by negotiation or formula related to carcass quality. As a result, a significant amount of the risk is shifted from the buyer to the seller, especially for pricing based on quality.

The USDA also reports both a daily carcass price and a daily cutout value for lamb (USDA, 2007c). The cutout value is a composite price that is an average of the wholesale value of all muscle cuts. The daily lamb cutout value is actually a rolling average for five days to account for the fact that the market for some cuts is thin on some days of the week. The USDA also reports prices for imported carcasses as well as for pelts. Imported carcasses represent a competing but separate market with American lamb because imported carcasses tend to be smaller and leaner than American carcasses.

Mandatory price reporting (MPR), legislated in 1999 and fully implemented in 2001 (Schroeder et al., 2002), requires prices for all "boxed lamb and lamb carcasses reported on form LS-128 and LS-129 must be reported on an FOB plant price basis" as well as live lamb purchase prices to be reported to the USDA Agricultural Marketing Service (AMS). Because the lamb market is quite thin, prices for specific time periods may not be reported because of confidentiality reasons (E. Rosa, personal communication, 2007). The MPR system has resulted in more price reporting for live lambs and lamb meat and has generated more information on imported lamb carcass prices. For example, the weekly direct slaughter sheep prices reported by AMS (USDA, 2007c) can be both domestic and imported prices, but often do not report imported prices (E. Rosa, personal communication, 2007). Again, the biggest single problem with the MPR system for lamb at this point is missing data resulting from confidentiality issues.

Formula pricing has become a major pricing method for lambs. Packers may also work on a grid system that offers higher prices for carcasses whose quality exceeds a set base and lower prices for carcasses below the base quality (see Table 2-9). The base price is typically determined from either the USDA carcass or cutout value. McDonnell (personal communication, 2007) indicated that 94 percent of the variation in lamb prices can be explained by changes in USDA carcass and pelt prices, a conclusion also reached by Greer and Ward (2000).

Although slaughter lambs are procured by packers in a number of different ways, formula pricing is one of the most important methods (Figure 2-15). The fact that over 80 percent of lambs are procured for slaughter through formula pricing or auctions indicates that both methods are important. McDonnell (personal communication, 2007) emphasizes the importance of the USDA carcass and cutout values both in formula pricing, contracts, and cash negotiations.

In September 2006, the Risk Management Agency of USDA established a price protection program for lamb called Livestock Risk Protection for

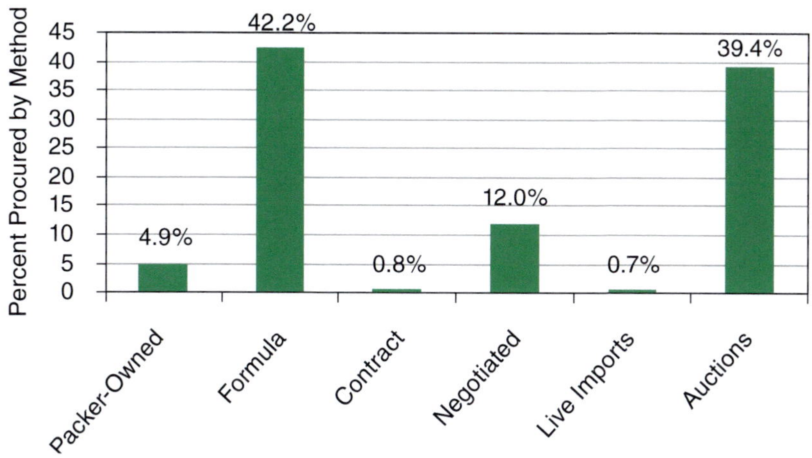

FIGURE 2-15 Percentage of lambs procured by packers by method of procurement, 2006.
Source: RTI (2007).

Lamb (LRP). This program is available in more than half of the states and is designed to allow producers to protect against declines in national slaughter lamb prices below a selected coverage price. This coverage is available through crop insurance agents. The LRP program is a positive development for many sheep producers because it offers the potential for reducing some of the risk associated with lamb production.

LIVE SHEEP TRADE

The United States exports a substantial number of live sheep (mostly cull ewes) to Mexico. However, sheep exports to Mexico have only begun to recover from the BSE border closure in 2004 and are still far below levels of exports experienced in 2003 (Figure 2-16). In 2006, live sheep exports (slaughter ewes) to Mexico increased 57 percent to 124,343 head as compared with the year earlier.

SHEEP RESEARCH, INSTRUCTION, AND EXTENSION/OUTREACH

Most of the sheep research in the United States is conducted by three groups: (1) state land-grant universities, (2) the USDA, and (3) private companies. Although there is no accurate estimate of the amount of sheep research conducted by private companies, the amount is small relative to

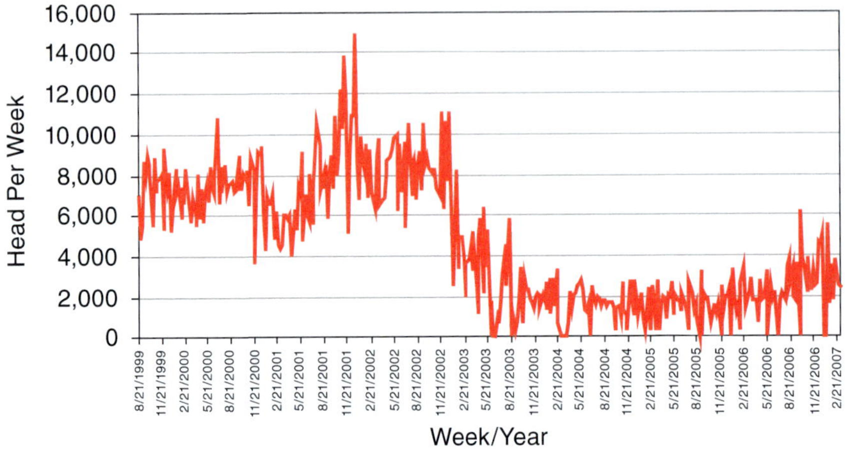

FIGURE 2-16 Weekly U.S. live sheep exports to Mexico, August 1999–March 2007.
NOTE: These data are not necessarily comparable to official final (monthly) numbers.
Source: Data from USDA (released by Agricultural Marketing Service [AMS] based on Animal Plant Health Inspection Service data collection) and compiled by LMIC. Used with permission.

the amount conducted by the public sector. Most private companies cannot justify large expenditures on sheep research programs because of the small potential domestic market for products developed by such programs.

Land-Grant Universities

The primary research and extension activities of the sheep industry as well as all sectors of U.S. agriculture are carried out by the land-grant universities. According to Campbell (1995), most universities, both private and public, focused their teaching programs on the basic sciences and the arts prior to 1860. The majority of their students were the children of the wealthy and professional classes. In 1862, the U.S. Congress passed the Morrill Act for the "donating [of] public lands to the several states and territories which may provide colleges for the benefit of agriculture and the mechanic arts." Under this act, the federal government granted land to each state with the income generated from these lands to be used for the establishment of universities whose primary purpose was the education of the children of farmers and the working classes. These universities became known as land-grant universities and continue today to offer instruction in agriculture and engineering. Land-grant universities have grown to include comprehensive instruction in the basic sciences, arts, and humanities as well.

Many "normal schools" or colleges were established from the early to mid-1800s to train primary school teachers in the existing western and northern states. From 1866 to 1890, several southern states established normal schools to train African American teachers. Although many of these institutions were similar to the land-grant universities established by the Morrill Act of 1862, the federal government was unable to gain cooperation from the southern states in the provision of land-grant support to the African American institutions. The passage of the Second Morrill Act by the U.S. Congress in 1890 expanded the 1862 system of land-grant universities to include historically African American institutions. Many of the African American schools were incorporated into this system and became known as "1890 Institutions." Each of the southern states that did not have an African American college by 1890 established one later under the Second Morrill Act. Several of these 1890 universities have developed small ruminant research programs, primarily with meat and dairy goats, but some also include sheep.

The Hatch Act of 1887 provided for the continual funding of agricultural research at land-grant universities "to aid in acquiring and diffusing among the people of the United States useful and practical information on subjects connected with agriculture" (Campbell, 1995). Federal funds continue to come to each land-grant university each year to support agricultural research. The distribution of Hatch Act funds among various research initiatives at each university is largely at the discretion of each university, within general guidelines from the federal government.

Extension services for farmers were added to the mission of land-grant universities when the U.S. Congress passed the Smith-Lever Act of 1914. The act established the Cooperative Extension Service at land-grant universities to disseminate the results of agricultural research to farmers. The extension service in each state is cooperatively funded by federal, state, and county governments. Most land-grant universities have at least one faculty or staff member with full- or part-time responsibility for sheep extension activities in the corresponding state. The sheep extension specialist organizes and conducts educational programs for sheep producers in the state with county livestock or agricultural extension agents located in each county of the state. The agents in county extension offices are an expansion of the land grant university into every county of every state.

The U.S. Department of Agriculture

The USDA, through its Cooperative State Research, Education, and Extension Service (CSREES) agency, is a major funding source for agricultural programs at land-grant universities through (1) joint funding with state and county governments of the cooperative extension services in each state, (2) funding state Hatch Act allocations for agricultural research, and (3) funding special and competitive grant programs in priority research areas. In addition to the cooperative and supportive programs with land-grant universities through CSREES, USDA has its own in-house research organization known as the Agricultural Research Service (ARS). The ARS has major sheep research activities at three of its agricultural research stations: (1) the U.S. Sheep Experiment Station in Dubois, Idaho; (2) the Roman L. Hruska Meat Animal Research Center in Clay Center, Nebraska; and (3) the Dale Bumpers Small Farms Research Center in Booneville, Arkansas. Although some sheep research is also conducted at several other ARS stations, those research activities are quite small compared to the sheep research programs at the Idaho, Nebraska, and Arkansas ARS sites.

Interaction and collaboration among sheep research and extension efforts at different land-grant universities and sheep research efforts at ARS stations are accomplished through regional research committees. Regional agricultural research, more recently designated multistate research, was authorized in 1946 to fund cooperative research on problems important to multiple states and to avoid duplication of research efforts. A portion of USDA Hatch Act funds going to land-grant universities is designated for multistate research. In addition to providing funding and a valuable infrastructure for research coordination, multistate research leverages additional funding (often four to five times depending on the project) from other sources such as state appropriations, private industry, and other federal agencies.

There are two types of regional research committees: (1) technical committees that receive funds to conduct a regional research project and to hold an annual meeting for coordination of the project and (2) coordinating committees that do not have a funded regional project but receive funds for an annual meeting to discuss and coordinate individual projects funded from other sources. While there are currently no technical committees dealing with sheep, there are two coordinating committees. Western Education and Research Activity 39 (WERA-39: Coordination of Sheep and Goat Research and Education Programs for the Western States) is composed of sheep research and extension personnel from 11 universities.[4] North Central Education and Research Activity 190 (NCERA-190: Increased Efficiency of Sheep Production) is composed of representatives from 14 universities and 3 ARS stations.[5] These two committees coordinate their activities by meeting together every few years, given that five universities and one ARS

[4]More information on WERA-39 can be found online at http://nimss.umd.edu/homepages/home.cfm? trackID=6936

[5]More information on NCERA-190 can be found online at http://nimss.umd.edu/homepages/home.cfm? trackID=3851

station are members of both groups. The universities and the ARS stations represented in these coordinating committees account for the majority of the sheep production research activity in the United States.

The Role of Sheep Producer Organizations

A limited amount of sheep research is also funded by various national sheep producer organizations, including the American Lamb Board (ALB), ASI, National Lamb Feeders Association (NFLA), Dairy Sheep Association of North America (DSANA), and many national purebred sheep breed societies. In addition, there are numerous state and county sheep producer organizations and state purebred breed societies. These organizations exist primarily to service the needs of their members and, thus, are funded through dues of members or fees for services in the case of the national breed societies. Some of these national and state sheep organizations receive funding from producers through mandatory or voluntary deductions on sheep and/or wool sales within that state, commonly referred to as "checkoff" funds. Because most national, state, and regional organizations have limited monetary resources, there is relatively little direct monetary support of sheep research and extension efforts from these organizations. A mandatory national producer checkoff on slaughter sheep sales supports activities of the ALB. While the legislation that created the ALB allows for the expenditure of funds on sheep research, the vast majority of the lamb checkoff funds are allocated to lamb promotion activities.

The ASI and its predecessor, the American Sheep Producers Council (ASPC), have played a greater role in sheep research and education activities than have the other sheep organizations. Because of budgetary constraints, research support has never been an important part of the ASI agenda. The main efforts of ASI have been in the area of sheep producer education and legislative/regulatory concerns. Before 1996, ASI spent approximately $500,000 annually on research and education activities (Thomas and Miller, 2001). Since that time, however, annual ASI support for similar programs has declined to only about $250,000 (P. Rodgers, personal communication, 2007).

The ASI carries on a number of important functions in the education arena, including the development, updating, and publishing of the *Sheep Production Handbook* (ASI, 2002). The handbook is used by sheep producers and is the most widely used textbook in university sheep production courses. The ASI also established and continues to publish the *Sheep & Goat Research Journal*, an applied research journal. In addition, ASI regularly organizes and sponsors numerous national conferences and symposia on topics of current interest to the sheep industry. Every 4 years, ASI publishes a Research and Education Priority List for the U.S. Sheep Industry which is used by a variety of individuals and organizations as decisions are made at the national and state levels on agricultural research and extension funding.

In recent years, the National Institute for Animal Agriculture has been a major source of print and Internet-based information for producers on the disease of scrapie and the national scrapie eradication programs (NIAA, 2008).

Research and Extension Funding and Activities

The Current Research Information System (CRIS), a public database maintained by USDA, contains information on public agricultural research conducted by USDA agencies (primarily at the ARS laboratories) and universities, predominantly the land-grant universities (USDA, 2007d). The USDA agencies receive funding via appropriations from the U.S. Congress, whereas land-grant universities derive funding from multiple sources, including appropriations from state legislatures, Hatch Act funding through USDA, and grant funding from federal agencies (e.g., USDA, the National Institutes of Health, the Department of Energy, the EPA, and the Department of Defense) and private sources, including commodity organizations, direct industry support, and research endowments. Often these multiple sources of funding are combined to support an individual research project at a land-grant university providing valuable flexibility for research management.

Between 1998 and 2005, average annual funding for all public agricultural research was $3.9 billion, with average annual increases of 5.6 percent (Figure 2-17). During the same period, average annual funding for livestock (beef cattle, dairy cattle, swine, and sheep) and poultry research was $657 million (16.7 percent of all agricultural research funding) with average annual increases of 2.7 percent, less than half the annual increase for all agricultural research. Funding for livestock research actually declined by 1.5 percent between 2004 and 2005.

During that period, sheep research expenditures were the lowest among the livestock sectors at approximately $44.5 million per year or 6.8 percent of total livestock research expenditures (Table 2-10). Funding for beef, dairy, and poultry research increased by an average annual 3.5–4.1 percent over that period, while funding for sheep research increased a modest 0.7 percent per year on average. Funding for swine research actually decreased by –0.2 percent per year during that same period.

The CRIS includes summaries of all USDA-funded research by ARS and at universities. A search of the CRIS system in May 2007 using the keywords of Sheep, Wool, or Lamb identified 1,260 individual active projects in 2005 (USDA, 2007d). Although the search may have missed a few projects with application to the sheep industry, some projects using sheep as a biological model or projects with a very small portion of sheep-related activity may also have been identified. There are likely some additional projects con-

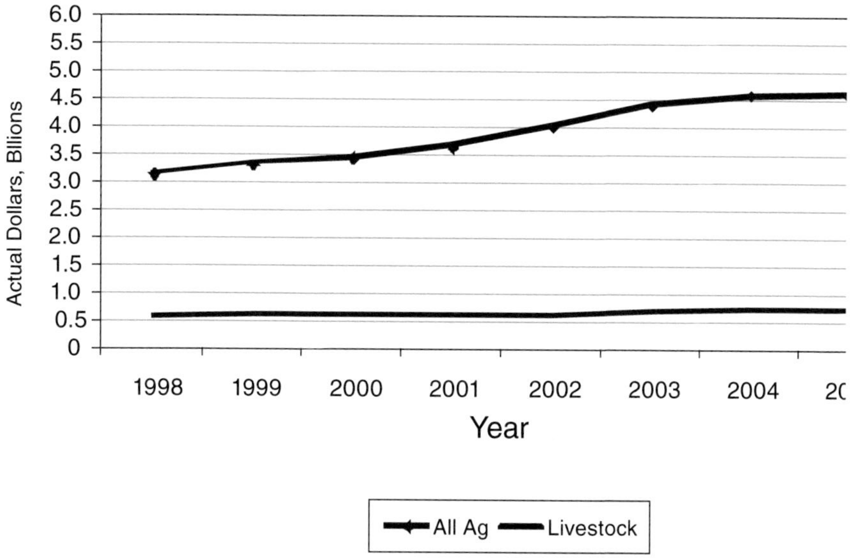

FIGURE 2-17 Public funding for all agricultural and livestock research, 1998–2005.
Source: USDA (2007d).

TABLE 2-10 Average Annual Public Funding of Livestock Research by Type of Livestock, 1998–2005

Livestock Sector	Expenditure ($1,000)	Share of Total (%)	Average Annual Change (%)
Beef cattle	205,939	31.3	+3.5
Dairy cattle and milk	168,998	25.7	+3.7
Pigs	118,517	18.0	–0.2
Poultry and eggs	119,258	18.1	+4.1
Sheep and wool	44,540	6.8	+0.7
Total livestock	657,250	99.9	+2.7

Source: USDA (2007d).

ducted at universities with state or private funds that do not appear in this information system if the university did not require the reporting to CRIS of projects funded from nonfederal sources. Therefore, a reasonable estimate of the number of sheep research projects at public institutions with application to the sheep industry in 2005 is 1,300 to 1,500.

According to USDA (2007d), 92.4 full-time equivalent (FTE) ARS and university scientists were working on sheep research projects in 2005, the lowest number of FTEs associated with sheep research between 1998 and 2005. The highest number of FTEs involved in sheep research during that period was 157.3 and the average was 113.4. The average annual number of scientist FTEs working on other types of livestock and their products between 1998 and 2005 was much higher, including 459.5 for beef, 409.8 for dairy cattle, 324.1 for poultry, and 295.8 for swine.

In addition to sheep research efforts reported in CRIS by USDA-ARS and at universities, the Economic Research Service of USDA also conducts economic analyses and compiles market reports important to the sheep industry. Unlike sheep research, quantitative data on expenditures, activities, and personnel involved in sheep extension are not readily available. Although virtually every state has someone designated as the sheep extension specialist, their appointment for these duties may vary from 100 percent to 5 percent or less depending upon the importance of sheep to the agricultural economy of the state. Likewise, county agricultural, livestock, and youth extension agents may spend a majority, a minority, or no time at all dealing with sheep extension activities. In counties where sheep are an important agricultural commodity, the county extension agent may be quite knowledgeable about sheep and be able to service most needs of local producers. In counties where sheep are not important, however, sheep producers may be referred to the state sheep specialist with their problems. In states where there are very few sheep, the person designated as the state sheep specialist may refer more technical questions to his or her more knowledgeable colleagues at other universities. Nevertheless, the extension network is in place in every county in the United States so that sheep producers can access the system to get answers to their questions usually at no cost. Most state sheep specialists maintain computer websites that provide both basic and technical information on sheep production.

Development of Sheep Research and Extension Programs

The individual lead scientist on a project, for the most part, determines the direction of sheep research and extension programs within the corresponding university or ARS station. There is no one body or organization that assigns scientists to specific research or extension programs. Some states and some ARS research stations have sheep advisory committees composed of representatives of the sheep industry that make recommendations to individual scientists on the types of research and extension programs they feel would be most valuable to the industry. If a formal advisory committee does not exist to advise scientists on priority needs, there are still generally good informal communications between sheep producers and extension and research scientists so that research needs of the industry are well known by scientists.

Neither of the two regional coordinating committees for sheep research (WERA-39 or NCERA-190) has authority to set research agendas at individual institutions. Participation by scientists in these committees, however, results in their familiarization with the strengths, interests, and current programs of scientists at other institutions. This knowledge results in collaborative projects and prevents unnecessary duplication of research among institutions.

Of course, scientists applying for competitive grants at the national level must develop proposals that meet the research objectives of the particular grant program if they expect to be successful. The objectives of most grant programs reflect priorities expressed by the corresponding livestock industries. Even scientists seeking funding for their own research projects from within their organizations (e.g., ARS scientists seeking support for a project within their own research station and university scientists seeking support from their respective university's Hatch Act allocation) must write a project proposal that is peer-reviewed for scientific worth and potential value to the sheep industry before it is funded.

NATIONAL SHEEP ORGANIZATIONS

Several national organizations work to strengthen the U.S. sheep and lamb industry. The roles they play in the industry are discussed in some detail here.

American Sheep Industry Association, Inc. (ASI)

The American Sheep Industry Association (ASI) is the primary national organization representing the interests of sheep producers throughout the United States. A federation of 44 state sheep associations, as well as individual members and other sheep-related organizations, ASI is organized with a board of directors, an executive board, councils, committees, and officers. ASI and its constituent groups define, develop, and execute policy on issues affecting the sheep industry. Councils and committees include Animal Health, Legislative Action, Predator Management, Public Lands, Research and Education, and Resource Management. The goals of ASI are to:

- Develop an industry vision for the future;
- Be an advocate of public policy to protect, promote, and support the economical viability of the industry;
- Create strong national and international markets for wool through advertising, promotion, and marketing;

- Advance and coordinate science and technology of production and marketing; and
- Promote communication and cooperation between all segments of the industry, related businesses, and government agencies.

Funding for ASI work on legislative and membership issues comes from member dues and individual donations. Individual and state member dues are $0.03 per stock sheep and $6 per member and must be renewed annually.

American Lamb Board

The American Lamb Board (ALB) was authorized by the U.S. Secretary of Agriculture to administer the Lamb Promotion, Research, and Information Order established under the Commodity Promotion, Research, and Information Act of 1996 [7 U.S.C. § 7411-7425; P.L. 104-127]. The ALB works to strengthen the domestic lamb industry's position in the marketplace through advertising, public relations, culinary education, and retail promotions. The 13-member board represents all sectors of the lamb meat industry including six producers, three feeders, one seedstock producer, and three first handlers (harvesters and processors of lambs).

In effect, the mission of ALB is to increase the demand for U.S. lamb through consumer advertising and promotion programs. The promotional activities of the ALB are funded by an assessment of $0.005/lb ($0.011/kg) on sheep sold by producers, seedstock producers, feeders, and exporters. Also, an assessment of $0.30/head is paid by first handlers. The ALB annual budget for 2005 was approximately $2.7 million. The majority of the funds (75 percent) were spent on promotion. The effectiveness of the ALB in promoting increased consumption of lamb is considered in some detail in Chapter 4.

National Lamb Feeders Association

The National Lamb Feeders Association (NLFA) is a nonprofit organization whose primary purpose is to initiate, sponsor, and carry out plans, programs, policies, and activities that promote, encourage, and improve the production of lambs and sheep. Membership is open to all persons engaged in promoting and improving the production and marketing of lambs and sheep. Funds for NFLA activities come primarily from membership dues of $25 for associate, nonvoting membership, $100 plus $0.10/head for full voting membership, and $500 for industry associate membership (such as packers and breakers).

National Sheep Improvement Program

The National Sheep Improvement Program (NSIP) is a computerized, performance-based program designed to identify sheep of high genetic merit for economically important traits. The NSIP uses the most modern, scientifically proven technology to transform production records into estimates of genetic merit called EPDs. This technology has been used extensively in the dairy, beef cattle, and swine industries for many years, and is only now being implemented in the sheep industry.

National Sheep Industry Improvement Center/ American Sheep and Goat Center

The USDA National Sheep Industry Improvement Center (NSIIC) was approved by the U.S. Congress as part of the 1996 Farm Bill. Its mission is to assist the U.S. sheep and goat industries in enhancing the production and marketing of sheep, goats, and their products and to strengthen the infrastructure of the U.S. sheep and goat industries through grants and low-interest loans. The NSIIC was privatized in 2006 and became the American Sheep and Goat Center (ASGC). The ASGC manages funds that are used in direct loans, loan guarantees, cooperative agreements, equity interest, investments, repayable grants, and grants to eligible entities either directly or through an intermediary.

National Livestock Producers Association—Sheep and Goat Fund

The National Livestock Producers Association (NLPA) maintains the Sheep and Goat Fund, a revolving fund established within NLPA to encourage innovation and efficiency in the sheep and goat industries by providing credit to eligible and qualified entities to make capital available for enhancing production methods and services; improve marketing efficiency, product quality, and industry infrastructure; and create opportunities for adding value to sheep and goat products. The fund was established as a result of a joint effort between NLPA and NSIIC. The ASGC now has oversight responsibility for the Sheep and Goat Fund, including budget approval and membership on the NLPA Board.

Dairy Sheep Association of North America

The Dairy Sheep Association of North America (DSANA) represents the interests of dairy sheep producers and sheep milk processors in the United States, Canada, and Mexico. A major activity of the DSANA is sponsorship of the annual Great Lakes Dairy Sheep Symposium, which rotates between the three major areas of dairy sheep concentration in North America: (1)

Wisconsin and Minnesota; (2) New York, Vermont, New Hampshire, and Maine; and (3) southern Ontario and Quebec. The organization is funded through member dues.

North American Hair Sheep Association

The North American Hair Sheep Association (NAHSA) is composed of both purebred and commercial hair sheep producers, with the majority of the membership in Texas. Some goals of the organization are to develop high-value markets for meat and hides from hair sheep and to sponsor production sales of hair sheep and educational forums for the membership. The NAHSA has received a number of marketing and product development grants to further assist hair sheep producers.

National Sheep Breed Associations

Most of the more than 50 breeds of sheep in the United States are represented by a breed association or society whose purpose is to maintain breed purity through the maintenance of pedigree records of registered animals and to promote the benefits of their breed to the industry.

POLICIES AND REGULATIONS RELATED TO THE LIVE SHEEP INDUSTRY

A number of policies directly impact the production of live sheep. Many other policies affect live sheep production and feeding indirectly through their effects on lamb, wool, and other lamb product markets. This section focuses on those policies directly impacting the live sheep industry. Other policies affecting the industry through their effects on other segments of the sheep and lamb supply chain are discussed in Chapter 4.

Policies Related to Eradication of Scrapie

Scrapie, a transmissible spongiform encephalopathy (TSE), has been a major concern in the U.S. sheep industry for over 50 years. Although different from bovine spongiform encephalopathy (BSE), the emergence of BSE as a possible human health hazard has renewed emphasis on eradicating scrapie as well as BSE. Information on scrapie, its clinical signs and transmission, and testing methods and strategies for eradication are discussed in depth in Chapter 3. The U.S. Congress has committed substantial funding to the USDA Animal and Plant Health Inspection Service (APHIS) to eliminate scrapie in the United States by 2010. Concerns relating to scrapie transmission have resulted in restrictions on live sheep exports from the

United States, nonuse of rendered byproducts from sheep for use in certain animal feedstuffs, and a national scrapie initiative through APHIS. Requirements of the national scrapie program include: (1) routine testing of sheep at slaughter, (2) official identification (ID) to facilitate traceback of positive sheep found at slaughter, (3) movement restrictions and the removal of certain risk animals from infected and source flocks, and (4) a certified scrapie-free flock program that requires unique flock and animal ID and pedigree information, as well as annual inspection of the flock for 5 years, recording all causes of death, and postmortem diagnosis of any suspect cause of death. Certified Scrapie Free flock status may be achieved in 5 years through this program. Although some countries are requiring more than 5 years for certification, certified status may permit the export of live animals for breeding purposes. Since research studies have documented that scrapie is not transmitted through semen, several countries have accepted the import of semen from rams that meets their protocols and testing requirements. The ASI and other sheep organizations support the scrapie eradication program. Chapter 3 provides more details on the current status of scrapie and other disease issues for the U.S. sheep industry.

National Animal Identification System (NAIS)

Currently a USDA APHIS-administered voluntary program, the National Animal Identification System (NAIS) was initiated as a result of concerns regarding TSEs and other diseases, quality assurance for safety and security of the U.S. food chain, and compliance with international requirements for the export of meat and other animal byproducts. The sheep ID requirements for scrapie are currently accepted as meeting the NAIS goal of enhancing animal traceability. The long-term goal of NAIS is to require individual animal ID that will comply with international standards utilized by all major livestock-producing nations. The sheep industry currently supports the NAIS goals.

Predator Management Programs

Administered at the federal level by the USDA APHIS Wildlife Services Agency, predator management programs are financed by federal appropriations; by state departments of agriculture; by sheep, goat, and cattle producers; and by state departments of wildlife. The extent of economic costs and animal losses due to predation is discussed in depth in a previous section in this chapter titled Predation Management. Animal damage control programs funded and implemented by Wildlife Services, and supported by state agencies and producers, are vital to the survival of the U.S. sheep industry.

Endangered Species Act (ESA)

The general public, including most livestock producers, support the protection and recovery of threatened and endangered species (TES). The Endangered Species Act (ESA) is administered by the USDI Fish and Wildlife Service (FWS). Specific examples of issues of concern that affect sheep production, especially on public lands, include the designation of the Sierra Nevada Bighorn sheep as a unique distinct population segment protected by the ESA, as well as other bighorn populations in Idaho, Wyoming, and Montana considered to be "threatened" by the presence of domestic sheep; the ESA protection of grizzly bear populations in the U.S. Intermountain States in the past; and the release of ESA-protected wolf subspecies in several regions of the northern, intermountain, and southwestern regions of the United States. The major concerns of the livestock industry with respect to the ESA are threefold. First, the survival of TES should be critical if the risks to survival of a species are caused by humans (e.g., hunting of whooping cranes to near extinction for their plumage). The wisdom of TES designation is in question, however, if a species decline is caused by natural processes. Second, although ESA mandates that TES designation and recovery plans must be determined on the basis of best available science, some scientists and livestock producers question whether the FWS has been objective in selecting the science on which decisions are based. This concern was discussed in more detail in the previous section titled Wildlife Interactions. Finally, TES designation often results in management for a single species, which is not consistent with established principles of ecosystem management.

Minor Use and Minor Species Act of 2004

The Minor Use and Minor Species (MUMS) Act of 2004 (P.L. 108-282, 108th Congress) facilitates the development, approval, and use of animal drugs intended for less common animal species or those with less common conditions (AVMA, 2005). Given the critical shortage of such drugs in the United States, veterinarians, animal owners, and livestock producers have had limited options for treating these animals if they become ill. The shortage of approved drugs results in animal suffering, loss of animal life, and financial loss to those who raise the animals.

The designation of sheep as a minor species under the MUMS Act makes more medications legally available to veterinarians and sheep producers to treat sheep, particularly for uncommon diseases. The MUMS Act provides pharmaceutical companies innovative options for overcoming the financial roadblocks they face in bringing limited-demand animal drugs to the market. Before this legislation, pharmaceutical companies could rarely afford to bring to market such drugs because the markets were too small to generate an adequate financial return.

The law modifies provisions of the Federal Food, Drug and Cosmetic Act (21 U.S.C. § 9) to provide sponsors of a veterinary drug for sheep at least three new ways of bringing their products to market, including (1) "conditional approval" to make and keep the product on the market for up to 5 years while collecting the required effectiveness data; (2) requesting the Food and Drug Administration (FDA) to add the drug to an index of legally marketed but unapproved new animal drugs when the potential market for the drug is too small to support the costs of the drug approval process, even under a conditional approval; and (3) approval as a "designated" drug for which 7 years of marketing exclusivity are granted so that the sponsor faces no competition in the marketplace for the approved use of the drug for that time.

Public Land Policies, Regulations, Fees, and Management

With an estimated 25–30 percent of U.S. sheep grazing on public land allotments, public land grazing permits are critical to the sheep industry. Public lands are managed primarily by the USDI BLM and the USDA Forest Service, as well as the Department of Defense, the USDI National Park Service, and various state agencies that manage state-owned lands. Grazing allotments are normally for a single livestock species. Grazing fees are determined on an animal unit basis[6] annually by a congressionally mandated formula. Grazing allotment permits are generally renewed for 10 years but may be reviewed earlier if noncompliance or other environmental concerns (such as fire, floods, and TES designation) arise.

The National Environmental Policy Act (NEPA) of 1969 (P.L. 91-190, 42 U.S.C. §§ 4321-4347, January 1, 1970, as amended by P.L. 94-52, July 3, 1975, P.L. 94-83, August 9, 1975, and P.L. 97-258, § 4(b), Sept. 13, 1982), was landmark legislation passed by the U.S. Congress to ensure that public lands were protected and conserved for future generations. The act requires environmental assessments (EA) and more in-depth environmental impact analysis (EIA) based on best available science in more sensitive situations. NEPA embraces multiple use concepts, including appropriately managed livestock grazing, and public inputs to the decision-making process. Although the legislative intent of NEPA is sound, certain public interest groups opposed to livestock grazing have used NEPA to legally challenge agency decisions that permit continued livestock grazing when permits are reviewed for renewal. When bighorn sheep were released in the northern Sierra Nevada mountain range over 20 years ago, commitments were made to sheep producers by the U.S. Forest Service and California Fish and Game that their grazing permits would not be affected. A lawsuit filed against the

[6]An animal unit is defined as one cow and her calf or five ewes and their lambs.

U.S. Fish and Wildlife Service (FWS) in 2004 by the San Francisco-based Center for Biological Diversity challenging all grazing permits adjacent to the designated Sierra Nevada bighorn habitat has resulted in reduction of grazing allotments for sheep producers, and the FWS has now responded to this lawsuit by designating over 400,000 acres as Sierra Nevada bighorn habitat for the current population of less than 100 bighorn sheep. The Center for Biological Diversity and the Idaho-based Western Watersheds Project recently filed a lawsuit against the USDA ARS U.S. Sheep Experiment Station over grazing sheep on federal land. The research station has been grazing federal lands for over 90 years and is frequently used as a model for land stewardship by the U.S. Forest Service. The lawsuit was settled to the mutual agreement of USDA ARS and the plaintiffs in February 2008, and sheep grazing will continue on experiment station lands and lands with grazing permits.

Even though a primary use of public land is for wildlife habitat, federal agency land managers have little or no control over this use since wildlife on public lands is managed and controlled by state wildlife agencies. If wildlife numbers, such as elk, become overpopulated in a multiple use area with livestock grazing and the state wildlife agency is unwilling to manage the elk population, then the only option available to the public land manager is to reduce livestock grazing. Properly managed grazing systems on private lands optimize livestock and wildlife in harmony with the forage and habitat resources.

Although invasive and nonnative plant species are widely acknowledged as the most serious threat to rangeland health on public lands, public management or control of invasive species is limited. Congress has recognized this concern through legislation but has not appropriated sufficient funds to begin even limited control measures on public lands. As discussed in an earlier section, targeted sheep grazing can be an effective part of the land manager's efforts to control many of the invasive plant species.

Mandatory Price Reporting

Before 2001, information on livestock and meat market transactions reported by USDA was based on data voluntarily submitted by market participants. However, an increasing number of livestock transactions were being made under long-term marketing arrangements (LMAs) that set sales terms well before delivery of the animals for slaughter (Perry et al., 2005). Because the terms of LMAs were not reported during that period, USDA livestock price and volume data were increasingly based on a declining number of transactions. Concerns emerged that the cash market prices reported by USDA did not reflect an increasing share of livestock sales. Along with growing concentration in the meat packing industry at the time, the increasing lack of transparency in livestock transactions fueled concerns of packer manipulation of markets.

The Livestock Mandatory Reporting (LMR) Act of 1999 (P.L. 106-78) was the legislative response to these concerns. The LMR Act required major meatpackers to report all transactions covering hog, cattle, and lamb purchases and commitments to USDA. The act also required packers to report the details of fresh wholesale beef and lamb transactions. In implementing LMR, the intent was to facilitate price discovery in the industry. Initial producer response to the LMR Act was negative primarily because of implementation problems that severely reduced the amount of price and volume data reported. At the same time, the data did not show that contract prices were higher than cash prices as many in the industry expected. The recent study by RTI International (2007) found that for the lamb industry, the primary effect of the LMR Act has been to reduce price risk rather than to influence the level of the price paid for slaughter lambs. The study found that implementation of the LMR Act in 2001 increased the slaughter lamb price by only 0.129 percent. Further, the study concluded that LMAs have had only a small effect on slaughter lamb prices. They found that a 10 percent increase in slaughter lamb purchases through formula pricing increases the slaughter lamb price by an estimated 2.54 percent. In contrast, a 10 percent increase in slaughter lamb purchases through cash markets increases slaughter prices by an estimated 2.68 percent. On the other hand, they found that a 10 percent increase in packer ownership reduces slaughter lamb prices by only an estimated 0.23 percent.

Federal Slaughter Inspection and Food Safety

The federal regulatory agencies that administer inspection and food safety programs include the USDA APHIS, Food Safety and Inspection Service (FSIS), FDA, and the EPA. The FSIS is responsible for ensuring that packing plants adhere to the Humane Methods of Slaughter Act (USDA, 2007e), and veterinarians employed by FSIS enforce the requirements for humane treatment of all animals prior to slaughter. Employees of FSIS inspect all sheep on the harvest floor of a processing plant. The FSIS is also responsible for governing the Pathogen Reduction/Hazard Analysis and Critical Control Point (HAACP) final rule (USDA, 2007f), which enforces safety, sanitation, handling, and processing requirements of all federally inspected slaughter and meat processing plants. Federally inspected plants are inspected and evaluated prior to the start of a production day and during all hours of production, and a follow-up inspection takes place after the closing clean-up is completed. All plants that ship meat and byproducts across state lines (interstate commerce) are federally inspected. State inspection is implemented in smaller local operations; however, their products are currently restricted to intrastate markets. An increasing number of state-inspected facilities are used by direct marketers providing lamb to local markets and private customers. The HAACP is further discussed in Chapter 4.

Animal Welfare Regulations

There are no specific animal welfare or humane handling regulations for sheep. Sheep, however, are covered under two federal humane/welfare regulations: (1) the Humane Methods of Livestock Slaughter Act and (2) the 28-Hour Law. Various state regulations also address animal welfare issues.

Humane Methods of Livestock Slaughter Act

The Humane Methods of Livestock Slaughter Act (7 U.S.C. §§1901–1907) provides the requirements for the humane handling of livestock before and during the slaughter process. Included are provisions for holding and transportation of certain animals prior to slaughter and a separate section for the handling and holding of nonambulatory or "down" livestock. The regulations promulgated under the act outline requirements for the specific method of slaughter such as captive bolt and electrical current. The regulations emphasize the minimization of "excitement and discomfort" to the livestock prior to slaughter.

Under this act, only two methods of livestock slaughter are deemed humane and, therefore, legal: (1) rendering the animal "insensible to pain by a single blow or gunshot or an electrical, chemical or other means that is rapid and effective, before being shackled, hoisted, thrown, cast, or cut"; and (2) slaughtering "in accordance with the ritual requirements of . . . any other religious faith that prescribes a method of slaughter whereby the animal suffers loss of consciousness by anemia of the brain caused by the simultaneous and instantaneous severance of the carotid arteries with a sharp instrument and handling in connection with such slaughtering. . . ."

Twenty-Eight-Hour Law of 1877

The 28-hour law (49 U.S.C. § 80502) addresses the transportation of animals, including those raised for food or in food production, across state lines. The statute provides that animals cannot be transported by "rail carrier, express carrier or common carrier" (except by air or water) for more than 28 consecutive hours without being unloaded for 5 hours for rest, water, and food. Sheep may be confined for an additional 8 consecutive hours without being unloaded when the 28-hour period of confinement ends at night.

State Welfare Regulations

Few states have specific regulations regarding farm animal welfare. New Jersey enacted regulatory standards for the humane handling, housing, treatment, and care of livestock and poultry. It is the first and only state currently to have such broad provisions related to livestock and poultry welfare.

Ewe Lamb Replacement and Retention Program

The Ewe Lamb Replacement and Retention Payment Program (ELRRPP) operated through the Farm Service Agency (FSA) of the USDA provided producers direct payments in 2004 and 2005 to help them replace and retain ewe-lamb breeding stock. The ELRRPP intended to strengthen the lamb industry by providing the financial ability to invest in larger and genetically improved breeding stocks. The program paid eligible operations $18 for each qualifying ewe lamb retained or purchased for breeding purposes. The ELRRPP legislation continues to exist, but funding has not been provided since 2005.

Wool Price Supports and Incentive Payments

The National Wool Act of 1954 established an incentive payment program for growers to be paid out of wool tariff revenues (USDA, 1999). The incentive portion included the establishment of a target price for raw wool and a payment to producers on the basis of the percentage difference between the national average market price and the target price. When the National Wool Act was repealed in 1992, wool production declined markedly and the wool share of total revenues from sheep production fell to only about 10 percent compared to 20 to 25 percent in years when the Wool Act was in force (calculated from data in Meyer et al., 2006). The Farm Security and Rural Investment Act of 2002 reinstated wool price supports through marketing assistance loans (MALs) and loan deficiency payments (LDPs) for the 2002 to 2007 crop years. The wool LDP program is designed to support wool prices approximately at comparable price differences for similar grades of wool in the international marketplace. Current LDP payments to producers for nongraded wool, for example, are $0.41/kg. The only payment differentiation from nongraded wool is for less than 18.6-micron staple length wool, which has varied from $1.76/kg to over $2.20/kg. Although the market price varies substantially for graded wools, the LDP does not provide an incentive to growers for marketing graded, table skirted, and classed wools other than the superfine grade of less than 18.6 microns. Chapter 5 has more details on the current wool policy.

MAJOR ACCOMPLISHMENTS, OPPORTUNITIES, AND CHALLENGES OF THE U.S. LIVE SHEEP INDUSTRY

The U.S. sheep industry is often described as an industry in decline. More appropriately, perhaps, the U.S. sheep industry could be described as being in "transition." Without question, U.S. sheep inventories have declined dramatically since the 1940s along with the number of sheep producers, sheep ranches, investments in research and extension, and government support. A closer look at the industry, however, shows that myriad forces of change—from the growing competition of lamb and textile imports to the changing dietary and fabric preferences of consumers—are honing and transforming the industry toward a potentially more efficient and competitive future. The signs of the transition taking place are already beginning to appear, such as the recent slowing of the long-term decline in inventories in many range sheep states and the modest growth in many farm flock states. Other signs of the industry's transition include the growth in direct marketing, the growth in importance of hair sheep breeds in several regions of the United States, and the emergence of the dairy sheep industry. The transformation is the result of numerous accomplishments of the industry in adjusting to the forces of change. Even so, the future survival of the industry will depend critically on the extent to which the industry recognizes and embraces the available opportunities for growth and change and how the industry chooses to respond to the many challenges it continues to face.

Major Accomplishments of the U.S. Live Sheep Industry

As the scale and competitiveness of the industry have diminished over time, the sheep industry and the supporting federal and state research and extension network have worked together to enhance the productivity and profitability of sheep production. A few of the more important accomplishments in this regard include the following:

- *Increased productivity*. The number of kilograms of lamb weaned per ewe has increased approximately 120 percent in the last 60 years. This has been achieved through genetic progress and improved nutrition, health, and management practices that have increased both the number of lambs weaned per ewe and harvest weight of lambs. The genetic potential exists to further increase productivity within the various production systems. A number of industry programs have facilitated these productivity increases. The land-grant tradition of research programs to develop new technologies combined with extension programs to provide producer education on uses of existing and new technologies was adopted by the sheep industry over 30 years ago through the Sheep Industry Development (SID) program, which is now a part of ASI. With the assistance of university and industry leaders and industry financial support, SID published the *Sheep Production Handbook*, which is now in its seventh edition and is the standard textbook for education programs and producer reference handbook throughout the United States. In addition, SID, and now ASI, have sponsored numerous symposia and published their proceedings to highlight new technologies and to identify continuing research and education needs. The industry continues to seek increased funding to support sheep research and education programs that will further increase lamb and wool productivity.

- *Improvements in lamb processing and merchandising.* Over 60 percent of lamb carcasses are further processed to retail-ready or HRI (hotel, restaurant, and institution)-ready products at the slaughter plant level, which is higher than for other red meats. Oxygen-free packaging systems have increased the shelf life of fresh lamb. These breakthroughs have increased merchandising opportunities, especially in low-volume markets. Shelf-ready lamb products historically included legs, shoulders, loins and racks or loin and rib chops, and lamb shanks, with remaining portions of the carcass often used for pet food products. Product development programs initiated by ASI representatives, meat research scientists, and lamb meat industry teams have developed value-added products that have improved consumer acceptance of lamb. Examples include sirloin chops from the traditional leg of lamb that are merchandised at more than twice the price of legs, dividing the remainder of the leg into butt and shank portions to reduce product package size to more desirable consumer amounts, further processing lamb shoulders into steaks and shish kebab products, further processing rib cages into products such as Denver ribs and lamb riblets, processing the remainder of the carcass trim into ground lamb, and more. New product developments have been promoted with industry checkoff funds and cooperative promotions with lamb processors and retailers.
- *Increased cull ewe market value.* Cull ewes historically were sold for \$8–\$12 per ewe. Emergence of the demand for cull ewes for slaughter in Mexico has increased their sale price to \$40–\$60 per ewe, depending on quality and weight. Domestic demand for cull ewes (mutton) has increased in recent years with the increased Hispanic and Asian immigrant populations in the United States. With an average 5- to 6-year productive life, reduced depreciation costs have increased profit potential for sheep producers. The marketing of cull ewes to Mexico has largely been developed by producers and industry entrepreneurs. The domestic market has been developed by cull ewe processors and marketers that serve growing U.S. ethnic markets.
- *The use of sheep and goats in vegetation management.* The use of

sheep and goats for targeted or prescribed grazing has been well documented in the last 10 years by researchers and practitioners. Sheep are used, for example, to control invasive and nonnative weed, grass, and small shrub species. Goats are also effective for weed control and are the preferred animal species for heavily infested brush control. Sheep are also used to graze firebreaks to protect ecologically sensitive areas and residential areas near the urban/wild land interface and to control competing vegetation in new tree plantations. The American Sheep Industry Association has published a handbook on targeted grazing (Launchbaugh, 2006).

- *Mapping of the sheep genome.* Scientists in U.S. and international laboratories are seeking gene markers for the major economically important traits of sheep such as larger litter size, lean meat production, superfine wool production, internal parasite resistance or tolerance, reduced seasonal breeding for year-round lambing, and foraging behavior. Although these state-of-the-art genetic technologies are in their infancy, future potential could accelerate genetic progress.
- *Sheep breed and genetic improvements.* The number of available sheep breeds has increased substantially in the last 40 years. Several of the new breeds perform at higher levels than breeds that were available before their arrival. Some notable examples are the high litter sizes of the Finnsheep and Romanov, the high milk production of the East Friesian and Lacaune, and the increased muscle mass of the Texel. The relatively recent increase in breeders of hair sheep offers a genetic resource for expansion of the U.S. sheep industry in the southeastern United States, where wool sheep are poorly adapted and sheep numbers have been historically low. Also, the NSIP has developed one of the most advanced programs for genetic improvement in the world. It utilizes state-of-the-art genetic technology to help producers genetically improve their flocks for reproduction, growth, and wool production. The NSIP calculates across-flock EPD based on the animal's own performance and that of its relatives, which producers use in their selection, mating, and culling decisions. The increased use of artificial insemination (AI) practices has also enhanced the availability of genetically superior sires and improved the market potential for superior germplasm.
- *Sheep research and extension support.* The generation of sheep research results and the transfer of these results into practice on U.S. sheep farms is accomplished through the cooperative funding of sheep research and extension by federal, state, and local governments and the integration of extension service and research scientists within land-grant universities. Over $40 million is spent annually on 1,300 to 1,500 sheep research projects conducted at ARS stations and at land-grant universities. State sheep extension specialists and county extension agents interact directly with research scientists and assist sheep producers in the transfer of research results into practice. Although sheep producers provide few funds for sheep research,

their national organizations such as ASI, support valuable educational functions and influence sheep research and extension activities conducted by public institutions.

- *Improvements in sheep-environment interface.* Research on forage feeding has developed new types of forages that use the land resources wisely. For example, research on the use of brassicas in a cropping system as a second crop is allowing producers to plant a cover crop that may reduce soil erosion and will provide nutritious feed for ewes and lambs when pastures normally are dormant. The discussion on targeted grazing earlier in this chapter emphasizes the current and future potential of sheep to contribute to environmental and rangeland health.
- *Improvements in lamb feeding and nutrition.* The National Academy of Sciences (NAS) recently published a revised edition of the nutrient requirements for sheep included in the report, *Nutrient Requirements of Small Ruminants* (NRC, 2007). A cursory review of the historical editions clearly demonstrates that nutrient requirements have increased to meet the nutrient demands of the significant increases in reproductive efficiency and lamb growth rates in the last 50 years. Research reported in the current edition of the NAS publication documented the increased needs of both macro- and micronutrients to support current levels of productivity. Educators have developed and producers have implemented new feeding regimes for both forage-based and concentrate-based feeding regimens that recognize the current nutrient demands for reproduction, lamb growth, and wool production. Producers now have a better understanding of how the nutritive value of certain feedstuffs will contribute to reproduction, lamb growth and feed efficiency, and product quality to the consumer.

Major Opportunities and Challenges of the U.S. Live Sheep Industry

Key opportunities for enhanced efficiency and competitiveness of the U.S. sheep industry include the following:

- *Continued productivity improvements.* The potential clearly exists to continue the momentum to improve productivity and production efficiency through continued improvements in the genetic, nutrition, animal health, and management programs that have contributed to the dramatic increases in kilograms of lamb marketed per ewe in the last 60 years. Realizing the potential, however, will require continued public support of research and education programs, as well as industry leadership and producers who are prepared to meet the challenges of change.
- *Targeted or prescribed grazing.* Using sheep and goats for specific vegetation management practices is an important tool available to land managers. Chemical and mechanical means, controlled burns (fire), and manual

labor are other options, all of which are expensive and often not acceptable to the public. Producer practitioners have been paid to implement targeted grazing or have been provided grazing at no cost, suggesting that this could be a potential supplemental income source or a means of reducing production costs for sheep and goat producers, and at the same time providing a public good for communities and society.

- *Forage finishing for competitive advantage.* Lambs can be grown postweaning on high-quality forage pasture or crop residues to slaughter weight or to weights that minimize grain feeding in feedlots. With current high feed grain costs and the likelihood of continued increases, forage finishing could enhance lamb's competitive position relative to other red meats. Forage-feeding systems have historically been lower cost than concentrate feeding, suggesting that they could be more price-competitive with imported lamb products in the retail meat case.
- *Emerging ethnic markets for lighter-weight lambs.* The emerging and rapidly expanding ethnic markets for lamb present a particularly important opportunity for greater direct marketing of lambs. Ethnic markets for high-quality, lighter-weight lambs at current market premiums could potentially revolutionize lamb production, marketing, feeding, and processing systems. Lambs from the hair sheep breeds are generally smaller framed and lighter weight at slaughter, and fit well with the demands of many ethnic markets. Highest priority should be to invest in research to better understand this market and its implication for the U.S. sheep industry.
- *The sheep genome and gene biotechnology.* Completion of the mapping of the sheep genome is a major breakthrough in the potential for genetic improvements. Australia and New Zealand are already merchandising gene markers for superfine wool production and internal parasite resistance, and continue to invest in the development of additional gene markers for genetic improvement in lamb and wool production. Although there are currently only limited funding resources for sheep genomics research in the United States, the potential use of these new technologies for genetic improvement clearly suggests that this should be a high-priority investment. Gene biotechnology, including molecular genetics and gene therapy, have the potential of providing more effective vaccines and medications to protect against animal diseases and to correct gene deficiencies and dysfunctions. Research using sheep as the biological model for human biomedical research has developed the technologies for direct use of gene therapies and molecular genetics to address sheep genetic deficiencies and dysfunctions. However, little support is available for this research opportunity.

Despite the opportunities, the U.S. sheep industry faces a number of key challenges that must be overcome to achieve sustainable growth and economic competitiveness, including the following:

• *Critical mass.* A major challenge facing the industry is the decline in sheep numbers, which has resulted in the decline of industry infrastructure, including lamb and wool market outlets, sheep slaughter and processing facilities, low volumes of product for available markets, lack of qualified sheep shearers and wool classers, reduced market providers of the supplies and equipment essential for sheep production, and reduced state and federal support for sheep research and education faculty. While these changes in infrastructure and markets have been forced by the reduction in the scale of the industry, the consequence is that the sheep industry today has a weak platform from which to maintain the transition underway, let alone to launch significant and rapid industry growth and development.

• *Predation.* The continued management (control) of predator populations by USDA Wildlife Services, as well as state and producer-supported programs, is critical to the survival of the U.S. sheep industry. Increased predation problems in wildlife populations in many states indicate the need for sheep industry alliances with wildlife agencies and interest groups.

• *Livestock, wildlife, and threatened and endangered species management.* Management programs will need to recognize the importance of ecosystem management based on the best available science rather than specific species management plans or practices that exclude appropriate multispecies management systems.

• *Price determination/price discovery.* The Mandatory Price Reporting (MPR) system has resulted in more price reporting for live lambs and has generated more information on imported lamb carcass prices. Even so, the lack of information on prices due to confidentiality issues continues to be a primary challenge in feeder and slaughter lamb markets. Also, relatively little public data are available for assessing relative prices between domestic lamb carcasses and imported lamb carcasses. At the same time, most slaughter lambs are now being priced using formulas or are packer fed so that the lamb price discovery process is now based largely on carcass or cutout values. Consequently, price is determined by negotiation or formula related to carcass quality. As a result, a significant amount of the risk is shifted from the buyer to the seller, especially for pricing based on quality.

• *Adjustment to emerging markets for lamb.* While continued growth of ethnic-based markets for lighter-weight lambs may have a potentially positive impact on lamb demand and prices, the existing industry infrastructure must respond sufficiently to accommodate the potentially major changes this market opportunity may require. Changes in sheep industry leadership policies and programs may be required.

• *Adoption of genetic improvement technology.* Compared to breeders of dairy cattle, beef cattle, swine, and poultry, purebred breeders of sheep lag behind in the adoption of genetic improvement technology, with decisions often based on show ring performance rather than commercial efficiency criteria. Only a small number of sheep breeders are using accurate estimates of genetic merit and performance-based criteria for economically important traits in their selection decisions. The result is less than optimum genetic improvement in the entire U.S. sheep population, which may ultimately result in sheep being less competitive with other livestock species. Increased enrollment of purebred flocks in the NSIP could be helpful in this regard.

- *Sheep and wool research funding.* Sheep and wool research receives the smallest amount of funding of any of the livestock sectors, resulting in less development of new technology and educational support for the sheep industry relative to other livestock sectors. The lack of new technology for the sheep industry relative to other livestock species will likely result in the sheep industry being less competitive, leading to further decreases in sheep numbers, further reductions in sheep research and extension efforts, even less new technology, and continued negative pressure on the competitiveness of the industry.
- *Sheep research and extension at land-grant universities.* A decreased emphasis on sheep research is already evident at land-grant universities. For example, in most departments of animal science there is at least one research scientist working in each of the core areas of nutrition, reproduction, genetics, and product (meat, milk, or eggs) with each of the livestock sectors of dairy cattle, beef cattle, pigs, and poultry. However, there are few departments that have three or more scientists working in the sheep area. Most departments have two or fewer. The main focus of the sheep research program at a particular university or ARS station most often is in the core area of the sheep scientist's training, resulting in a deficiency of work in other areas. Some sheep scientists try to become generalists and cover all areas with highly applied research efforts. This approach is effective in the short term in generating useful information for producers but does not result in the generation of new basic knowledge for the development of new technologies for application in the future. A possible solution to this downward spiral is the formation of several sheep research and extension consortia among land-grant universities and ARS stations. Three or more neighboring states with similar types of sheep production systems could form a consortium. Land-grant universities and ARS stations within those states could each agree to cover certain specific sheep research areas so that all major areas important to sheep production in the region are covered by at least one effective research program. The results of all research programs would be readily available to sheep extension personnel in all states for application to sheep producers. By necessity, several informal arrangements of sharing expertise across states have evolved among sheep scientists. There is now a need, however, to formalize such arrangements in order to gain the most benefit out of limited sheep research and extension resources.

REFERENCES

ALBC (American Livestock Breeds Conservancy). 2007. Conservation priority list: Livestock breeds. Pittsboro, NC. Online at: http://www.albc-usa.org/cpl/wtchlist.html. Accessed May 5, 2008.

Arnold, A. M., and H. H. Meyer. 1988. Effects of gender, time of castration, genotype and feeding regimen on lamb growth and carcass fatness. J. Anim. Sci. 66:2468–2475.

ASI (American Sheep Industry Association). 2002. Sheep Production Handbook. Centennial, CO: American Sheep Industry Association.

AVMA (American Veterinary Medical Association). 2005. Minor Use and Minor Species Animal Health Act. Schaumburg, IL: American Veterinary Medical Association. Online at: http://www.avma.org/issues/drugs/mums/default.asp. Accessed May 5, 2008.

Bastian, C., and G. Whipple. 1998. An historical overview of lamb marketing in the United States and considerations for the future. Sheep Goat Res. J. 14:4–15.

Berger, P. 1991. Long-term effects of feeding during calfhood on subsequent performance in beef cattle (a review). Livest. Prod. Sci. 28:179–201.

Bodenchuk, M. J., J. R. Mason, and W. C. Pitt. 2002. Economics of predation management in relation to agriculture, wildlife, and human health safety. Proceedings of the 3rd National Wildlife Research Center Special Symposium. Fort Collins, CO: National Wildlife Research Center.

Boland, M. A., A. Bosse, and G. W. Brester. 2007. The Mountain States Lamb Cooperative: Can vertical integration keep lamb producers from being fleeced? Rev. Agric. Econ. 29:157–169.

Bouton, P. E, P. V. Harris, D. Ratcliff, and D. W. Roberts. 1978. Shear force measurements on cooked meat from sheep of various ages. J. Food Sci. 43:1038–1039.

Brethour, J. R. 1994. Estimating marbling score in live cattle from Ultrasound images using pattern recognition and neural network procedures. J. Anim. Sci. 72:1425–1432.

Bunge, R., D. L. Thomas, and T. G. Nash. 1995. Performance of hair breeds and prolific wool breeds of sheep in southern Illinois: Lamb production of F1 adult ewes. J. Anim. Sci. 73:1602–1608.

Campbell, J. R. 1995. Reclaiming a Lost Heritage. Ames, IA: Iowa State University Press.

Cook, C. W. 1985. Biological efficiency from rangelands through management strategies. Pp. 54–64 in Proceedings of the Conference on Multispecies Grazing, F. H. Baker and R. K. Jones, eds. Morrilton, AR: Winrock International.

Crouse, J. D., R. A. Field, J. L. Chant, C. L. Ferrell, G. M. Smith, and V. L. Harrison. 1978. Effect of dietary energy intake on carcass composition and palatability of different weight carcasses from ewe and ram lambs. J. Anim. Sci. 47:1207–1218.

Deakin, G. A., ed. 2007. Report of registrations by purebred sheep associations. The Banner 30(2):20.

Ely, D. G. 1994. The role of grazing sheep in sustainable agriculture. Sheep Goat Res. J. (Spec. Iss.):37–51.

EPA (U.S. Environmental Protection Agency). 2002. Concentrated animal feeding operations: Clean water act requirements. What are the federal requirements for horse and sheep CAFOs? Washington, DC. Online at: http://purl.access.gpo.gov/GPO/LPS45185. Accessed May 5, 2008.

Falxa, K., L. W. Van Tassell, and J. P. Hewlett. 2002. Advantages of multispecies grazing: Perceptions of Idaho and Wyoming producers. Sheep Goat Res. J. 17(1):1–8.

Foreyt, W. J., K. P. Snipes, and R. W. Kasten. 1994. Fatal pneumonia following inoculation of healthy bighorn sheep with *Pasturella hemolytica* from healthy domestic sheep. J. Wildlife Dis. 2:137–145.

Gaskin, J. F., and B. A. Schaal. 2002. Hybrid *Tamarix* widespread in U.S. invasion and undetected in native Asian range. Proc. Nat. Acad. Sci. 99:11256–11259.

Glimp, H. A. 1988. Multispecies grazing and marketing. Rangelands 10:275–278.

Glimp, H. A., and L. Haug. 2004. Prescribed grazing for vegetation management. Paper presented at Symposium on Prescribed Grazing, University of California, Davis and at ASI Symposium on Prescribed Grazing for Vegetation Management, Phoenix, AZ.

Goll, D. E., R. W. Bray, and W. G. Hoekstra. 1963. Age-associated changes in muscle composition. The isolation and properties of collagenous residue from bovine muscle. J. Food Sci. 28:503–509.

Greer, H. C., and C. E. Ward. 2000. Regional differences in slaughter lamb marketing and prices. Sheep Goat Res. J. 16(2):52–57.

Houghton, P. L., and L. M. Turlington. 1992. Application of ultrasound for feeding and finishing animals: A review. J. Anim. Sci. 70:930–935.

Jahnke, L. C., S. H. Phillips, S. H. Anderson, and L. L. McDonald. 1987. A methodology for identifying sources of indirect costs of predator control. Pp. 159–169 in Vertebrate Pest Control and Management Materials, S. A. Shumake and R. W. Bullard, eds. Vol. 5. ASTM STP 974. Philadelphia, PA: American Society for Testing and Materials.

Knowles, D., and A. Rink. 2006. Outline of concerns relating to perception of disease transmission issues at the livestock wildlife interface in the western United States. Invited paper prepared for American Sheep Industry Association, the Agricultural Research Service (USDA), and the Nevada Department of Agriculture.

Koch, D. W., C. Kercher, and R. Jones. 2002. Fall and winter grazing of Brassicas—a value-added opportunity for lamb producers. Sheep Goat Res. J. 17(2):1–13.

Kuber, P. K. 2005. Youth food animal quality assurance curriculum guide. Columbus: The Ohio State University Extension.

Launchbaugh, K., ed. 2006. Targeted Grazing: A Natural Approach to Vegetation Management and Landscape Enhancement. Englewood, CO: American Sheep Industry Association.

LMIC (Livestock Marketing Information Center). 2008. Livestock Marketing Information Center website. Online at: http://www.lmic.info/. Accessed April 8, 2008.

Martin, K. D., T. Schommer, and V. L. Coggines. 1996. Literature review regarding the compatibility between bighorn and domestic sheep. Biennial Symposium of the Northern Wild Sheep and Goat Council 10:72–77.

McClure, K. E., R. W. Van Keuren, and P. G. Althouse. 1994. Performance and carcass characteristics of weaned lambs either grazed on orchardgrass, ryegrass, or alfalfa or fed all concentrate diets in drylot. J. Anim. Sci. 72:3230–3237.

Mead, L. E. 1967. 1966 Purebred sheep review. Sheep Breeder and Sheepman 87(3):24–32.

Meyer, L., S. MacDonald, and R. Skinner. 2006. Cotton and Wool Situation and Outlook Yearbook. Report No. CWS-2006. U.S. Department of Agriculture. Economic Research Service, November. Online at: http://usda.mannlib.cornell.edu/usda/ers/CWS-yearbook//2000s/2006/CWS-yearbook-11-21-2006.pdf. Accessed May 5, 2008.

Miller, D. C. 1998. Accuracy and application of real-time ultrasound for evaluation of carcass merit in live animals. Online at: http://www.cals.ncsu.edu/an_sci/extension. Accessed May 5, 2008.

Mosier, N. A. 2006. Cellulosic ethanol—biofuel beyond corn. Purdue University Extension Publication ID-335. West Lafayette, IN: Purdue University.

Murphy, M. A., H. N. Zerby, and F. L. Fluharty. 2003. The effects of energy sources and ionophore supplementation on lamb growth, carcass characteristics, and tenderness. Sheep Goat Res. J. 18:89–95.

Murphy, T. A., S. C. Loerch, K. E. McClure, and M. B. Solomon. 1994. Effects of grain or pasture finishing systems on carcass composition and tissue accretion rates of lambs. J. Anim. Sci. 72:3138–3144.

Nada, T., K. Fujita, M. Fujii, M. Yoshida, T. Haryu, S. Misumi, and T. Okumura. 2005. Development of X-ray computed tomography for live standing cattle. Anim. Sci. 76:513–517.

NIAA (National Institute for Animal Agriculture). 2008. Eradicate scrapie! Online at: http://www.animalagriculture.org/scrapie/Scrapie.htm. Accessed February 26, 2008.
Notter, D. R., R. F. Kelly, and B. W. Berry. 1991. Effects of ewe breed and management system on efficiency of lamb production: III. Meat characteristics. J. Anim. Sci. 69:3523–3532.
NPB (National Pork Board). 2007. Des Moines, IA. Online at: http://www.pork.org. Accessed May 2, 2008.
NRC (National Research Council). 2007. Nutrient Requirements of Small Ruminants: Sheep, Goats, Cervids, and New World Camelids. Washington, DC: The National Academies Press.
NSIP (National Sheep Improvement Program). 2007. Ovine EPD. Online at: http://www.members.aol.com/ovineepd/ovineepd.htm. Accessed May 5, 2008.
Oltjen, J. W., and H. A. Glimp. 1992. Standardized performance analysis for 6 Nevada range sheep operations. (Presentation to American Sheep Industry Association), Davis, CA.
Perry, J., J. MacDonald, K. Nelson, W. Hahn, C. Arnade, and G. Plato. 2005. Did the mandatory requirement aid the market? Impact of the Livestock Mandatory Reporting Act. Outlook Report No. LDPM13501. Economic Research Service, U.S. Department of Agriculture, Washington, DC, September. Online at: http://www.ers.usda.gov/publications/ldp/sep05/ldpm13501. Accessed May 5, 2008.
Pimentel, D. L., L. Lach, R. Zuniga, and D. Morrison. 2000. Environmental and economic costs of non-indigenous species in the United States. BioScience 50:53–65.
RTI (Research Triangle Institute International). 2007. Livestock and Meat Marketing Study. Volume 5: Lamb and Lamb Meat Industries. RTI Project Number 0209230. Final report prepared for the Grain Inspection, Packer and Stockyard Administration, U.S. Department of Agriculture, Washington, DC, February.
Rector, B. S. 1983. Diet selection and voluntary forage intake by cattle, sheep and goats grazing in different conditions. Ph.D. Dissertation, Texas A&M University: College Station.
Roeber, D. L., K. Belk, S. B. LeValley, J. A. Scanga, J. N. Sofos, and G. C. Smith. Undated. Producing consumer products from sheep: The sheep safety and quality assurance program. Report for the American Sheep Industry Association. Colorado State University, Ft. Collins. Online at: http://www.colostate.edu/programs/SSQA/. Accessed May 5, 2008.
Schoenian, S. 2007. Sheep 201: Feeding lambs. Online at: http://www.sheep101.info/201/feedinglambs.html. Accessed May 5, 2008.
Schroeder, T. C., S. Grunewald, and C. E. Ward. 2002. Mandatory price reporting in fed cattle markets: Motivations and implications. Council on Food, Agricultural, and Resource Economics (C-FARE) Annual Symposium, Public Information and the Food and Agricultural System, Washington, DC. Online at: http://www.cfare.org/updates/2002Symposium TedSchroeder.pdf. Accessed May 21, 2007.
SGRJ (Sheep and Goat Research Journal). 2004. Special Edition: Predation. American Sheep Industry Association. Englewood, CO. Online at: http://www.sheepusa.org/index.phtml?page=site/researchopener&nav_id=601e0a31bbf6a0ef56f1f0591aa0dc78. Accessed May 5, 2008.
Shanks, B. C., P. G. Hatfield, R. A. Field, J. A. Hopkins, and G. S. Lewis. 2000. Influence of winter backgrounding systems on subsequent lamb finishing performance, body composition, carcass traits, and palatability. Sheep Goat Res. J. 16(3):88–93.
Shelton, M. 2004. Predation and livestock production perspective and overview. Sheep Goat Res. J. 19:1–5.
Suster, D., B. J. Leury, E. Ostrowska, K. L. Butler, D. J. Kerton, J. D. Wark, and F. R. Dunshea. 2003. Accuracy of dual energy X-ray absorbtiometry (DXA), weight and P2 back fat to predict whole body and carcass composition in pigs within and across experiments. Livest. Prod. Sci. 84:231–242.
Talley, B. 2008. Hair contamination still an issue in U.S. wool clip. Sheep Industry News 12(2):9.

Tatum, J. D., B. J. Klein, F. L. Williams, and R. A. Bowling. 1988. Influence of diet on growth rate and carcass composition of steers differing in frame size and muscle thickness. J. Anim. Sci. 66:1942–1954.

Thomas, D. L. 2007. Budget for a 50-ewe farm flock sheep enterprise: 2006/07 operating costs and income. Presented at Wisconsin Sheep and Wool Festival, Jefferson, WI.

Thomas, D. L., and L. R. Miller. 2001. Organization and funding of sheep research and extension in the United States—from research to practice. Livest. Prod. Sci. 72:65–73.

USDA (U.S. Department of Agriculture). 1992. United States standards for grades for slaughter lambs, yearlings, and sheep: Effective date July 6, 1992. Agricultural Marketing Service. Washington, DC. Online at http://www.ams.usda.gov/LSG/ stand/standards/sl-lamb.pdf. Accessed March 16, 2007.

USDA. 1994. Sheep and lambs on feed. National Agricultural Statistics Service. Washington, DC: USDA.

USDA. 1999. Economic Impact of the Elimination of the National Wool Act. Economic Research Service. Report prepared for the U.S. House of Representatives Committee on Appropriations, as requested in House Report 105–588, p. 16, 1998. Online at: http://www.ers.usda.gov/Briefing/Sheep/WoolActStudy.pdf. Accessed November 10, 2007.

USDA. 2001. Sheep 2001 Part IV: Baseline Reference of 2001 Sheep Feedlot Health and Management. Animal and Plant Health Inspection Service, National Animal Health Monitoring System, Washington, DC. Online at: http://www.aphis.usda.gov/vs/ceah/ncahs/nahms/sheep/index.htm. Accessed May 5, 2008.

USDA. 2002. Agricultural Atlas of the United States: Sheep and Lambs—Inventory. Map 02-M160. Maps and Cartographic Resources, National Agricultural Statistics Service, Washington, DC. Online at: http://www.nass.usda.gov/research/atlas02/. Accessed May 5, 2008.

USDA. 2003. Sheep and lambs on feed. National Agricultural Statistics Service. Washington, DC.

USDA. 2005. Sheep and goats death loss. National Agricultural Statistics Service. Washington, DC, May 6. Online at: http://usda.mannlib.cornell.edu/usda/current/sgdl/sgdl-05-06-2005.pdf. Accessed March 16, 2007.

USDA. 2007a. Farm income data, Table 5: Cash receipts. Economic Research Service Washington, DC, August. Online at: http://www.ers.usda.gov/ Data/FarmIncome/FinfidmuXls.htm. Accessed May 5, 2008.

USDA. 2007b. Mandatory price reporting frequently asked questions. Agricultural Marketing Service, Livestock and Seed Division, Washington, DC. Online at: http://www.ams.usda.gov/lsmnpubs/mpr/Q&A08.htm.

USDA. 2007c. Summary livestock and grain market news reports. Agricultural Marketing Service. Online at: http://www.ams.usda.gov/LSMNpubs/NatSum.htm. Accessed May 5, 2008.

USDA. 2007d. Current Research Information System (CRIS). Cooperative State Research, Education, and Extension Service, Washington, DC. Online at: http://cris.csrees.usda.gov. Accessed May 5, and July 19, 2007.

USDA. 2007e. Fact Sheets: Production and Inspection. Online at: http://www.fsis.usda.gov/Factsheets/Key_Facts_Humane_Slaughter/index.asp. Accessed May 5, 2008.

USDA. 2007f. About FSIS: Agency History. Food Safety and Inspection Service. Online at: http://www.fsis.usda.gov/About_Fsis/index.asp. Accessed May 5, 2008.

USDA. 2007g. Quick Stats: National Agricultural Statistics Service. Online at: http://www.nass.usda.gov/Data_and_Statistics/Quick_Stats/index.asp. Accessed May 5, 2008.

USDA. 2008. National Animal Germplasm Program, Agricultural Research Service, Washington, DC. Online at: http://www.ars-grin.gov/animal/. Accessed Feb. 26, 2008.

USITC (U.S. International Trade Commission). 1999. Lamb Meat Industry Adjustment Plan. Investigation No. TA-201-68. USITC Publication No. 3176, Washington, DC.

Van Dyne, G. M., N. R. Brockington, Z. Sxocs, J. Duek, and C. A. Ribic. 1980. Large herbivore sub-systems. Pp. 269–537 in Grasslands, Systems Analysis, and Management, A.I. Breyemer and G.M. Van Dyne, eds. Cambridge, UK: Cambridge University Press.

Walker, J. W. 1994. Multispecies grazing: The ecological advantage. Sheep Goat Res. J. (Spec. Iss.):52–64.

Ward, A. C., D. L. Hunter, M. D. Jaworski, P. J. Benolkin, M. P. Dobel, J. B. Jeffress, and G. A. Tanner. 1997. *Pasturella spp.* in sympatric bighorn and domestic sheep. J. Wildl. Dis. 3:544–557.

Weir, C. E. 1960. P. 212 in The Science of Meat and Meat Products. New York: Reinhold Publishing Co.

Westbrooks, R. 1998. Invasive Plants, Changing the Landscape of America: Fact Book. Washington, DC: Federal Interagency Committee for the Management of Noxious and Exotic Weeds (FICMNEW).

Wildeus, S. 1997. Hair sheep genetic resources and their contribution to diversified small ruminant production in the United States. J. Anim. Sci. 75:630–640.

Williams, G. W., and E. E. Davis. 1998. Lamb market structure. Sheep Goat Res. J. 14(1):16–34.

WRA (Western Range Association). 2007. Online at: http://www.allbusiness.com/ agriculture/livestock-except-dairy-poultry/3886111-1.html. Accessed May 5, 2008.

Yambayamba, E., and M. A. Price. 1991. Growth and performance and carcass compositions in beef heifers undergoing catch-up (compensatory) growth. Can. J. Anim. Sci. 71:1021–1029.

3

Sheep Health Issues

Maintaining the health of a flock is imperative for a successful sheep operation because disease imposes both apparent as well as hidden costs on the operation. Obvious expenses include the price of treatments and death loss. Hidden costs can be attributed to disease-related production inefficiencies. For example, a ewe that is lame or one that suffers from ovine progressive pneumonia (OPP) will produce less milk than her genetics or nutrition should allow and, therefore, may raise lambs with a reduced rate of gain. There are little or no data, however, on the economic impacts or the true prevalence of most of the disease conditions affecting sheep in United States. The lack of this information makes it quite difficult to make decisions regarding the allocation of resources as well as the determination of research and policy priorities.

Some of the more common costs associated with ill or unthrifty sheep include but may not be limited to:

- Treatment;
- Mortality;
- Premature culling;
- Lower number of live lambs born as a result of (1) failure of the ram to settle ewes, (2) reduced conception rates, (3) failure to carry lambs to term, and (4) failure to give birth to live lambs;
- Reduced birth weights;
- Weak lambs (little or no colostrum);
- Reduced rate of gain either due to poor milk production by the ewe or poor health of the lamb;

- Disease that results in wool loss or lower quality wool; and
- Labor costs associated with treatment.

Maintaining the health of the national flock has even broader implications than for any given individual flock. An introduction of a foreign animal disease or an endemic disease that mutates may result in large-scale repercussions such as a food safety event that could result in human illness and decreased consumer confidence. In the past decade, > 70 percent of the emerging diseases have been zoonotic (Woolhouse and Gowtage-Sequeria, 2005). The U.S. sheep industry has been fortunate to avoid a significant crisis resulting from a highly contagious or zoonotic disease outbreak. To prevent complacency, the industry as a whole will need to systematically review worldwide conditions and disease reports as well as suspicions of diseases mutating to a more virulent strain. One of the greatest vulnerabilities is the risk of introducing a foreign animal disease. When introduced into a native population, foreign animal diseases may spread rapidly because the animals have no immunity to the disease. In many cases, the outbreaks result in a high death losses and/or severe production losses. To illustrate that the cost of prevention is usually much less expensive than control or elimination, the World Organization for Animal Health (OIE) commissioned a study to compare the cost of preparedness and prevention versus the cost of control for avian influenza. Even without considering the indirect financial impacts, the benefits of prevention far outweighed the potential outbreak costs and losses (Agra CEAS Consulting, 2007).

The intent of this chapter is to examine broad issues regarding health and how they may impact the economic viability of the sheep industry as a whole, as well as individual sheep operations. The chapter is not intended to serve as a comprehensive review of sheep diseases. Certain diseases are highlighted to illustrate different points. A detailed review of scrapie is included because it is the only sheep disease that has had a congressionally funded control program in place for over 50 years.

MANAGEMENT PRACTICES REGARDING HEALTH

Sound flock health management practices are the key to disease prevention and control. While it is important to have vaccination and treatment programs for specific diseases, a holistic approach founded on prevention can be the most effective tool for maintaining a healthy flock. This in turn increases productivity. The basic management tools include:

- Biosecurity, including maintaining a closed flock or restricting the number and source of replacements;
- Recordkeeping, including animal identification;

- Technical resources; and
- Availability of effective vaccines and treatments.

Biosecurity and Replacement Practices

Diseases gain entry into and spread within flocks by various routes. Most commonly, they are brought in by the introduction of new animals or by animal contact at points of concentration such as shows, sales, fairs, and sale barns. Disease can also be spread by vectors that include visitors, equipment, feed, and insects. Good hygiene and biosecurity measures are important methods of preventing the spread of disease into and within a flock. Although there are costs associated with biosecurity measures, they have the potential to halt the entry of pathogens that may result in disease. As stated previously, the costs associated with disease are both direct (treatment and mortality losses) and indirect (production losses) and usually are higher than the expense of prevention.

Biosecurity refers to measures taken to keep diseases out of populations, herds, or groups of animals or to limit the spread of diseases. Successful biosecurity measures must address isolation of new animals brought to the farm; isolation of sick animals; regulation of the movement of people, animals, and equipment; correct use of feed; and procedures for cleaning and disinfecting facilities. The responsibility for a successful biosecurity program falls on the owner (European Community, 2007). The American Sheep Industry Association (ASI) has a fact sheet on biosecurity that provides useful advice.[1]

The 2001 National Animal Health Monitoring System (NAHMS) study (USDA, 2002) found that more than one-half of the sheep operations with 100 or more head added replacements from the outside in 2000. This was also true for approximately 25 percent of operations with 1–24 sheep and close to 40 percent of the operations with 25–99 head. Approximately 80 percent of all operations added sheep in the last 9 years (USDA, 2003a).

New replacements from other flocks pose a risk of introducing a disease into an existing flock. Certain precautions may be taken to minimize this risk. Effective prevention measures may include obtaining a complete history of the flock of origin, testing for certain diseases prior to and after movement, isolation and quarantine on the new farm, vaccination, treatments such as medicated foot baths, and deworming. When introducing/purchasing animals from another farm a good practice is to investigate the health status of the flock of origin. This inquiry should include a discussion

---

[1]http://www.sheepusa.org/index.phtml?page=site/text&nav_id=3c081c2af5f98f1a054911d06824094f.

with the producer about biosecurity practices, as well as a review of vaccination and testing programs, veterinary visits, and laboratory reports.

Although quarantine is an effective way to monitor for and prevent disease introduction, only 33.9 percent of operations accounting for 16.4 percent of new additions administer any type of quarantine prior to commingling new animals with the existing flock. Contact with other sheep during shows, exhibitions, breeding, grazing, and other encounters also poses a risk of disease transmission. Over 50 percent of operations report such contacts, yet less than 30 percent made an attempt to decrease nose-to-nose contact (USDA, 2003a).

On most operations, the major treatments performed on the new additions either prior to or upon arrival include deworming (75.2 percent) and vaccination (61.8 percent). A larger percentage of operations performed these two treatments after arrival than before movement. Another effective method of keeping certain diseases out of flocks is to test for diseases prior to an animal leaving a flock and then retest again in quarantine. This does not appear to be a common practice. Only 6.3 percent of operations test for OPP and 3.4 percent for Johne's disease prior to or upon arrival into the new flock (USDA, 2002).

Visitors who borrow equipment, feed, rendering, and other service-related vehicles can also bring disease agents onto a farm. This is especially true with highly contagious diseases such as foot and mouth disease (FMD). Of operations that allowed visitors, only 22.6 percent had any biosecurity requirements for visitors.

A proactive management practice is to prevent introduction of disease into a flock rather than reacting to an introduction or outbreak of disease. Prevention is also the safest and least costly method of protection. A plan of prevention requires each farm or ranch operator to carefully evaluate their management practices in order to identify specific practices that could present potential risks in their production unit and incorporate the common sense biosecurity practices necessary to reduce those risks. Biosecurity management plans are intended to minimize the factors that increase the risk of disease transmission into and within flocks and maximize the factors that decrease disease. The existing data suggest that certain producer behaviors and lack of effective disease management practices increase the vulnerability of many sheep operations to disease (USDA, 2003b).

There are a number of important endemic diseases or disease conditions that can enter a flock unless adequate biosecurity measures are in place. These are described in the following subsections:

Infectious Foot Rot

Infectious foot rot can quickly become a chronic problem. It is difficult to eliminate, requiring time, labor, and financial resources. Foot rot can be introduced by adding sheep from any source, borrowing rams for breeding, sharing infected pastures, sharing common holding areas such as at sales and shows, and sharing common hauling or shearing trailers. Warmth and moisture, which are conducive environmental conditions for the transmission of foot rot, can lead to a large proportion of a flock becoming infected within 1–2 weeks.

Foot rot in sheep has two forms, benign and virulent. The virulent form may be quite severe, causing an animal to limp or be unwilling to place weight on one foot. If more than one foot is affected an animal may graze on its knees or not get up at all. Economic losses may be considerable and consist of treatment costs for labor, drugs, vaccines, and equipment; decreased flock productivity (poor growth rates, reduced wool growth and poor wool quality and reduced fertility); and loss of sales.

Foot rot is the result of the action of two bacteria. *Fusobacteria necrophorum,* which is present in the environment, assists *Dichelobacter* (formerly Bacteroides) *nodosus,* the causal agent in its quest to invade the hoof. *D. nodosus* does not survive in the environment for more than a few days to 2 weeks. It can, however, persist in the feet of infected sheep for many years even under dry conditions.

Foot baths, parenteral antibiotics, topical treatments, vaccines, trimming of the feet, and rigorous culling, as well as determination and dedication may be required to eliminate foot rot from a flock. Some producers opt to depopulate and start over. Prevention is the best option (Seaman and Evers, 2006; Radostits et al., 2007).

Ovine Progressive Pneumonia (OPP)

Ovine progressive pneumonia (OPP), which is also known as Maedi-Visna in Europe, is caused by an ovine lentivirus that induces a persistent infection in sheep and may result in lymphoproliferative changes in the lung, mammary gland, brain, and/or joints. Because of the nature of the virus, the animal never develops complete immunity and fails to completely eliminate the virus. Lambs may become infected at birth or shortly thereafter by the ingestion of infected colostrum and/or milk. Lateral transmission via a respiratory route has also been documented in older sheep. The importance of each route may vary between flocks and management practices.

Even in flocks that have a significant rate of infection, the manifestation of clinical disease is low. Clinical disease is not apparent before 2 years of age and is most common after 3 years. The disease develops insidiously with the earliest sign being a loss of body condition, which is why this disease is sometimes called "thin ewe syndrome." Dyspnea develops later and may lead to exercise intolerance, causing a ewe to fall behind the rest of the flock. In some cases there is mammary gland or joint involvement. The disease is also known as "hard bag." The milk is normal in appearance but there is little to no volume.

Economic losses associated with OPP are primarily the result of premature culling, the loss of value for the mature clinically affected animals, mortalities, and the possible effects of subclinical infection on productivity. A review of other countries' import regulations indicates that these may be barriers to U.S. sheep.

Once OPP has been introduced into a flock it is difficult and costly to eradicate because there is no treatment. If a flock is free of OPP, considerable effort should be made to maintain this status and prevent introduction of the disease. Control may be attempted by segregating lambs at birth and hand raising them on bovine milk or milk replacers. Another strategy is to test and cull serologically positive animals and their offspring (Radostits et al., 2007).

Seroprevalence of OPP was measured during the 2001 NAHMS sheep study by testing sheep on randomly selected sheep operations. More than 21,000 samples collected at 682 operations were tested. The seroprevalence was calculated to be 24.2 percent. Overall, 36.4 percent of the operations had one or more positive animals (USDA, 2003c).

Contagious Ecthyma

Contagious ecthyma, also called soremouth or orf, is caused by the orf virus, genus *Parapoxvirus*. It is commonly introduced into a flock by the purchase of new animals or by contact with other sheep at fairs and shows. Soremouth most commonly affects young lambs but may also affect adult sheep, especially in previously unexposed flocks. The animals develop papules on the lips and mouth with occasional involvement of the udder and teats. The papules progress to vesicles and then scabs. The virus remains infectious in the scabs for long periods under dry conditions.

Once infected, the immunity in a sheep lasts for many years; however, this immunity is not transferred via colostrum, and hence newborn lambs are susceptible. In flocks with preexisting immunity the economic impacts are usually minor. More severe financial losses may occur if very young animals develop lesions severe enough to interfere with nursing. This would also be true if lesions develop on the teats of lactating ewes. The disease can be zoonotic, but most people recover uneventfully (Radostits et al., 2007).

Caseous Lymphadenitis (CL)

Caseous lymphadenitis (CL) can be of considerable economic importance to the sheep industry because it results in reduced growth weights, reduced fleece weights, and carcass condemnation. The disease affects both sheep and goats and is a chronic recurring disease. Sheep are usually exposed to the organism through skin or mucous membranes, but the bacteria usually enter through wounds or abrasions. The source of the bacterial agent (*Corynebacterium pseudotuberculosis*) can be abscesses on the sheep, the environment, or equipment (especially shearing equipment). The pus contains large amounts of bacteria that can survive in bedding and soil for many months.

Sheep infected with CL present with abscesses especially around the head and neck, and the number of abscesses usually increases with age. Most often, the abscesses are in surface lymph nodes but can become systemic and appear in internal nodes and organs (e.g., lungs and liver). An initial infection may result in no clinical signs; however, as the animal ages, this changes. The disease may contribute to "thin animal syndrome." Bronchopneumonia, abortion, arthritis, and central nervous system abscesses may occur with CL infection, but these are not common findings.

A diagnosis of CL is usually made using history and clinical signs. The organism may be cultured. Although there are serological tests available, they have limitations because infected animals may test negative if abscesses are walled off, young animals (< 6 months) may test positive because of colostrum titers, and vaccinated animals will test positive.

Treating animals infected with CL with antibiotics is not effective. Animals with signs should be culled. Because shearing is one of the primary means of within-flock transmission, older animals should be shorn last. The vaccine reduces incidence and prevalence, but it will not cure the disease and will not prevent new infections. Because eradication is extremely difficult, prevention is key (Radostits et al., 2007).

Johne's Disease

Johne's disease, which is caused by *Mycobacterium avium paratuburculosis*, causes a wasting condition in sheep for which there is no vaccine in the United States and no treatment. Like OPP, elimination of Johne's disease from a flock is difficult and expensive. Johne's disease also has trade implications for sheep exported from the United States. The disease is discussed in greater detail in sections below.

Ram Epididymitis (RE)

Ram epididymitis is more common in western range flocks and commonly introduced with the purchase or use of infected rams. The organism responsible for RE is *Brucella ovis*. Rams infected with *B. ovis* have reduced fertility or may be sterile. The disease may influence the number of rams required. Lambing percentages may be reduced by 30 percent in recently infected flocks and by 15–20 percent in flocks with endemic infection. In flocks that have a prevention/control program for RE, there is an additional return of $12.00 per ewe mated (Kimberling and Schweitzer, 1989). Ewes are somewhat resistant, but the disease may cause early embryonic death and, in some cases, abortion (Radostits et al., 2007).

Scrapie

Scrapie is an insidious, degenerative disease affecting the central nervous system of sheep, goats, and moufflon. Scrapie is discussed in detail later in this chapter.

Internal Parasites (Resistant Worms)

Internal parasites are addressed in the section on diseases of economic impact or concern.

Recordkeeping

To maximize profitability, it is imperative that production parameters including disease status are measured. Accurate assessments of production efficiency cannot be made without records and identification. Without records, there are no yardsticks by which to measure possible problems and improvements. Records provide a baseline and can assist in determining which management practices help improve the operation and which may add costs without return. Without records, many of the silent inefficiency losses caused by disease may go unnoticed. The NAHMS 1996 Sheep Health and Management Practice study found that almost 20 percent of producers kept no records (USDA, 1996). In 2001, the study found that slightly over 30 percent did not keep records (USDA, 2002). The study did not go into detail about which specific records were kept or what production parameters were recorded. Records that would assist in evaluating the overall health status of a flock include the number of ewes exposed to the ram, any abortions and cause, number of lambs born, number of live lambs born, reason for lamb deaths, birth weights, body and udder condition of the ewe, condition of the colostrum, any apparent clinical disease conditions, any serological or other evidence of disease, necropsy results, number of lambs at weaning, weaning weights, yearling weights, vaccinations given, treatments, cost of the treatment (including labor), and number of animals culled and reason for culling.

Identification

Another important component of flock management and disease control is identification. Identification is critical for timely tracing of disease spread and eventual elimination. In addition, if animals are not identified, measuring production efficiency is difficult especially in relationship to the cost of disease. The 2001 NAHMS study found that 27.4 percent of all operations participating in the survey used flock identification (all animals have the same identification), of which more than 87 percent of the operations with flocks over 1,000 head used flock identification (USDA, 2002).

Perhaps most important, the 2001 NAHMS study found that 80 percent of all operations used some form of individual identification, including 68 percent of the large flocks (> 1,000 head), emphasizing the importance of identification if scrapie is ever to be eliminated from the United States. Hence, in November 2001, a federal regulation requiring the identification of certain classes of sheep went into effect (*Federal Register*, 2001). The statistics mentioned above reflected industry practices prior to this regulation.

The sheep industry has a more complete ID system than exists for any other species in the United States at this time. Three national sheep industry programs have an identification component, including the National Scrapie Eradication Program (NSEP), the Scrapie Flock Certification Program (SFCP), and the National Animal Identification Program (NAIS). Only the NSEP is mandatory.

National Scrapie Eradication Program (NSEP) Regulations

Near the end of 2001, final regulations for the NSEP were published under Title 9, Part 79 of the Code of Federal Regulations (CFR; *Federal Register*, 2001). These regulations require official identification for sheep and goats moving interstate. The purposes of the regulations were to provide a more effective national program for scrapie surveillance and to facilitate the traceback of scrapie-positive animals. The lack of identification and records was found to be hampering scrapie control efforts

Animals required to be identified in 9 § CFR 79.3 must be officially identified to the flock of birth upon change of ownership. In cases where the flock of birth cannot be determined, the animal is to be officially identified to the flock of origin (flock in which an animal most recently resided for breeding).

Animals required to be officially identified include:

a. All breeding sheep;
b. All sexually intact animals for exhibition;
c. All sheep over 18 months of age;
d. All exposed and high-risk animals including all low-risk exposed animals, genetically susceptible exposed animals, genetically less susceptible exposed animals, and genetically resistant exposed sheep;
e. All suspect and test-positive animals;
f. Animals from noncompliant flocks;
g. Breeding goats, except low-risk commercial goats.

Animals not required to be individually identified include:

a. Slaughter sheep (sheep in slaughter channels) under 18 months. (Note: If a sexually intact sheep is sold at an unrestricted sale [any sale that is not a slaughter or feeding for slaughter sale], it must be identified. If these animals cannot be identified to flock of origin, they must be identified with slaughter only tags);
b. Wethers for exhibition and wethers under 18 months of age;
c. Slaughter goats (goats in slaughter channels);
d. Low-risk commercial goats;
e. Animals shipped directly to an approved slaughter facility or an approved market when all the animals in a section of a truck are from the same premises of origin and are accompanied by an owner's statement. (Note: The owner's statement must contain the information needed for the plant or market to identify the animals);
f. Animals moved for grazing or similar management reasons whenever the animals are moved from a premises owned or leased by the owner of the animals to another premises owned or leased by the owner of the animals.

Under NSEP rules, the flock is assigned a premise number and the identification and movement information must be recorded. These records must be retained for at least 5 years after the date of the movement. Lambs under 18 months of age moving to slaughter and animals not leaving the premise do not have to be identified. Some states have additional restrictions for sheep moving within the state for shows, sales, and other events.

The identification requirements are a part of the scrapie eradication effort. In the event that scrapie is eliminated from the United States and/or there is no longer a program, the identification requirements may not be applicable, potentially hampering other management practices. To some extent, this is what happened in the cattle industry where identification was linked to the brucellosis program. As brucellosis was eliminated, identification and records were no longer available for managing other diseases.

Scrapie Flock Certification Program (SFCP)

The SFCP monitors flocks over time and confers a certified status on those that do not have evidence of the disease after a minimum of 5 years complying with movement, identification, recordkeeping, and sampling requirements. The basis of this program is to provide a source of sheep having a negligible risk of scrapie for domestic and international markets.

The SFCP requires that the following sheep within a flock enrolled in the complete or export-monitored category be identified with a program-approved ID such as an official eartag:

1. All animals one year of age or older;
2. All acquired animals before commingling with the flock unless already identified with an approved device; and
3. All sexually intact animals < 1 year old with a change of ownership unless moving directly to slaughter or to a terminal feedlot.

National Animal Identification System (NAIS)

After the first case of bovine spongiform encephalopathy (BSE) was detected in the United States in 2003, there was a movement to develop and implement a nationwide identification and tracing (recordkeeping) system (now known as the National Animal Identification System or NAIS), which could trace animal movements in 48 hours. After the initial public and media reaction subsided, certain livestock species groups began to withdraw support for a mandatory identification program and urged the U.S. Department of Agriculture (USDA) to implement a voluntary national identification program. Other species groups still support the need for a mandatory system. Currently, USDA has taken the position that every aspect of the NAIS will be voluntary in nature.

The NAIS is an identification and information system consisting of three components, all of which are voluntary: (1) premises registration, (2) animal identification, and (3) tracing. An owner may choose only to register a premise and not participate with animal identification or tracing. The premises registration component of the NAIS ensures the availability of a nationwide communications network to assist livestock owners and animal health officials in the event of an animal disease emergency. Upon registering with a state or tribal animal health authority, a premise is assigned a unique premise identification number that corresponds to the contact information provided.

The identification component provides the owner with a nationally unique identification for the animals on a registered premise. The identification number stays with individual animals (animal identification number [AIN]) or with lots or groups of animals (group identification number [GIN]) throughout their lifetime. This number links the animal to its birthplace or premises of origin. When combined with animal tracing, the AIN/GIN also links the animal to each premises or location that has been reported for it.

The federal government, states, and industry groups actively encourage producers to at least register their premises. To accomplish the goal of a 48-hour trace, a high rate of voluntary participation in NAIS will be necessary. Sheep identification requirements for scrapie are currently accepted as meeting the NAIS goal of enhancing animal traceability. By developing consistent standards for the official identification of livestock species across USDA Animal and Plant Health Inspection Service (APHIS) programs, NAIS supports USDA and the industry's long-term goal of enhancing international trade and marketing opportunities.

The Sheep Working Group of NAIS (NAIS, 2006) recommended that:

- The existing mandatory identification should be used as a starting point for NAIS as it is more complete than what exists for any other species.
- The Working Group acknowledges that a purely visual identification system will not work to attain the goal of tracing within 48 hours and an electronic system is needed. However, a proven system of electronic identification and tracking does not exist for small ruminants.
- USDA should conduct research to develop an electronic system or test systems developed in other countries.

Educational Resources

To educate themselves on various issues regarding sheep production including animal health, producers utilize a multitude of resources. According to the 1996 NAHMS survey (USDA, 1996), the following are the top 10 resources used by producers, with the percentage of sheep operations that use the associated resource in parentheses:

Magazines/newsletters (70.5)
Neighbors/other sheep producers (69.4)
Use of a veterinarian (63.3)
Other books (50.5)

Shearer (49.5)
University/Extension (49.2)
Fairs and shows (42.3)
Meetings (35.4)
Feed and drug salespersons (26.5)
Sheep Industry Development (SID) *Sheep Production Handbook* (24.1)

In the 2001 NAHMS survey (USDA, 2002), producers were asked what sources of information are important. Their top responses and the percentage of operations reporting that the resource is very important are as follows:

1. Veterinarians, private practitioners, consultants (39.1)
2. Other sheep producers (30.0)
3. Shearers (29.3)
4. Magazines, newsletters (22.7)
5. University/Extension (22.0)
6. SID handbook (10.5)
7. Meetings (10.2)
8. Feed and drug salespersons (9.2)
9. Internet (7.3)

In the 1996 study, more than 35 percent of the operations did not list veterinarians as a source of sheep information (USDA, 1996). In the 2001 study, only 39.1 percent of all sheep operations reported that veterinarians are a very important source of sheep health information and 27.9 percent of operations listed veterinarians as not important (USDA, 2002).

In 2000, 46.1 percent of sheep operations consulted a veterinarian. Past NAHMS studies of other species revealed a significantly different picture. The percentages of operations involving other species that consulted a veterinarian were as follows (USDA, 2002):

- Swine (2000) — 78 percent
- Equine (1998) — 73.8 percent
- Dairy (1996) — 98.1 percent
- Feedlot (1999) — 97.4 percent
- Beef (1997) — 55 percent

Exploring this subject further may shed light on reasons for the difference between sheep and most other species. There do not appear to be data on whether producers feel they have sufficient information and assistance on prevention and control of disease, lambing issues, or nutritional problems. Do many sheep producers not use the services of veterinarians because of the lack of availability, lack of knowledge about sheep, cost, fewer disease and health issues, or something else? Would the increased use of veterinarians and veterinary services provide a benefit and added value or merely add a cost without adequate return?

Commonly accepted anecdotal information suggests that there is a shortage of veterinarians whose practices include small ruminant medicine or who are knowledgeable about sheep. In recent years, there has been a growing concern within the veterinary profession about a shortage of veterinarians in the area of food supply veterinary medicine (FSVM). The FSVM specialty includes private practice serving dairy, beef, swine, poultry, sheep, and goat operations, as well as federal and state animal health officials involved with food safety and animal health. The American Veterinary Medical Association (AVMA) and other organizations have studied the potential problem of shortages within FSVM. A study published in 1999 predicted that there would be a decrease in the demand for large-animal practitioners (Brown and Silverman, 1999).

A 13-panel Delphi study published in 2006 found that FSVM will face opportunities and challenges. This study confirmed the anecdotal information and conflicted with the 1999 report. A key finding was a predicted shortage of food supply veterinarians. Interestingly, both the poultry and small ruminant panels predicted a close match between supply and demand. However, the small ruminant panel predicted a slight shortage of practitioners (Prince et al., 2006). A 2007 ASI policy resolution declared that the U.S. sheep industry is in need of qualified veterinarians with knowledge of sheep diseases and management practices. The ASI supports a curriculum that includes an emphasis on related sheep diseases and applied sheep health management practices. The real problem may well be that there has been a shortage of knowledgeable and experienced veterinarians willing to do small ruminant work over such an extended period of time that the industry has found ways to manage without veterinarians.

Sheep Disease Pharmaceuticals

In the United States, there is a critical shortage of approved animal drugs intended for less common animal species (minor species) or a major species such as cattle suffering from an uncommon condition. Sheep and goats are considered minor species. Few animal drugs are specifically approved for sheep and goats. Consequently, veterinarians and livestock producers often have limited options for treating sheep if they become ill. The shortage of approved drugs may result in animal suffering, increased mortality, and financial loss to those who raise the animals.

Congress recognized that statutory changes might be needed to address the described shortage of approved animal drugs in the Animal Drug Availability Act (ADAA) (P.L. 104-250), passed in 1993 (see Chapter 2 for a further discussion of this federal policy). A section of the legislation recognized particular problems relating to the availability of approved animal drugs for minor uses in major species and for use in minor species. The law directed the Secretary of Health and Human Services to consider and announce proposals for legislative or regulatory change to the approval process for such drugs. The Food and Drug Administration (FDA) concluded that federal statutes should be amended. The FDA proposals provide the conceptual base for the Minor Use and Minor Species Animal Health Act of 2001.

The Minor Use and Minor Species (MUMS) Animal Health Act of 2001 is similar to the Human Orphan Drug Act of 1983. This act is intended as a mechanism to provide FDA-authorized drugs for less common animal species and health indications. Specifically, MUMS seeks to provide labeled drugs for needy minor species, including sheep, goats, game birds, emus, ranched deer, alpacas, llamas, deer, elk, rabbits, guinea pigs, pet birds, reptiles, ornamental and other fish, shellfish, wildlife, and zoo and aquaria animals. The MUMS Act is also designed to provide major species (cats, dogs, horses, cattle, swine, turkey, chickens) with needed therapeutics for uncommon indications, so-called minor uses.

Despite MUMS, however, there is still a shortage of therapeutics for sheep. For example, there is a tremendous need for additional dewormers as internal parasites have developed a resistance to the drugs currently available (see further discussion below). The ASI supports legislative and regulatory efforts that will restructure and expedite the drug approval process while maintaining product safety and efficacy.

DISEASES OF ECONOMIC IMPACT OR CONCERN

The two NAHMS studies (USDA, 1996, 2002) provide data on diseases that are present and are of concern to the sheep industry. Many of the conditions identified as of high or moderate concern, such as intestinal parasites, contagious ecthyma (soremouth), foot rot, and mastitis, are not usually fatal if the animals receive proper treatment. Nevertheless, these conditions are insidious and debilitating, affect production efficiency, and add to production costs. Unfortunately, there are no hard data as to the actual costs of the diseases, which would include production losses, treatments, and mortalities. Data regarding actual costs are essential to set priorities and make educated decisions on research, prevention, and control needs.

The most commonly reported disease or condition of concern in the 1996 NAHMS study was stomach/intestinal worms, reported to be of moderate or high concern by 62.1 percent of sheep operations (Figure 3-1). Nearly half of all operations surveyed reported problems with stomach/intestinal worms in the previous 5 years (Figure 3-2). In the 2001 NAHMS

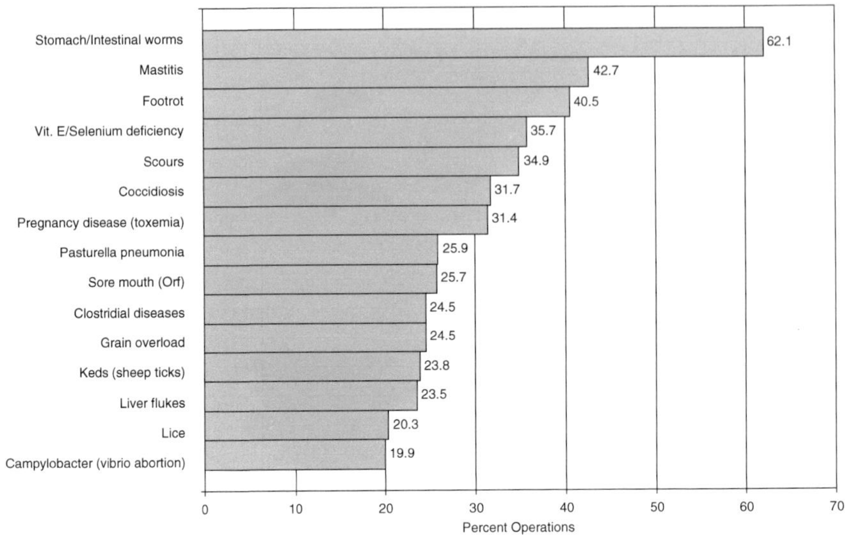

FIGURE 3-1 Percent of operations that reported moderate or high concern for the top 15 conditions of concern.
Source: Reproduced from USDA (1996a).

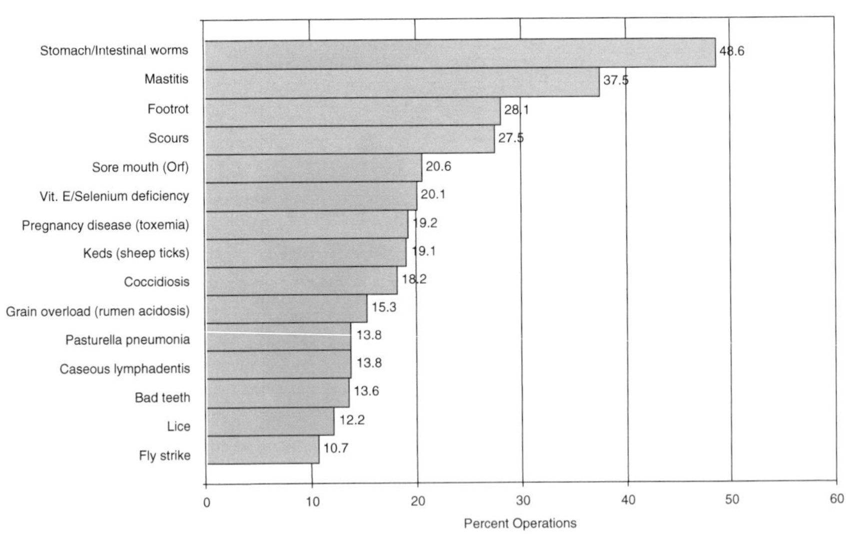

FIGURE 3-2 Percent of operations on which the top 15 conditions of concern were known to be present in the last 5 years.
Source: Reproduced from USDA (1996a).

study, 74 percent of sheep operations reported the presence of stomach or intestinal worms in the previous three years (Figure 3-3).

Although the purpose of this section is not to provide a technical review of the common diseases, it examines intestinal parasites as an example of the need for economic and other data. The most common internal parasites affecting sheep throughout the United States are roundworms or nematodes. The gastrointestinal nematodes of greatest importance in sheep are members of the order Strongylida and include *Haemonchus, Teladorsagia, Trichostrongylus, Cooperia, and Nematodirus.* These species have a direct life cycle, meaning that sheep are the only host. Adult worms in the gastrointestinal tract of sheep lay eggs that are passed in the feces and contaminate the environment. When conditions are suitable, the eggs hatch into a larval stage that matures. The first two larval stages are nonparasitic but the third stage is infective. Part of the parasite maturation process takes place on the pasture and part within the abomasum (stomach) or small intestine of the sheep. Sheep that graze an infected pasture ingest the third-stage larva, which finishes development into egg-laying adult worms.

During periods where environmental conditions are harsh, the parasite larvae in the stomach undergo a process called hypobiosis, which means their development is delayed. The delay extends until pasture conditions are favorable for survival of the parasite. Over time, sheep may develop some resistance to these worms. The development of resistance is complex and involves genetic components. Lambs are extremely susceptible as are older

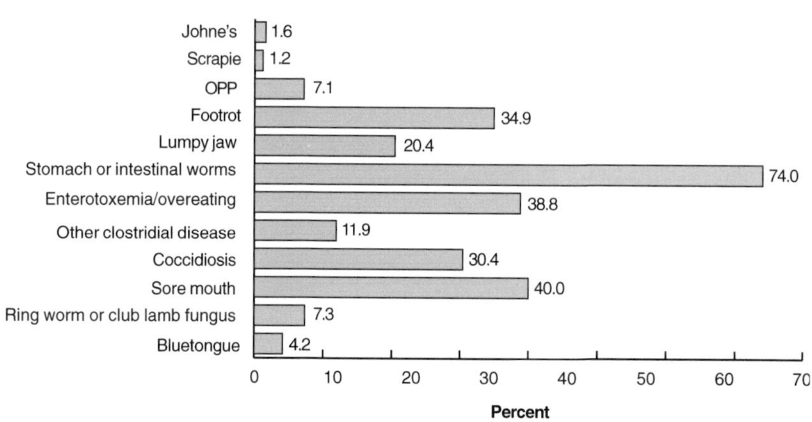

FIGURE 3-3 Percent of operations where diseases were present (suspected or confirmed) during the last 3 years.
Source: Reproduced from USDA (2003a). The survey was completed in early 2001.

sheep that have compromised immune systems from other diseases, from poor nutrition, or during lambing. Ewes usually experience a temporary relaxation of immunity slightly before and after lambing. Infected ewes may shed large numbers of worm eggs into the environment that develop into larvae and are ingested by the lambs.

The overall effect of internal parasites is known as parasitic gastroenteritis. Clinically, symptoms of internal parasitism may be insidious at the onset. Infected lambs grow more slowly and become unthrifty, food intake is often reduced, and, if the lamb is left untreated, it may die. Larger worm burdens may result in higher lamb mortalities. Clinical disease is not restricted to lambs. Adult sheep already in poor condition, stressed by milk production, or with existing disease conditions may also suffer clinical manifestations such as weight loss, reduced milk production, and even death.

Haemonchus contortus, the barberpole or wireworm can be particularly severe. The fourth-stage larva and adult *Haemonchus* are known to be vigorous bloodsuckers. The worms can deplete volumes of blood large enough to result in anemia and death. *H. contortus* is very prolific and a female may produce thousands of eggs each day, which may then contaminate grazing pastures. Unlike the other gastrointestinal parasites, sheep burdened with *H. contortus* do not usually have diarrhea as a primary sign, thus owners may not appreciate the extent of the infection until it is too late to save the animal (Zajac, 2006).

Damage caused by *Teladorsagia circumcincta (*formerly called *Ostertagia circumcincta*) is targeted at the abomasal gastric glands. Heavy infection with *T. circumcincta* can cause diarrhea, and hypoproteinemia, which may result in death. More commonly, moderate infections cause diarrhea with poor weight gain or weight loss (Zajac, 2006).

In the United States, the third common nematode group to contribute to parasitic gastroenteritis is *Trichostrongylus*. Severe infections of *Trichostrongylus* can result in diarrhea and weight loss (Zajac, 2006). There have been two human infections of *Trichostrongylus* reported in Australia. Vegetables grown using fresh goat manure as fertilizer were eaten raw and apparently transmitted *Trichostrongylus* (Ralph et al., 2006).

Drugs used to treat internal parasites in sheep are called anthelmintics. The anthelmintics available in the United States belong to three major chemical classes. These are:

- Benzimidazoles—albendazole and fenbendazole are sold in the United States for small ruminants. Albendazole is labeled for sheep but not goats, and fenbendazole is not labeled for sheep.
- Macrocyclic lactones—ivermectin and moxidectin are approved for oral use in sheep. Currently ivermectin is the only anthelmintic approved for limited use in nonslaughter organically raised sheep in the United States.

• Nicotinic anthelmintics—this group includes levamisole, morantel, and pyrantel. Only levamisole is approved for sheep.

The limited availability of anthelmintics for sheep is significant because internal parasites have developed, or have begun to develop, resistance to the treatments. Some of the past treatment regimens actually assisted the development of anthelmintic resistance. Failure of the overall parasite control program on sheep operations is often due to mismanagement (Radostits et al., 2007).

A proportion of all worm populations have the genetic capacity for resistance to some or all classes of anthelmintics. This allows their survival when others die. Frequent use of anthelmintics as the major way of controlling parasites has allowed the development of resistant populations that have become a major problem, especially in the South. Although frequent use of anthelmintics is perhaps the major selection force for the development of resistance, underdosing a treatment also provides a powerful selection mechanism. When a population of worms becomes resistant to one member of a class of drugs, it effectively becomes resistant to other members of that class. Unfortunately, when resistance of a worm population to an anthelmintic becomes evident, it will likely remain even if the use of that compound is discontinued for several years. Resistant worms can be introduced onto a new farm with the introduction of new animals (Craig, 2006).

Current treatment strategies to increase the efficacy of the anthelmintics involve the judicious use of drugs combined with pasture management and monitoring the sheep for fecal egg counts and other disease indicators, such as anemia. A summary is provided by Craig (2006):

• Delivery of treatment—dividing the dose of an anthelmintic over several days or fasting the animal before treatment may enhance the effectiveness.

• Use of drug combinations—this may be the only choice in places where resistance to multiple classes of anthelmintic has developed. Although certain drugs may not be effective on their own, these may work in combination with others. They should be used at full dosage and concurrently.

• Drug rotations—the practice of changing drugs within a grazing season actually selects for resistance to all the drugs used and can accelerate the process of resistance.

• Targeted or selective treatment—this approach involves treating only sheep with signs of disease (e.g., high egg counts or pale mucous membranes). The untreated animals will carry worms that are susceptible to treatment. Worms that are not treated are called "in refugia." Having some worms in refugia (not exposed to an anthelmintic) slows the development of resistance by diluting the frequency of resistant genes. Consequently, when a dewormer is required, it will be effective because the worms will be susceptible to treatment (Kaplan, 2004). The FAMACHA system uses a patented color chart to allow producers to identify animals with moderate or severe anemia (*H. contortus* infections) for treatment, leaving up to 80 percent of the animals untreated. Other strategies include only treating thin animals or leaving 10–20 percent of a group untreated when changing pastures to maintain a level of "refugia."

- Culling of more susceptible animals—certain sheep may consistently have high fecal egg counts or clinical signs of parasitism. Culling these animals can improve the overall resistance of the flock (Barger and Dash, 1987).
- Use of breeds that show resistance to parasites—some breeds have a degree of genetic resistance to parasites and may carry a smaller worm burden. It is possible to identify these animals and select for a flock carrying a higher level of resistance to parasitism. However, even in these breeds there are differences among individual animals that influence the course of disease.
- Pasture management practices—pasture rotation, alternative grazing or cograzing with different species such as cattle or horses, or using pastures that have been cropped may lower the worm burden when sheep are returned to the pastures.
- Nontraditional treatments—the use of copper oxide wire particles or certain botanical products may be useful alternatives to anthelminthics, but most need more scientific scrutiny before they can be adapted for widespread use.

As discussed above, the newer control strategies require a more detailed level of sheep and pasture management.

There are no data that capture the total economic loss to the sheep industry because of internal parasites despite it being the disease condition of highest concern and highest reported incidence. Such data are important as they could be used to encourage research for additional control measures that may include the development of new anthelmintics and evaluating the effectiveness of other management practices. Inadequate data is a problem for most other sheep disease conditions such as mastitis, foot rot, and Johne's disease, not just internal parasites.

Causes of Death

A USDA (2005) study by APHIS examined the nonpredator causes of death for lambs and adult sheep in the United States. Since 1994, nonpredator causes of death have been responsible for 62 percent of lamb and adult sheep losses, more than from predators nationally and in all regions of the country. In the 2004 NAHMS study of mature sheep losses, old age was reported as the leading cause of death (26.8 percent of adult sheep death loss) followed by lambing problems (13.4 percent); digestive problems such as internal parasites, bloat, scours, and acidosis (12.9 percent); and cause unknown (12.1 percent). The primary nonpredator causes of death in lambs were respiratory problems such as pneumonia, and shipping fever (22.8 percent); digestive problems, such as internal parasites, bloat, scours, and acidosis (19.8 percent); weather-related causes such as chilling, drowning, and lightning (14.8 percent); lambing problems (14.7 percent); and cause unknown (13.3 percent).

Loss estimates for the Pacific (California, Oregon, and Washington) and West Central (Arizona, Colorado, Idaho, Montana, New Mexico, Nevada, Texas, Utah, and Wyoming) regions were made after docking, marking, or branding, while estimates for the Central, Northeast, and Southeast regions were made from birth. In fact, in the western states, the National Agricultural Statistics Service (NASS) defines lamb crop as those lambs marked, docked, or branded. The exclusion of predocking losses from these regions is because lambs are usually born on the range and less likely to be counted. Five states conduct surveys that include these early losses: Colorado, Idaho, Montana, Utah, and Wyoming. Other than Colorado, which shows predocking losses at 32 percent of all lamb losses, the other four states report predocking lamb losses over 50 percent of all lamb losses. (USDA, 2005).

The lack of data regarding lamb numbers during the predocking period in states that have some of the highest sheep numbers is significant. An accurate estimate of the actual lamb crop in the United States cannot be developed without death loss data for the predocking period. More important, efforts to reduce such losses are encumbered without an accurate baseline that could provide answers to various questions relating to possible reasons for lower numbers of lambs:

- Are the ewes giving birth to multiple lambs?
- Are lambs born dead?
- Are lambs dying shortly after birth from insufficient colostrum or poor mothering?
- Are lambs dying at a few weeks of age from scours, pneumonia, or other reasons?
- Are predators responsible for the majority of deaths?

Scrapie

As a disease entity for individual flock owners in the United States, scrapie did not make the list of the top 15 conditions of moderate or high concern in the 1996 NAHMS study (USDA, 1996), but it is the only sheep disease that has had a congressionally funded program aimed at control and/or eradication since 1952. In fact, federal regulations mandating an accelerated scrapie eradication program went into effect in 2001. When publishing this proposed rule, APHIS estimated that it would cost approximately $100 million over 7 years to accomplish the task of eliminating scrapie outbreaks (*Federal Register*, 2001).

During the 1980s, scrapie was considered "a minor disease in a minor species." In 1987–1988, APHIS published a proposed rule asking for comments on whether to discontinue the scrapie program. The publication of the proposed rule came shortly after the United Kingdom's announcement that it had identified a new disease entity that affected cattle and appeared to be related to scrapie. Comments on the proposed rule overwhelmingly requested that APHIS not discontinue the scrapie program but work together with the sheep and allied industries to develop a new program. Many comments cited BSE as the reason for a renewed interest in control. In 1990, the spread of BSE to other species, including cats, highlighted the hidden dangers associated with these diseases. A new program was developed through the process of negotiated rulemaking and went into effect in 1992. The new program was multifaceted in that it consisted of the establishment of a national scrapie flock certification program (SFCP) and interstate regulations to identify and restrict the movement of certain sheep from scrapie-infected and source flocks. The basis of the certification program is to provide a source of sheep having a negligible risk of scrapie (Detwiler et al., 1997). The foundation of this program is still in effect although it has been modified based on new scientific findings. As stated above, there is a concerted effort to eliminate scrapie from the United States by 2010. The UK announcement in 1996 that 10 deaths (adolescents and young adults) were thought to be attributed to BSE brought the transmissible spongiform encephalopathies (TSE) to the world's center stage.

Classical Scrapie

Scrapie is an insidious, degenerative disease affecting the central nervous system of sheep, goats, and moufflon. The disease is also called La tremblante (French: trembling), Traberkrankheit (German: trotting disease), or Rida (Icelandic: ataxia or tremor). Scrapie has been reported worldwide and affects most sheep-producing regions with few notable exceptions. Australia and New Zealand are commonly accepted to be scrapie free. The disease has been recognized for over two centuries in England, Wales, and Germany (Parry, 1983).

Scrapie is the prototype of the group of the TSE diseases. Other TSE include Creutzfeldt-Jakob disease (CJD), kuru, Gerstmann-Sträussler-Scheinker disease (GSS), fatal familial insomnia (FFI), and variant Creutzfeldt-Jakob disease (vCJD) in humans and transmissible mink encephalopathy (TME), chronic wasting disease (CWD), bovine spongiform encephalopathy (BSE), atypical BSE, BSE in a goat (Eloit et al., 2005), feline spongiform encephalopathy (FSE), and atypical scrapie in sheep. These diseases are caused by transmissible agents yet to be fully characterized, all of which share a number of common characteristics:

- Prolonged incubation period of months or years;
- Progressive debilitating neurological illness that appears to always be fatal;
- Pathological changes confined to the central nervous system (CNS); and
- The transmissible agent elicits no detectable specific immune response in the host, which has inhibited the development of a live animal diagnostic test and vaccines.

The clinical disease occurs primarily in sheep of breeding age because the minimum incubation period is usually between 18 and 24 months. Scrapie occurs most frequently in sheep of either sex between 2 and 5 years of age. Although cases of the disease are not common before 18 months, a few cases of natural scrapie have been reported in sheep at approximately 1 year of age and it may occur in animals over 5 years of age (Detwiler and Baylis, 2003).

Sheep usually become exposed to scrapie through an oral route. Once the agent enters the body, it replicates and infects most tissues, including tonsils, spleen, lymph nodes, brain, and spinal cord (Hadlow et al., 1979, 1982; van Keulen et al., 1996, 1999); blood (Hunter et al., 2002); and peripheral nerves (Groschup et al., 1996). There is also evidence that scrapie-infected sheep may harbor infectivity in actual muscle cells apart from the peripheral nerves (Pattison and Millson, 1962; Andréoletti et al., 2004; Casalone et al., 2005). In the past, milk had not been considered a risk factor for scrapie. However, in 2005, a published study began to suggest milk as a possible route of transmission. This research found that sheep concurrently infected with scrapie and a chronic inflammatory disease such as OPP have prions in the mammary gland and may shed these into milk (Ligios et al., 2005). Recently, it was found that lambs fed milk from scrapie-affected ewes became infected with scrapie (Konold et al., 2008). Certain sheep may shed the scrapie agent via the placenta and likely the birthing fluids. Thus, scrapie is thought to spread from an infected ewe at or near the time of lambing to susceptible lambs or other susceptible sheep that may be exposed to the placenta (see Pattison et al., 1972, 1974; Onodera et al., 1993; Race et al., 1998).

Over the decades, there has been debate about the nature of scrapie and whether it is a genetic disease or is transmitted between sheep (Dickinson et al., 1974; Parry, 1983). Research in the 1990s found that scrapie was influenced by genetics, specifically the sequence of the gene coding for prion protein (Westaway et al., 1994; Hunter et al., 1994, 1996; Hunter, 1997). There is also evidence indicating that scrapie is not a genetic disease (Hunter et al., 1997). Most classical scrapie cases have historically been found in sheep carrying the VRQ and ARQ alleles, and it was thought that sheep having two ARR alleles were resistant to scrapie. Consequently, many countries shifted their scrapie control program to a genetics-based program or at least included a genetic (breeding) component (Detwiler and Baylis, 2003). Newer diagnostics as well as extensive testing programs in Europe have begun to call into question whether there truly are genotypes that are fully resistant to TSE infection (Buschmann et al., 2004; Le Dur et al., 2005; Groschup et al., 2007).

There have been many different approaches to scrapie control and eradication. Countries such as Australia and New Zealand, which have detected the disease after introduction but before widespread transmission, have apparently been successful in eradicating the disease within a short period of time. Other countries, in which the disease has become endemic, such as Canada, Iceland, and the United States, have implemented various strategies in an effort to eliminate the disease. The emergence of BSE in 1986 and the experimental transmission of BSE to sheep and goats have been the impetus for a significant increase in research and have prompted a number of other scrapie-endemic regions such as the European Union to initiate programs to eliminate the disease

The United States has tried many options to eradicate scrapie (see the Appendix for a chronology of U.S. control efforts). Historically, none has been very successful, which may be attributed to a number of factors. First and foremost, the lack of scientific advancements, especially in the area of preclinical diagnostics and understanding the exact nature of the pathogenesis, has hampered control programs. For example, when during the incubation period an animal starts to shed the agent and by what route(s) are not known. Such information is necessary to identify animals that may be exposed. All of the tests have limitations. None have adequate sensitivity to be able to declare an individual animal free of scrapie (O'Rourke et al., 2000; González et al., 2005). Only recently have live animal tests become available to diagnose an infected animal before the onset of signs, which is usually years after the animal has become infected and may be a risk to other sheep. Despite some of these limitations, recent scientific advancements have finally provided the necessary tools for the current control program to demonstrate that its measures have resulted in a decrease in the prevalence of classical scrapie.

The properties of the causative agent are unusual because it is able to survive conditions that normally destroy other agents such as bacteria and viruses. The causative agent can survive conditions such as boiling, dry heat up to 600°C, formalin fixation, and treatments with many common disinfectants (WHO, 1999). The trait of survivability also allows the agent to remain in the environment for extended periods of time. How long the agent may persist and pose a risk to other animals if shed into the environment is unknown. Two different studies have shown that scrapie infectivity can survive in the environment at least 2–3 years and possibly a great deal longer (Brown and Gajdusek, 1991; Seidel et al., 2007). The most recent studies have demonstrated infectivity not only in the soil itself, but also in the aqueous soil extract (Seidel et al., 2007).

Despite more than 250 years of potential for human exposure to scrapie, there is no scientific evidence to date indicating that scrapie poses a public health risk (Harries Jones et al., 1988). During a consultation held in 1999, the World Health Organization reviewed existing evidence, including a 15-year epidemiological study (Brown et al., 1987), and came to the same conclusion (WHO, 1999). Nevertheless, with the detection of atypical manifestations of certain TSE such as scrapie and BSE, public health officials must remain vigilant for any potential change in virulence and host susceptibility.

Prevalence of Classical Scrapie

Before the Scrapie Ovine Slaughter Surveillance (SOSS) study, the estimated prevalence of scrapie in the United States was 0.07 percent. However, this estimate was based on results obtained from the 1996 NAHMS sheep study, which relied on data from a producer-generated mail-in survey (USDA, 2003d,e). The prevalence in this study was estimated from those producers reporting confirmed or suspected scrapie in their flocks over a period of 5 years.

More recently, estimates of prevalence derived from the results of the SOSS study were actually based on positive tests. The vast differences between the methodologies of the NAHMS and SOSS studies prevent any accurate comparison.

Of the 12,508 samples collected in the SOSS study, 12,491 valid test results were obtained (99.9 percent). The study concluded that the overall weighted national prevalence of scrapie in mature sheep was 0.20 percent. On a regional basis, scrapie was found to be most prevalent in the Eastern (0.52 percent), followed by the Central (0.21 percent) and Mountain (0.14 percent) regions (Figure 3-4). Even though the small sample size from the Western region prohibited an estimate for the region itself, the results for the region were included in the national estimate (USDA, 2003d,e).

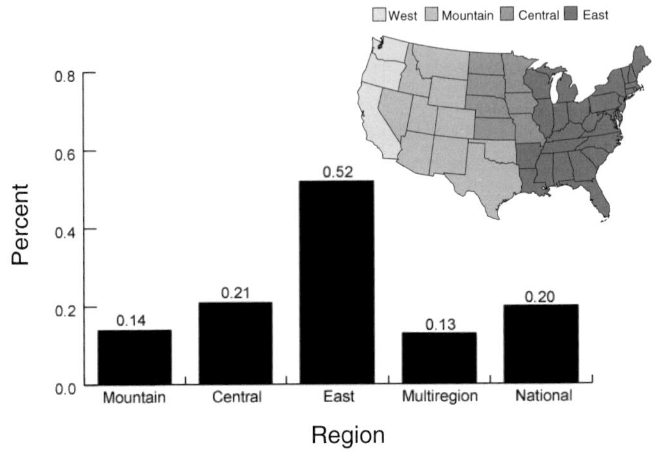

FIGURE 3-4 Percent of that sheep tested positive for scrapie by region.
Note: Because of the low number of samples obtained in the Western region, results for the Western region are included in the national estimates but are not listed individually.
Source: Reproduced from USDA (2003d). Samples were collected between April 1, 2002, and March 31, 2003.

About 84 percent of the positive samples were from black-faced sheep and less than 0.01 percent from white-faced sheep. Mottled-faced sheep represented 0.12 percent of the positives (USDA, 2003d,e). The breed results are similar to those that have been obtained over the years through passive surveillance. The study found that black-faced sheep were significantly more likely to test positive even when adjusting for other factors. The fact that scrapie appears to be predominant mostly in black-faced and mottled-faced sheep and extremely limited in white-faced sheep is important because the majority of sheep (67 percent) in the United States are white-faced (USDA, 1996). All of the sheep that tested positive were homozygous for glutamine (QQ) at codon 171. This genotype has been characterized as one of the most susceptible for classical scrapie (USDA, 2003d,e).

Current Scrapie Program

The current National Scrapie Eradication Program (NSEP) is comprised of active testing of targeted mature sheep and goats at slaughter with trace-back of positive animals to the flock of origin. Additional infected flocks are identified by tracing exposed animals out of these flocks. Effective tracing can be done as a result of the identification requirements implemented in 2001. As infected or source flocks are identified, government veterinarians work with producers to develop a plan to remove animals determined to be most susceptible. In addition, the premise may have to be cleaned and disinfected. Most flocks use a genetics-based plan to clean up. Indemnity funds are available for animals that must be removed to comply with the clean-up plan. In addition, those exposed animals that have been sold from the flock will be traced and tested if still alive.

There are three basic steps in the current NSEP:

1. When an infected flock has been identified, the sheep are genotyped to determine disease risk.
2. Susceptible genotypes are either removed or their movement restricted.
3. The identified flock is placed under surveillance for 5 years.

In most cases, producers will be able to keep many more of their sheep with the genetics-based plan. This plan allows owners to retain or sell without restrictions nearly all sheep that are AA RR, AA QR, and most AV QR from infected or source flocks once owners have met certain conditions (see NIAA, 2003, for more details). On average, an estimated 60 percent of a flock can be preserved when using a genetics-based plan compared to 25 percent when using a traditional plan.

Another component of the scrapie eradication effort is the Scrapie Flock Certification Program (SFCP). Given that the live animal testing for scrapie cannot yet guarantee absence of scrapie infectivity, the SFCP monitors flocks over time and confers a certified status on those animals that do not have evidence of the disease after a minimum of five years of complying with movement, identification, recordkeeping, and sampling requirements. The basis of this program is to provide a source of sheep that would have a negligible risk for having scrapie.

Most important, the accelerated eradication program requires an identification and recordkeeping system, which allows diseased, exposed, and high-risk animals to be traced back to their flock/herd of origin so that the spread of scrapie within and from these flocks/herds can be prevented. The scrapie eradication regulations published in 2001 also established a program to recognize states that conducted an active scrapie control program consistent with federal requirements. Additional restrictions were placed on the movement of sheep from states that do not require that scrapie be a reportable disease and/or do not quarantine infected and source flocks.

Congressional funding for the NSEP has increased from $15 million in 2003 to $18.4 million in 2007. When publishing the proposed rule in 2000, APHIS estimated the total cost to eradicate scrapie over a 7-year period to be 100 million dollars (USDA, 2000). In a letter to Undersecretary of Agriculture Bruce Knight, ASI requested that the USDA increase its scrapie budget request to $28.6 million for fiscal year 2009.

In Fiscal Year (FY) 2006, the NSEP made significant progress as shown by the reduction in the percentage of sheep of all face colors that tested positive at slaughter and the decrease in the number of positive animals found at slaughter. The result was fewer newly infected and source flocks and fewer animals indemnified in comparison to FY 2005. For example, the number of scrapie-positive, black-faced sheep dropped from slightly under 0.9 percent in FY 2004 to < 0.5 percent in FY 2006 (USDA, 2007a).

Economic and Trade Effects of Scrapie

The USDA has estimated that producers annually incur losses of $20 million from scrapie (USDA, 2000). The costs are associated with decreased productivity of infected flocks, diminished potential for exports (live animals, germplasm, and byproducts), and increased costs of disposal (USDA, 2000).

Because scrapie is always fatal, there are two considerations when examining the direct economic impact of scrapie on domestic flocks. These are the overall prevalence of the disease across all U.S. flocks and the within-flock prevalence of infected flocks. The overall weighted prevalence of scrapie in the United States (0.20 percent) is low, especially when compared to diseases such as OPP which had a seroprevalence calculated at 24.2 percent. Seroprevalence of OPP was measured during the 2001 NAHMS sheep study by testing sheep on randomly selected sheep operations. Over 21,000 samples collected at 682 operations that agreed to participate were tested. Overall, 36.4 percent of the operations had one or more positive animals (USDA, 2003c).

Although uncommon, flocks heavily infected with scrapie and that contain a high percentage of susceptible animals may experience significant mortality losses. In flocks where scrapie is endemic, the number of infected animals increases and the age at onset of clinical signs decreases over a period of several years making these flocks economically unviable. Surveys of farmers in Great Britain and the Netherlands have found mean within-flock incidences of 0.37 percent and 1.2 percent, respectively (Schreuder et al., 1993; Hoinville et al., 2000). Other observations have included within-flock incidences ranging from 1 percent to 20 percent (Young et al., 1964; Sigurdarson, 1991; Hoinville et al., 2000).

With the promotion of breeding programs that protect sheep from the clinical signs and subsequent death from scrapie, most U.S. sheep producers are not likely to experience a significant within-flock incidence. Hence, for many U.S. producers, the direct animal losses from scrapie may be insignificant. This conclusion is reflected in the 1996 NAHMS survey, which found that scrapie was not among the top 15 diseases of concern (USDA, 1996).

The epidemiological finding that BSE was most likely spread by the feeding of ruminant meat and bone meal (MBM) to ruminants and the theory that the species of origin was sheep prompted voluntary action by U.S. renderers in the late 1980s. Many independent renderers imposed a voluntary ban on the transportation and processing of sheep offal and/or heads, which had an impact on ovines slaughtered at small and very small packing plants. For those plants, the alternatives were to either send the head and offal back home with the producer or pass the cost of disposal back to the producer. If the producers had to dispose of the material themselves, the choices were to dispose of the materials on their own properties or through another mechanism such as a landfill at a cost. One study estimated that the costs for such disposal would average $150/ton (U.S. HHS, 1997). In that study, large lamb slaughterers were still able to use the byproducts for pet food. Another study reported value of the byproducts at $3.00 per carcass (Seitzinger et al., 2006).

In 1997, the FDA enacted a regulation prohibiting the feeding of most mammalian protein (both cattle and sheep) to ruminants. The diversion of cattle offal created a separate stream for ruminant MBM in which sheep materials could be included without the extra costs of disposal. Sheep offal were then included with the bovine materials in meals to be fed to species such as pigs and poultry. This regulation actually assisted the sheep industry in that ovine offal could be combined with the larger pool of bovine offal and utilized, instead of having to be destroyed.

Scrapie and BSE impact the trade of various sheep and lamb products including live animals, germplasm (semen and embryos), meat, and byproducts such as MBM especially for pet food. Worldwide, only New Zealand and Australia are commonly recognized as scrapie-free and, thus, are currently the nations that can freely sell breeding stock to producers in most other countries. During the 1990s, actual statistics on the value of Australian breeding stock sold indicates the amount was $29.2 million.

The Terrestrial Animal Health Code (TAHC) of the World Organization for Animal Health (OIE) includes provisions to recognize a country free of scrapie, which would open up trade for the United States if the current eradication program is successful (OIE, 2007). The TAHC also includes provisions for individual flocks to be recognized scrapie-free, thus allowing the export of breeding stock. As of June 2007, APHIS has modified the SFCP to allow for the monitoring and recognition of flocks to be certified for export, which will allow trade at least from some flocks without having the country declared scrapie-free. The modifications for the export-certified flock category are in accordance with the TAHC.

Scrapie may also be a reason for regulatory restrictions on the export of lamb MBM or the export of pet food to other countries. The actual value of the lost trade is not known and may not be significant. According to the Pet Food Institute, the U.S. supply and consistency of ovine MBM is not adequate to meet the demands of the domestic market so that supplemental byproducts are imported from Australia and New Zealand (N. Cook, personal communication, 2007). Even if scrapie were to be eradicated immediately, restrictions on the trade of any ruminant MBM would still apply. As recommended by the TAHC, once a country has detected BSE in domestic cattle, other countries usually prohibit the importation of many ruminant products, especially ruminant MBM even for use in pet food (OIE, 2007).

Scrapie is not the only disease that may impede or restrict the exportation of U.S. sheep to other countries. In addition, OPP, also known as Maedi-Visna, and ovine Johne's disease (OJD) are both present in the U.S. sheep population. A number of countries have restrictions or requirements on imports of sheep from countries with these diseases. The TAHC includes a chapter on Maedi-Visna that outlines the provisions for trade. In addition to being recognized as scrapie-free, both Australia and New Zealand have been recognized as free of OPP. Australia and New Zealand both have a control program for Johne's disease (Seitzinger et al., 2006).

Unlike scrapie, there are no U.S. government programs in place to eliminate or control OPP or OJD. A producer-driven effort to test for and control OPP is operated by the OPP Concerned Sheep Breeders Society. An attempt by the state of New Jersey to combine enrollment in the scrapie certification program and monitoring for OPP has been discontinued. Currently, Minnesota has initiated a pilot program that combines monitoring for OPP with the scrapie certification program. Although there is a voluntary control program for bovine Johne's disease, there is no organized program for OJD.

In 2006, USDA examined the market and economic effects of eliminating scrapie alone versus eliminating scrapie, OPP, and OJD (Seitzinger et al., 2006). The study found that eliminating scrapie increased annual revenues to U.S. sheep and lamb producers by a total of $10.8 million dollars. The least increase in revenue from eliminating scrapie alone was from exports. The presence of OPP and OJD in the United States did not allow for a significant increase in revenue by the elimination of scrapie alone, as countries also have restrictions for OPP and OJD.

Atypical Scrapie

The term "atypical scrapie" has been adopted to describe forms of scrapie in which the findings in tested animals (including animals that appear to be clinically normal) were different from those that were previously recognized as classical scrapie. Norwegian scientists published reports about the first cases of nonclassical scrapie in 2003. They called this Nor 98 scrapie (Benestad et al., 2003). Hence, atypical scrapie cases are also called Nor 98 or Nor 98-like.

Both types of scrapie (classical and atypical) involve the presence of abnormal prion protein. Most atypical cases differ from classical scrapie in that the lesions found in the brain as well as the deposits of prions have been concentrated mostly in the cerebrum and cerebellum rather than the brain stem. The patterns seen by Western Blot testing also differ from classical scrapie. The average age of sheep with atypical scrapie is greater than that of classical scrapie. Research has found that over 58 percent of the cases of atypical scrapie in Germany were 6 years or older and 26.7 percent were older than 10 (Lühken et al., 2007).

Classical scrapie identified in clinically affected sheep usually occurs in certain genotypes of sheep.[2] Most classical scrapie cases have traditionally been in sheep carrying the VRQ and ARQ alleles (homozygous or heterozygous), only occasionally in combination with ARR and AHQ. Atypical scrapie has been found predominantly in sheep carrying the $AF_{141}RQ$, AHQ (Moum et al., 2005) ARQ, and ARR alleles. Atypical cases of scrapie have been found in ARR/ARR sheep in a number of European countries (European Union, 2006). Investigations of sheep that had difficulty walking (ataxia) led to the identification of Nor 98 scrapie in Norway, which was the first of the atypical scrapie variations to be characterized (Benestad et al., 2003). Subsequently, a number of countries have identified this type of scrapie or cases that have similar characteristics (De Brosschere et al., 2004; Gavier-Widen et al., 2004; Epstein et al., 2005).

Before the fall of 2006, there had been no cases of a Nor 98-like scrapie found in U.S. sheep. When the SOSS study was conducted, a number of the characteristics associated with atypical scrapie had not been identified and some atypical cases of scrapie may have been missed.

In November 2006, however, a sample was collected from an apparently healthy mottled-face sheep at slaughter (Wyoming Livestock Board, 2007). The characteristics found by the various diagnostic tests conducted revealed a Nor 98-like scrapie case. The sheep was traced back to a flock in Wyoming that was subsequently depopulated. Since then, four additional cases of Nor 98-like scrapie have been identified in the United States (USDA, 2007b). One of these cases of Nor 98-like scrapie (atypical) has been found in an ARR/ARR Suffolk sheep in the United States (D. Sutton, personal communication, 2007). This case is significant because many countries, including the United States, have adopted control programs that promote breeding

[2]For more information, see the USDA information sheet "The Genetics of Scrapie Susceptibility" online at http://www.aphis.usda.gov/animal_health/animal_diseases/scrapie/downloads/scrapie_genetics.pdf.

for ARR animals. These programs were initiated when it was thought that arginine (R) at codon 171 conferred complete resistance.

There has been speculation that atypical scrapie may be a spontaneous disease, similar to CJD in humans. This theory has originated because most flocks have not been found to have additional cases (Hopp et al., 2006; Lühken et al., 2007). While a spontaneous origin is definitely a possibility, if this "strain" of scrapie is predisposed to long incubation periods where sheep usually are culled or die from other causes, the disease may go undetected. The inability to diagnose such long incubation cases may lead to the conclusion that the disease is spontaneous. In regard to control programs, it will be important to determine whether the disease occurs sporadically or has a low rate of transmission.

Many characteristics of "atypical" strains of scrapie are still unknown. For example, whether the disease transmits naturally from sheep to sheep is unknown, along with which tissues are infected and the potential host range. Research on these and other issues related to "atypical" scrapie is underway.

The public health risk from atypical scrapie is also unknown. There is no basis for attributing a greater or lesser risk to humans from atypical scrapie than from BSE or classical scrapie. If atypical scrapie is not a new phenomenon and has simply been discovered recently, then the lack of epidemiological association between prion diseases in humans and sheep or in the consumption of sheep products suggests that atypical scrapie does not represent a risk to humans. This is not, however, a demonstration of absolute safety.

In February 2007, the UK Chief Medical Officers sent a letter to neurologists to remind them to remain vigilant and refer unusual human neurological cases through the established national arrangements. The UK Spongiform Encephalopathy Advisory Committee (SEAC) considered the reminder important in part because of the unknown human health implications of atypical scrapie (SEAC, 2007).

Considerations for atypical scrapie include the following:

- Monitoring of the epidemiological situation and related research to ensure timely and effective policy changes are made if needed;
- Monitoring the occurrence of scrapie in the so-called "resistant" genotypes. If these genotypes have extended incubations or are subclinical carriers, the problem may go unnoticed.

These are important considerations for the United States because even if classical scrapie is eliminated from the U.S. flock, atypical scrapie may continue to occur in a subset of the sheep population.

THREAT OF EMERGING OR EXISTING DISEASES

Threat from an Existing Disease

In the past decade, over 70 percent of the emerging diseases have been zoonotic (Woolhouse and Gowtage-Sequeria, 2005). The U.S. sheep industry has been fortunate to avoid a significant crisis resulting from a disease outbreak. To prevent complacency, the industry as a whole will need to systematically review worldwide conditions and disease reports as well as suspicions of diseases mutating to a more virulent strain. One such disease that should be reviewed is Johne's disease.

Johne's disease is caused by *Mycobacterium avium* subspecies *paratuberculosis* (MAP), which is a small bacterium related to *Mycobacterium avium* (avian TB). The organism affects cattle, sheep, and goats by causing a chronic inflammation of the intestines. The inflammation progresses to a severe state affecting the absorption of nutrients, which causes the animal to lose condition and eventually progress to death. Crohn's disease is a human condition that is also characterized by a chronic inflammation of the intestines. There is continuing research and debate as to whether the Johne's agent is the causative agent or contributing factor for Crohn's (Radostits et al., 2007) To date, there has been no definitive causal link between the two diseases. If a definitive link between Johne's disease and Crohn's disease, a chronic debilitating human disease, were made in the future, lamb meat and sheep milk and sheep milk products may come under scrutiny.

The MAP organism has a waxy cell wall, tends to clump, and appears to be quite resistant to inactivation. Studies have shown the bacterium to survive pasteurization temperatures (Millar et al., 1996; Grant et al., 2002). The agent is shed in large amounts in the feces. Hence, contaminated feed, bedding, water, and soiled udders serve as sources of infection. Animals less than 6 months of age are thought to be most susceptible to infection. The incubation period for OJD is long, during which time the animal can be shedding the agent and be a source of infection for other sheep. Available diagnostic tests suffer from a number of problems. Infection is often difficult to confirm during the incubation period. OJD may be confused with parasitism, chronic malnutrition, caseous lymphadenitis, and OPP. Control of the disease in infected flocks is currently difficult because there is no reliable live animal test that can detect animals shedding very low numbers of MAP (Radostits et al., 2007).

Ovine Johne's disease has been documented on all continents. In the United States, OJD is estimated to be present on 4.7 percent of sheep operations (USDA, 2004). Currently there is no national, state, or producer-driven control program for OJD. For Johne's disease in cattle, there is a voluntary program conducted by states with federal support. In the United States, surveys have indicated that the primary problem is in the dairy cattle sector but it is also increasing in the nation's beef cattle herd.

If there were ever a definitive link made between Johne's and Crohn's or if the public perceived there was a risk, a number of food safety questions would need to be answered:

- Given that the Johne's agent is shed in large numbers (millions of organisms) in the feces, will fecal contamination of the carcass result in meat contaminated by this agent? If so, will the current microbial intervention steps be effective for this agent?
- There is evidence that the Johne's organism is present in lymph nodes. It would be difficult to process meat totally free of lymph nodes. Would this also provide a source of the organism in food for humans?
- Can the agent survive cooking?
- If the organism survives cooking, what interventions would be necessary to state that lamb and mutton are safe in the event a definitive link is made?
- Given that studies have shown that the organism survives pasteurization, can sheep shed the agent in milk? Would sheep milk and sheep milk products be a risk?

The National Advisory Committee on Microbial Criteria for Foods (USDA) has assigned a subcommittee to conduct an assessment of the food safety importance of MAP (USDA, 2006). The committee has been asked to limit its deliberations to the consideration of a very specific set of questions. It has not been asked to consider the question of whether or not MAP is a human pathogen. The committee was asked to consider the following questions:

- What food, water, or environmental sources are of most concern with respect to exposure of humans to MAP?
- What are the frequencies and levels of MAP contamination found in the above-referenced sources?
- What is the efficacy of the current methods of detection for MAP?
- What processing interventions are available for the foods of concern to eliminate or reduce the levels of MAP contamination to an acceptable level or to ensure that MAP does not enter the food supply?
- What are the research needs to determine additional sources of MAP; the frequencies and levels of MAP contamination in specific sources of concern; potential processing interventions to eliminate or reduce the levels of MAP contamination; and potential processing interventions to prevent MAP from entering the food supply?

Threat from a Foreign Animal Disease

The findings of the NAHMS 2001 sheep study (USDA, 2003a) indicate that certain behaviors and lack of certain practices increase the vulnerability of many sheep operations to disease. One of the biggest vulnerabilities is the risk of introducing a foreign animal disease. When introduced into a native population, foreign animal diseases may spread rapidly as the animals have no immunity to the disease. In many cases, the outbreaks result in a high death loss and/or severe production losses.

Foot-and-mouth disease is an example of such a disease. It is a highly contagious disease that affects cloven-hoofed animals, including pigs, cattle, sheep, and goats. The virus that causes the disease can spread rapidly over large distances, mainly from animal-to-animal contact or from contact with other contaminated sources. While the United States is currently free from this disease, the risk of introduction is real. The introduction could be accidental through the illegal movement of animals or animal products, or intentional. The disease would have devastating consequences for the animal and allied industries. Initial outbreaks in an FMD-free country may warrant a full livestock movement stop for at least 1 week. A full movement stop of livestock prohibits the movement of all susceptible species (cattle, sheep, goats, swine, and other cloven-hoof species) from moving for any purpose including slaughter. Given the susceptibility of a broad range of species, the impact would affect not only the farm level, but also marketing and slaughtering channels, as well as meat and milk distribution. As an example, economic losses from the 2001 FMD epidemic in the United Kingdom exceeded $10 billion (Thompson et al., 2002). More than 6 million animals were slaughtered. The outbreak lasted 7 months. A more recent outbreak that occurred in August 2007 was not widespread and was contained in a relatively short period of time, yet is still expected to result in millions of dollars of loss.

In sheep, FMD results in signs that are often so subtle that they go unnoticed or the signs are mistaken for common sheep diseases, namely soremouth and foot rot. This is an important aspect of the disease, as infected sheep may be moved to many different locations and serve as a source of virus not only to other sheep and goats, but also spread the disease to pigs and cattle. In the 2001 UK outbreak, sheep were thought to play a major role in the widespread undetected initial spread of the disease.

Other foreign animal diseases of concern are peste despetits ruminants, sheep and goat pox, exotic strains of bluetongue, and Rift Valley Fever. The threat of these diseases further emphasizes the need for a good biosecurity program as well as identification and records for traceability.

MAJOR ACCOMPLISHMENTS, OPPORTUNITIES, AND CHALLENGES IN SHEEP HEALTH

Disease adversely impacts viability, overall growth, rate of gain, immunity, and reproductive performance, which in turn reduce income. Flock health programs should be an essential component of the overall management scheme of raising sheep. Other than scrapie, there is no other disease where a national effort has been made to reduce or eliminate the problem. In the United States, making and prioritizing recommendations regarding research needs and control measures is difficult because the data related to the costs of disease and the effect on productivity and profitability of the sheep industry are inadequate.

Major Accomplishments in Sheep Health

The sheep industry and the USDA, with support from congressional appropriations, have made substantial progress in the effort to eradicate scrapie. The insidious nature of this disease in combination with the lack of definitive scientific understanding has made this a great challenge. A few of the more important accomplishments in sheep health include the following:

- A reduced percentage of sheep of all face colors that test positive for scrapie at slaughter and a decrease in the number of positive animals found at slaughter, resulting in fewer new infected and source flocks and fewer animals indemnified in recent years.
- A more complete animal identification system for sheep than exists for any other species in the United States, providing records of movements and a framework that would allow for the tracing of other diseases and animal movements.
- The sheep industry was instrumental in obtaining designation of sheep within the MUMS Act, which has maintained the availability of critically needed drugs, anthelmintics, and vaccines.

Major Opportunities and Challenges for Sheep Health

Over the years, the U.S. sheep industry has not suffered the effects of a major disease outbreak or a food safety crisis. Consequently, consumers do not currently associate lamb and mutton with *Salmonella*, *E. coli*, or other food-borne pathogens. From an animal welfare aspect, the practice of raising sheep extensively has protected the industry from the severe criticism of those industries accused of having factory farms. The markets for lamb and other sheep products, such as milk, cheese, mutton, and wool, have the opportunity for growth unhampered by factors that may negatively influence public opinion and purchases. Other opportunities include:

- *Use of the identification system for animal movement as the foundation for an overall flock health program.* The scrapie-certification program could be used as the framework for other disease control programs. This is being done in Minnesota with OPP.
- *Improvement in biosecurity practices.* Over time, the vast majority of sheep operations add sheep to their flocks (i.e., they do not maintain closed flocks). Despite these additions, most producers do not implement any type of quarantine, require premovement testing, or take precautions to prevent nose-to-nose contact at shows, or other encounters. The industry has the opportunity to improve biosecurity practices that will help reduce the spread of endemic diseases and prevent the entry of highly contagious diseases.
- *Genomics, gene biotechnology, gene markers.* Although in its infancy in production applications, completion of mapping the sheep genome is likely to have major impacts on improving animal health and disease control. For example, New Zealand research has identified multiple gene markers for internal parasite resistance in sheep. New techniques are being used to provide improved medicines and vaccines.

Despite the accomplishments and opportunities, a number of key challenges in the area of sheep health still face the U.S. sheep industry, including the following:

- *Lack of data on the economic impacts or the true prevalence of most of the diseases or disease conditions affecting sheep in the United States.* The lack of this information makes it difficult for the industry as a whole to make decisions regarding the allocation of resources as well as the determination of research and policy priorities. Without knowing the economic impacts and prevalence of certain diseases, individual producers may not be able to maximize profits by implementing the necessary on-farm preventive or control measures.
- *Lack of approved drugs.* The unavailability of approved drugs will continue to adversely impact the sheep industry in efforts to prevent and control certain diseases. There will still be significant challenges of providing the necessary incentives for pharmaceutical companies to engage in research and development for small ruminant products. One of the biggest challenges will be the increasing drug resistance of intestinal parasites.
- *Growing shortage of large-animal veterinarians.* The veterinarian shortage may add to the already significant number of sheep operations that do not use the service of a veterinarian, which, in turn, may affect sheep welfare. As the niche industries grow, many of the new sheep owners

have not raised livestock before. The lack of practitioners to give advice, administer treatments, perform surgeries, and euthanize animals may lead to prolonged and unnecessary suffering.

- *Lack of death loss data.* An accurate estimate of the actual lamb crop in the United States cannot be developed without death loss data for the predocking period in some of the states with the highest sheep numbers. More important, efforts to reduce such losses are encumbered without an accurate baseline that could provide answers to various questions, such as the following relating to possible reasons for lower numbers of lambs:

—Are the ewes giving birth to multiple lambs?
—Are lambs born dead?
—Are lambs dying shortly after birth from no colostrum or poor mothering?
—Are lambs dying at a few weeks of age from scours, pneumonia, and other causes?
—Are predators responsible for the majority of deaths?

- *Monitoring research on atypical scrapie.* With the tremendous progress the industry and USDA have made in reducing the prevalence of classical scrapie, it will be important for all parties to continue to monitor the changing science and epidemiology surrounding atypical scrapie, especially the theories on origin, routes of transmission, and the effects of genetics. What is almost certain is that the current genetic approach as used around the world will not work for atypical scrapie. It may be that other genotypes could be considered but even that is unknown.
- *Research on the link between MAP and Crohn's disease.* If a scientific consensus is reached or if a definitive link is found between MAP and Crohn's, the sheep industry may find itself in a precarious position on a food safety issue. In a crisis, the unknowns are considered one of the biggest enemies. Currently the industry does not know the prevalence of OJD, if cooking kills the agent, or if pasteurization of sheep's milk is more effective than pasteurization of cow's milk. It is important to begin research in these areas.

REFERENCES

Agra CEAS Consulting. 2007. The World Organization for Animal Health (OIE) Prevention and control of animal diseases worldwide economic analysis, prevention versus outbreak costs. Final Report, Part I. Online at: http://www.oie.int/eng/OIE-WB_Conference_1007/OIE%20Economic%20Studies/OIE%20-%20Cost-Benefit%20Analysis%20(Part%20I).pdf. Accessed April 6, 2008.

Andréoletti, O., S. Simon, C. Lacroux, N. Morel, G. Tabouret, A. Chabert, S. Lugan, F. Corbiere, P. Ferre, G. Foucras, H. Laude, F. Eychenne, J. Grassi, and F. Schelcher. 2004. PrP^{Sc} accumulation in myocytes from sheep incubating natural scrapie. Nat. Med. 10:591–593.

Barger, I. A., and K. M. Dash. 1987. Repeatability of ovine faecal egg counts and blood packed cell volumes in Haemonchus contortus infections. Int. J. Parasitol. 17:977–980.

Benestad, S. L., P. Sarradin, B. Thu, J. Schönheit, M. A. Tranulis, and B. Bratberg. 2003. Cases of scrapie with unusual features in Norway and designation of a new type Nor98. Vet. Rec. 153:202–208.

Brown, J. P., and J. D. Silverman. 1999. The current and future market for veterinarians and veterinary medical services in the United States. J. Am. Vet. Med. Assoc. 215:161–183.

Brown, P., F. Cathala, R. F. Raubertas, D. C. Gajdusek, and P. Castaigne. 1987. The epidemiology of Creutzfeldt-Jakob disease: Conclusion of a 15-year investigation in France and review of the world literature. Neurology 37:895–904.

Brown, P., and D. C. Gajdusek. 1991. Survival of scrapie virus after 3 years' interment. Lancet 337:269–270.

Buschmann, A., A.-G Biacabe, U. Ziegler, A. Bencsik, J.-Y Madec, G. Erhardt, G. Lühken, T. Baron, and M. H. Groschup. 2004. Atypical scrapie cases in Germany and France are identified by discrepant reaction patterns in BSE rapid tests. J. Virol. Methods 117:27–36.

Casalone, C., C. Corona, M. I. Crescio, F. Martucci, M. Mazza, G. Ru, E. Bozzetta, P. L. Acutis, and M. Caramelli. 2005. Pathological prion protein in the tongues of sheep infected with naturally occurring scrapie. J. Virol. 79:5847–5849.

Craig, T. M. 2006. Anthelmintic resistance and alternative control methods. Vet. Clin. N. Am. Food Anim. Pract. 22:567–581.

De Brosschere, H., S. Roels, S. J. Benestad, and E. Vanopdenbosch. 2004. Scrapie case similar to Nor98 diagnosed in Belgium via active surveillance. Vet. Rec. 155:707–708.

Detwiler, L. A., and M. Baylis. 2003. The epidemiology of scrapie. Rev. Sci. Tech. Office International des Epizooties (OIE) 22(1):121–143.

Detwiler, L. A., A. L. Jenny, R. Rubenstein, and N. E. Wineland. 1997. Scrapie: a review. Sheep Goat Res. J. 12(3):111–131.

Dickinson, A. G., J. T. Stamp, and C. C. Renwick. 1974. Maternal and lateral transmission of scrapie in sheep. J. Comp. Pathol. 84:19–25.

Eloit, M., K. Adjou, M. Coulpier, J. J. Fontaine, R. Hamel, T. Lilin, S. Messiaen, O. Andréoletti, T. Baron, A. Bencsik, A. G. Bicae, V. Beringue, H. Laude, G. LeDu, J. L. Vilotte, E. Comoy, S. P. Deslys, J. Grassi, S. Simon, F. Lantier, and P. Sarradin. 2005. BSE agent signature in a goat [5]. Vet. Rec. 156:523–524.

Epstein, V., S. Panting, and S. Halfacre. 2005. Atypical scrapie in the Falkland Islands. Vet. Rec. 157:667–668.

European Community. 2007. A new animal health strategy for the European Union (2007–2013) where "Prevention is better than cure." Online at: http://ec.europa.eu/food/animal/diseases/strategy/animal_health_strategy_en.pdf. Accessed April 29, 2008.

European Union. 2006. Report on Monitoring and Testing of Ruminants for Transmissible Spongiform Encephalopathy (TSE) in the EU in 2006. Online at: http://ec.europa.eu/food/food/biosafety/bse/annual_report_tse2006_en.pdf. Accessed November 28, 2007.

Federal Register. 2001, August 21. Rules and Regulations, 9 CFR Parts 54 and 79.66(162): 43963–44003.

Gavier-Widen, D., M. Noremark, S. Benestad, M. Simmons, L. Renstrom, B. Bratberg, M. Elvander, and C. H. Segerstad. 2004. Recognition of Nor98 variant of scrapie in the Swedish sheep population. J. Vet. Diagn. Invest. 16:562–567.

González, L., M. Jeffrey, S. Sisó, S.Martin, S. J. Bellworthy, M. Stack, M. Chaplin, L. Davis, M. Dagleish, and H. Reid. 2005. Diagnosis of preclinical scrapie in samples of rectal mucosa. Vet. Rec. 156:846–847.

Grant, I. R., H. J. Ball, and M. T. Rowe. 2002. Incidence of *Mycobacterium paratuberculosis* in bulk raw and commercially pasteurized cows' milk from approved dairy processing establishments in the United Kingdom. Appl. Environ. Microbiol. 68:2428–2435.

Groschup, M. H., F. Weiland, O. C. Straub, and E. Pfaff. 1996. Detection of scrapie agent in the peripheral nervous system of a diseased sheep. Neurobiol. Dis. 3:191–195.

Groschup, M. H., C. Lacroux, A. Buschmann, G. Lühken, J. Mathey, M. Eiden, S. Lugan, C. Hoffman, J. C. Espinosa, T. Barron, J. M. Torres, G. Erhardt, and O. Andréoletti. 2007. Classic scrapie in sheep with the ARR/ARR prion genotype in Germany and France. Emerg. Infect. Dis. 13:1201–1207.

Hadlow, W. J., R. E. Race, R. C. Kennedy, and C. M. Eklund. 1979. Natural infection of sheep with scrapie virus. Pp. 3–12 in Slow Transmissible Diseases of the Nervous System,Vol. 2, S. B. Prusiner and W. J. Hadlow, eds. New York: Academic Press.

Hadlow, W. J., R. C. Kennedy, and R. E. Race. 1982. Natural infection of Suffolk sheep with scrapie virus. J. Infect. Dis. 146:657–664.

Harries Jones, R., R. Knight, R. G. Will, S. N. Cousens, P. G. Smith, and W. B. Mathews. 1988. Creutzfeldt-Jakob disease in England and Wales, 1980–1984: A case-control study of potential risk factors. J. Neurol. Neurosurg. Psych. 51:1113–1119.

Hoinville, L. J., A. Hoek, M. B. Gravenor, and A. R. McLean. 2000. Descriptive epidemiology of scrapie in Great Britain: Results of a postal survey. Vet. Rec. 146:455–461.

Hopp, P., M. K. Omer, and B. T. Heier. 2006. A case-control study of scrapie Nor98 in Norwegian sheep flocks. J. Gen. Virol. 87:3729–3736.

Hunter, N. 1997. Molecular biology and genetics of scrapie in sheep. Pp. 225–240 in The Genetics of Sheep, L. Piper and A. Ruvinsky, eds. Wallingford, UK: CABI Publishing.

Hunter, N., W. Goldmann, G. Smith, and J. Hope. 1994. The association of a codon 136 PrP gene variant with the occurrence of natural scrapie. Arch. Virol. 137:171–177.

Hunter, N., J. D. Foster, W. Goldmann, M. J. Stear, J. Hope, and C. Bostock. 1996. Natural scrapie in a closed flock of Cheviot sheep occurs only in specific PrP genotypes. Arch. Virol. 141:809–824.

Hunter, N., D. Cairns, J. D. Foster, G. Smith, W. Goldmann, and K. Donnelly. 1997. Is scrapie solely a genetic disease? Evidence from scrapie-free countries. Nature 386:137.

Hunter, N., J. D. Foster, A. Chong, S. McCutcheon, D. Parnham, S. Eaton, C. MacKenzie, and F. Houston. 2002. Transmission of prion diseases by blood transfusion. J. Gen. Virol. 83:2897–2905.

Kaplan, R. M. 2004. Drug resistance in nematodes of veterinary importance: a status report. Trends Parasitol. 20:477–481.

Kimberling, C. V., and D. Schweitzer. 1989. Brucella ovis infection and its management in ovine reproduction. Agri-Practice 10:36–39.

Konold, T. 2008. Evidence of scrapie transmission via milk. BMC Vet. Res. 4:14.

Le Dur, A., V. Béringue, O. Andréoletti, F. Reine, T. L. Laï, T. Baron, B. Bratberg, J. L. Viloe, P. Saradin, S. L. Benestad, and H. Laude. 2005. A newly identified type of scrapie agent can naturally infect sheep with resistant PrP genotypes. Proc. Nat. Acad. Sci. U. S. A. 102:16031–16036.

Ligios, C., C. J. Sigurdson, C. Santucciu, G. Carcassola, G. Manco, M. Basagni, C. Maestrale, M. G. Cancedda, L. Madau, and A. Aguzzi. 2005. PrPsc in mammary glands of sheep affected by scrapie and mastitis. Nat. Med. 11:1137–1138.

Lühken, G., A. Buschmann, H. Brandt, M. Eiden, M. Groschup, and G. Erhardt. 2007. Epidemiological and genetical differences between classical and atypical scrapie cases. Vet. Res. 38:65–80.

Millar, D., J. Ford, J. Sanderson, S. Withey, M. Tizard, T. Doran, and J. Hermon-Taylor. 1996. IS900 PCR to detect *Mycobacterium paratuberculosis* in retail supplies of whole pasteurized cows' milk in England and Wales. Appl. Environ. Microbiol. 62:3446–3452.

Moum, T., I. Olsaker, P. Hopp, T. Moldal, M. Valheim, and S. L. Benestad. 2005. Polymorphisms at codons 141 and 154 in the ovine prion protein gene are associated with scrapie Nor98 cases. J. Gen. Virol. 86:231-235.

NAIS (National Animal Identification System). 2006. Sheep Working Group Report, August. Online at: http://www.sheepusa.org/?page=site/text&nav_id=bbe4ac2bced3271a5a8900 2b58b479de. Accessed November 27, 2007.

NIAA (National Institute for Animal Agriculture). 2003. The ABCs of genetic based flock clean-up and monitoring plans. National Scrapie Education Initiative. Bowling Green, KY.

OIE (World Organization for Animal Health). 2007. Terrestrial animal health code, Chapter 2. Paris, France. Online at: http://www.oie.int/eng/normes/mcode/en_partie_2.htm and at: http://www.oie.int/eng/normes/mcode/en_chapitre_2. 3.13.htm. Accessed April 29, 2008.

Onodera, T., T. Ikeda, Y. Muramatsu, and M. Shinagawa. 1993. Isolation of scrapie agent from the placenta of sheep with natural scrapie in Japan. Microbiol. Immunol. 37:311–316.

O'Rourke, K. I., T. V. Baszler, T. E. Besser, J. M. Miller, R. C. Cutlip, G. A. H. Wells, S. J. Ryder, S. M. Parish, A. N. Hamir, N. E. Cockett, A. Jenny, and D. P. Knowles. 2000. Preclinical diagnosis of scrapie by immunohistochemistry of third eyelid lymphoid tissue. J. Clin. Microbiol. 38:3254–3259.

Parry, H. B. 1983. Pp. 31–51 in Scrapie Disease in Sheep, D. R. Oppenheimer, ed. New York: Academic Press.

Pattison, I. H., and G. C. Millson. 1962. Distribution of the scrapie agent in the tissue of experimentally inoculated goats. J. Comp. Pathol. 72:233–244.

Pattison, I. H., M. N. Hoare, J. N. Jebbett, and W. A. Watson. 1972. Spread of scrapie to sheep and goats by oral dosing with foetal membranes from scrapie-affected sheep. Vet. Rec. 90:465–468.

Pattison, I. H., M. N. Hoare, J. N. Jebbett, and W. A. Watson. 1974. Further observations on the production of scrapie in sheep by oral dosing with foetal membranes from scrapie affected sheep. Br. Vet. J. 130 (4):lxv–lxvii.

Prince, J. B., D. M. Andrus, and K. P. Gwinner. 2006. Future demand, probable shortages, and strategies for creating a better future in food supply veterinary medicine. J. Am. Vet. Med. Assoc. 229(1):57–69.

Race, R., A. Jenny, and D. Sutton. 1998. Scrapie infectivity and proteinase K-resistant prion protein in sheep placenta, brain, spleen, and lymph node: Implications for transmission and antemortem diagnosis. J. Infect. Dis. 178:949–953.

Radostits, O. M., C. C. Gay, K. W. Hinchcliff, and P. D. Constable (eds). 2007. A Textbook of the Diseases of Cattle, Horses, Sheep, Pigs and Goats, 10th edition, New York: Saunders Elsevier.

Ralph, A., M. V. N. O'Sullivan, N. C. Sangster, and J. C. Walker. 2006. Abdominal pain and eosinophilia in suburban goat keepers. Med. J. Aust. 184:467–469.

Schreuder, B. E. C., M. C. M. de Jong, J. J. Pekelder, P. Vellema, A. J. M. Broker, and H. Betcke. 1993. Prevalence and incidence of scrapie in the Netherlands: A questionnaire survey. Vet. Rec. 133:211–214.

SEAC (Spongiform Encephalopathy Advisory Committee). 2007. Minutes of the 96th meeting held on February 20, 2007, London, England. Online at: http://www.seac.gov.uk/minutes/final96.pdf. Accessed April 29, 2008.

Seaman, J., and M. Evers. 2006. Foot rot in sheep and goats. Fact Sheet New South Wales Department of Primary Industries. Online at: http://www.dpi.nsw.gov.au/__data/assets/pdf_file/0015/102381/footrot-in-sheep-and-goats.pdf. Accessed April 6, 2008.

Seidel, B., A. Thomzig, A. Buschmann, M. H. Groschup, R. Peters, M. Beekes, and K. Terytze. 2007. Scrapie agent (Strain 263K) can transmit disease via the oral route after persistence in soil over years. PLoS ONE 2(5):e435.

Seitzinger, A. H., P. L. Paarlberg, and J. G. Lee. 2006. Economic impacts of eradicating scrapie, ovine progressive pneumonia, and Johne's Disease on U.S. sheep, lamb, sheep meat, and lamb meat markets. Pp. 193–206 in The Economics of Livestock Disease Insurance, S. R. Koontz D. L. Hoag, D. D. Thilmany, J. W. Green, and J. L. Grannis, eds. Wallingford, UK: CABI Publishing.

Sigurdarson, S. 1991. Epidemiology of scrapie in Iceland and experience with control measures. Pp. 233–242 in Sub-acute Spongiform Encephalopathies, R. Bradley, M. Savey, and B. Marchant, eds. Boston: Kluwer Academic Publishers.

Thompson, D., P. Muriel, D. Russell, P. Osborne, A. Bromley, M. Rowland, S. Creigh-Tyte, and C. Brown. 2002. Economic costs of the foot and mouth disease outbreak in the United Kingdom in 2001. Rev. Sci. Tech. Office International des Epizooties (OIE) 21:675–687.

USDA (U.S. Department of Agriculture). 1996. Reference of 1996 sheep health and management practices. National Animal Health Monitoring System. Animal and Plant Health Inspection Service. Fort Collins, CO, September.

USDA. 2000. Declaration of emergency because of scrapie in the United States. Federal Register, Vol. 65, No. 53, p. 14521, March 17.

USDA. 2002. Part 1: Reference of sheep management in the United States, 2001. USDA: APHIS:VS, CEAH, National Animal Health Monitoring System, Fort Collins, CO, No. N3560702.

USDA. 2003a. Part II: Reference of sheep management in the United States, 2001. USDA: APHIS:VS, CEAH, National Animal Health Monitoring System, Fort Collins, CO, No.N3780403.

USDA. 2003b. Part III: Lambing Practices, Spring 2001. APHIS:VS, CEAH, National Animal Health Monitoring System, Fort Collins, CO, No. N3790403.

USDA. 2003c. Fact Sheet on Ovine Progressive Pneumonia. Online at: http://www.aphis.usda.gov/vs/ceah/ncahs/nahms/sheep/sheep01/OPP.pdf. Accessed November 28, 2007.

USDA. 2003d. Phase II: Scrapie: Ovine Slaughter Surveillance Study, 2002–2003. APHIS: VS,CEAH, National Animal Health Monitoring System, Fort Collins, CO, No. N419.0104.

USDA. 2003e. Scrapie Ovine Slaughter Surveillance Info Sheet. Online at: http://www.aphis.usda.gov/vs/ceah/ncahs/nahms/sheep/SOSSphase1.pdf. Accessed November 28, 2007.

USDA. 2004. Ovine Johnes Disease: Awareness, Management and Seroprevalence. National Animal Health Monitoring System, Animal and Plant Health Inspection Service, Fort Collins, CO.

USDA. 2005. Sheep and lamb nonpredator death loss in the United States, 1999. National Animal Health Monitoring System, Animal and Plant Health Inspection Service, Fort Collins, CO, December.

USDA. 2006. 2004–2006 Subcommittee: Assessment of the Food Safety Importance of *Mycobacterium avium* subspecies *paratuberculosis* (MAP). National Advisory Committee on Microbiological Criteria for Foods, Food Safety and Inspection Service, Washington, DC. Online at: http://www.fsis.usda.gov/About_ FSIS/NACMCF_Subcommittee_MAP/index.asp. Accessed April 29, 2008.

USDA. 2007a. Scrapie Program Report, FY 2006, Veterinary Services, Animal and Plant Health Inspection Service, February 5. Online at: http://www.aphis.usda.gov/animal_health/animal_diseases/scrapie/downloads/yearly_report_2006.pps. Accessed September 20, 2007.

USDA. 2007b. Scrapie Program September Monthly Report, FY 2007, Veterinary Services, Animal and Plant Health Inspection Service, February 5. Online at: http://www.aphis.usda.gov/animal_health/animal_diseases/scrapie/downloads/monthly_scrapie_rpt.pps#264,1,Scrapie Program. Accessed November 28, 2007.

U.S. HHS (U.S. Department of Health and Human Services). 1997. Substances prohibited from use in animal food or feed: Animal proteins prohibited in ruminant feed. Regulatory impact analysis. U.S. HHS, FDA, and AEI-Brookings Joint Center for Regulatory Studies. RIN: 0910-AA91, Federal Register 62(June 5):30396.

van Keulen, L. J., B. E. Schreuder, R. H. Meloen, G. Mooij-Harkes, M. E. Vromans, and J. P. Langeveld. 1996. Immunohistochemical detection of prion protein in lymphoid tissues of sheep with natural scrapie. J. Clin. Microbiol. 34:1228–1231.

van Keulen, L. J., B. E. Schreuder, M. E. Vromans, J. P. Langeveld, and M. A. Smits. 1999. Scrapie-associated prion protein in the gastrointestinal tract of sheep with natural scrapie. J. Comp. Pathol. 121:55–63.

Westaway, D., V. Zuliani, C. Cooper, M. D. Costa, S. Neuman, A. L. Jenny, L. A. Detwiler, and S. B. Prusiner. 1994. Homozygosity for prion protein alleles encoding glutamine-171 renders sheep susceptible to natural scrapie. Genes Dev. 8:959–969.

WHO (World Health Organization). 1999. WHO consultation on public health and animal transmissible spongiform encephalopathies: Epidemiology, risk and research requirements, with the participation of the Office International des Epizooties, 1–3. Geneva, December. Online at: www.who.int/csr/resources/publications/bse/WHO_CDS_CSR_APH_2000_2/en/pdf. Accessed March 12, 2007.

Woolhouse, M. E. J., and S. Gowtage-Sequeria. 2005. Host range and emerging and reemerging pathogens. Emerg. Infect. Dis. 11:1842–1847.

Wyoming Livestock Board. 2007. NOR98-like strain of scrapie found in Wyoming. Press Release, March 17, 2007. Cheyenne, WY. Online at: http://wlsb.state.wy.us/NewReleases/07Mar16FINALNOR98LIKESCRAPIEPRESSRLS.pdf. Accessed April 29, 2008.

Young, G. B., J. T. Stamp, C. C. Renwick, and A. G. Dickinson. 1964. Field observations on scrapie incidence. Pp. 199–206 in Report of Scrapie Seminar (ARS 91–53), January 27–30, 1964. Washington, DC: ARS.

Zajac, A. M. 2006. Gastro-intestinal nematodes of small ruminants: life cycle, anthelmintics, and diagnosis. Vet. Clin. N. Am. Food Anim. Pract. 22:529–541.

APPENDIX: A CHRONOLOGY OF SCRAPIE CONTROL IN THE UNITED STATES

1947 – First case of scrapie was diagnosed in a Michigan flock. The sheep were of British origin imported from Canada over a period of years.

1952 – Scrapie diagnosed in a California flock. Insistence from the United States Animal Health Association (USAHA) prompted the U. S. Secretary of Agriculture to declare a state of emergency to handle the disease. The eradication program included laboratory confirmation, quarantine and depopulation of infected flocks, and tracing and slaughter of exposed animals sold from infected flocks. The federal indemnity paid at this time was 50 percent of the difference between the appraised value of the animal and salvage, but not to exceed $25.00 per head for grade animals and $75.00 for purebreds.

1953 – The Act of 1884 was amended to include scrapie, and the emergency order was rescinded.

1954 – Title 9 Code of Federal Regulations Part 54 was promulgated. These regulations covered animals destroyed because of scrapie.

1955 – Regulations were amended to include goats.

1957 – Program was broadened to include source flocks. These were defined as flocks from which an affected animal was removed within 18 months before showing signs of scrapie. The source flock was also quarantined and depopulated. Exposed animals sold from the source flocks were traced and slaughtered.

1964 – Scrapie Field Trials began at Mission, Texas.

1965 – The widespread eradication program was modified to allow a provision for bloodline slaughter. In the event the disease was limited to one bloodline, slaughter could be confined to that genetic line. The nonbloodline animals were placed under 2 year quarantine with sale to slaughter only. After the quarantine period, the animals were subject to 18 months of surveillance. In lieu of the bloodline option the owner could opt to depopulate the entire flock.

Note: Although the bloodline program was in effect for 10 years, the bloodline option was used in only 4 of some 71 cases.

1975 – Bloodline option was eliminated.

Exposed animals could no longer be slaughtered for human consumption due to a perceived public health risk.

Federal indemnity was increased to $40 for grades and $90 for purebreds.

1978 – Federal indemnity was paid in the amount of two-thirds of the appraised value of the animal not to exceed $300. This formula was used for both grades and purebreds.

1980 – Canada adopts a bloodline program.

1982 – On recommendations by USAHA and the National Woolgrowers Association, the Cooperative Scrapie Eradication Program was reviewed.

1983 – The Scrapie Eradication Program as outlined in Veterinary Services Memorandum 557.1, dated April 8, 1983, went into effect. The program involved diagnosing infected animals, tracing and euthanizing bloodline animals, and maintaining infected and bloodline flocks under surveillance. The program concentrated primarily on the elimination of bloodline animals on the maternal side. The rationale for this change was to reduce indemnity payments and preserve valuable bloodlines without supposedly reducing the effectiveness of the program.

1987–1988 – Scrapie Review meetings held. These reviews involved representatives from industry, researchers, state regulators, and USDA APHIS.

Advanced notice in the *Federal Register* of proposed rulemaking, soliciting comments on whether to discontinue the Scrapie Eradication Program.

Comments received in response to this rulemaking overwhelmingly asked APHIS not to discontinue efforts to control scrapie. The commentators did request that the government officials in cooperation with industry groups devise a new program for the control of scrapie.

1990 – Scrapie Negotiated Rulemaking Committee established.

The following organizations were represented on the Rulemaking Committee:

American Association of Small Ruminant Practitioners
American Farm Bureau
American Hampshire Association
American Meat Institute
American Polypay Association
American Sheep Industry, Inc.
American Suffolk Society
Animal and Plant Health Inspection Service
Continental Dorset Club
National Assembly of Chief Livestock Health Officials
National Renderers Association
National Suffolk Association
United States Animal Health Association

1991 – The Rulemaking Committee agreed upon a core program for the control of scrapie. This program consists of the following facets:

a. A voluntary scrapie flock certification program;
b. One-time indemnification for infected and source flocks; and
c. Regulations to establish identification of sheep from scrapie-infected and source flocks moving in interstate.

1992 – Voluntary Scrapie Flock Certification Program (VSFCP) established.

Interstate regulations to identify sheep from scrapie-infected and source flocks established (Detwiler et al., 1997).

1997 – VSFCP modified to make it more user friendly for commercial producers.

1999 – APHIS reviewed and revised the VSFCP which became the Scrapie Flock Certification Program.

2001 – APHIS approves the third eyelid test for the diagnosis of scrapie in live sheep.

APHIS makes official eartags available to producers, dealers, and markets.

Federal regulations that enacted the NSEP went into effect: (1) Records and identification required for sheep moving in interstate commerce, (2) indemnity was reinstated and (3) a program to recognize states that conducted an active scrapie control program consistent with federal requirements was established.

Phase 1 of the Scrapie Slaughter Surveillance Study (SOSS) began.

2002 – Phase 2 of the Scrapie Slaughter Surveillance Study began. This phase was carried out to determine the prevalence of scrapie in the United States. 12,491 valid test results were obtained. It was determined that the overall weighted national prevalence of scrapie in mature sheep was 0.20 percent. More details are provided below.

APHIS adopted a genetics-based flock clean-up plan as a standard method for cleaning up scrapie-infected flocks.

2003 – USDA received $15 million in appropriated funding to conduct NSEP.

SOSS concluded and the Regulatory Scrapie Slaughter Surveillance began.

Scrapie Eradication Uniform Method and Rules published.

APHIS approved the immunohistochemistry on lymphoid tissue as an official test.

National Identification Development Team (NIDT) Steering Committee created.

2004 – SOSS results published.

Sheep-specific animal identification plan was presented to the NIDT Steering Committee.

2007 – APHIS received $18.4 million in appropriated funds to conduct NSEP.

2010 – Goal to eliminate scrapie outbreaks in the United States.

2017 – Goal for the United States to be declared "scrapie-free" by the World Organization for Animal Health (OIE) (D. Sutton, personal communication, 2007).

4

The U.S. Lamb Industry

The largest downstream component of the U.S. sheep industry supply chain is the lamb industry. Lamb consumers represent the end users of the lamb industry component of the supply chain. Ultimately, production is driven by consumer demand. Signals provided by consumers determine in large part how much lamb moves through market channels. At the other end of the lamb industry component of the supply chain are a relatively few number of packers who transform the live animals to meat and byproducts. Between packers and consumers are breakers, further fabricators, wholesalers, retailers, and foodservice purveyors who transport and further transform the meat for sale to consumers.

Not surprisingly, given the historical contraction of U.S. sheep inventories as chronicled in Chapters 1 and 2, U.S. sheep and lamb slaughter, along with lamb production and demand, have also declined over the years. Lamb production has declined more rapidly than lamb consumption, falling by over 80 percent from a high of nearly 0.50 billion kilograms in 1945 to the low of 84.4 million kilograms in 2006 (USDA, 2007a). Buoyed by growing imports in recent years, however, U.S. lamb consumption hit a low of 134 million kilograms in 1996 and has grown slowly over the last decade. The resilience of lamb consumption in the face of declining domestic production has arrested the slow decline in per capita consumption, which has held steady at 0.50 to 0.55 kg since the mid-1990s. Even so, as a share of total U.S. red meat consumption, lamb has dropped from over 5 percent in the 1930s to just under 1 percent since 2000. While these numbers accurately characterize historical changes in the U.S. lamb industry, focusing on these numbers alone fails to recognize the important contribution of the lamb industry to the U.S. agricultural economy and the potentially industry-transforming changes currently in process, such as the emergence of direct marketing, growth in ethnic demand, and other market forces creating some optimism about the future of the industry.

This chapter takes a close look at the U.S. lamb industry, with particular interest in the current status of and changes taking place in lamb production, marketing, consumption, and trade, along with the market forces and government policies that influence their patterns of change. The discussion is based on a wide range of research on the lamb industry, including supply issues (e.g., Whipple and Menkhaus, 1989; Purcell et al., 1991; Van Tassell and Whipple, 1994), demand issues (e.g., Whipple and Menkhaus, 1989; Williams et al., 1991; Byrne et al., 1993; Purcell, 1998), marketing margin and packer concentration issues (e.g., Menkhaus et al., 1989; Brester and Musick, 1995; Capps et al., 1995; Viator et al., 2007); trade issues and foreign lamb markets (e.g., Richie, 1979; Reynolds and Gardiner, 1980; Babula, 1996, 1997; U.S. International Trade Commission (US ITC), 1999; Vere et al., 2000; Muhammad et al., 2007); and the welfare implications of government policies (e.g., Whipple and Menkhaus, 1990). The chapter concludes with a summary of the major accomplishments, opportunities, and challenges facing the lamb segment of the U.S. sheep industry.

LAMB SLAUGHTER AND PRODUCTION

Even as slaughter has declined over the years, the average live weight of slaughter lambs has grown, particularly since the mid-1990s, so that lamb production has declined somewhat more slowly than slaughter (Figure 4-1). From 10.5 million head in 1970, federally inspected sheep and lamb slaughter dropped by more than half to 5.0 million head only 9 years later in 1979. Following a brief upsurge over the next few years to nearly 6.8 million head in 1984, slaughter began to decline once again, reaching only 2.5 million head in 2006 (USDA, 2007b). Over the same period, however, the average live weight of slaughter lambs increased from 47.5 kg to around 63.5 kg and dressed weight from 23.45 kg to about 31.75 kg (ASI, 2007). The heavier weights of the slaughter lambs helped slow the decline in lamb production from 250 million kg in 1970 to 80 million kg in 2007.

One consequence of the decline in slaughter has been a decline in the number of packers buying sheep and a drop in public and non-public auctions by over 70 percent between 1980 and 2005 (Figure 4-2). In turn, the decline in the number of packers has led to both regional and structural concentration in sheep and lamb slaughter. Regionally, about three-quarters of the 2.5 million head slaughtered in 2006 were concentrated in the Midwest and Mountain states (Table 4-1). Another 9.5 percent occurred in the Northeast, with the rest scattered among a large number of other states

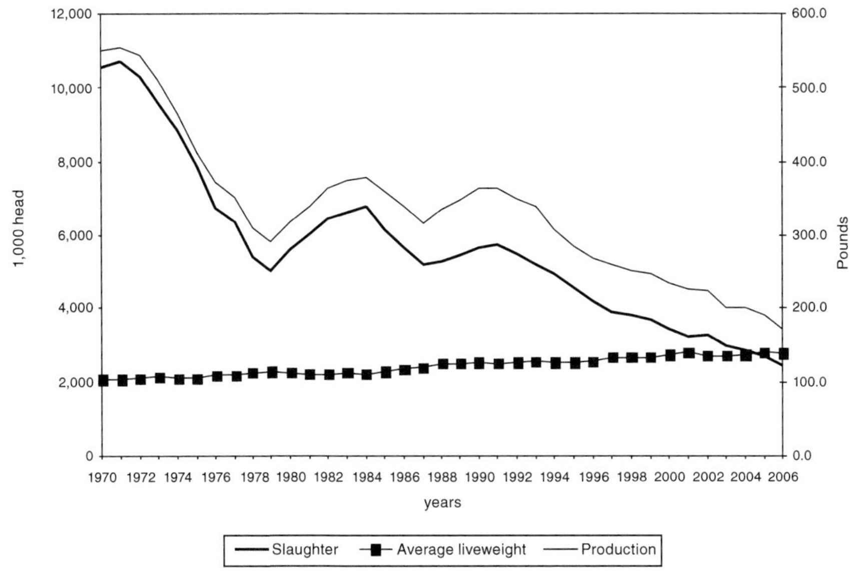

FIGURE 4-1 Lamb slaughter, production, and average liveweight, 1970–2004. (1 pound (lb) = 0.4536 kg.)
Source: USDA (2006).

around this country. Structurally, only a few firms slaughter a large share of the sheep produced in the country. In 2005, the four largest slaughtering firms accounted for 69.6 percent of the federally inspected lamb slaughter (USDA, 2007c). The percentage of lambs slaughtered by the four largest packers, termed the four-firm concentration ratio (CR4), however, has declined from a high of 75 to 80 percent in the early 1990s, following a rash of mergers and acquisitions in the meat packing industry (Williams et al., 1991; USDA, 2007c). Most packing facilities are located strategically near lamb feeders, consumers, or both. Of 205 lamb packing plants, one is classified by the USDA Food Safety Inspection Service (FSIS) as large (500 or more employees), while 42 are classified as small (10 to 499 employees) and another 162 as very small (fewer than 10 employees or less than $2.5 million in annual sales).

The majority of finished lambs (both grain-fed and grass-fed) are purchased by packers for slaughter. Packers separate the pelts and offal from the lamb carcasses, which are inspected by FSIS. Lamb carcasses are also usually quality graded by the USDA Agricultural Marketing Service (AMS). Packers have traditionally marketed their products as hanging carcasses

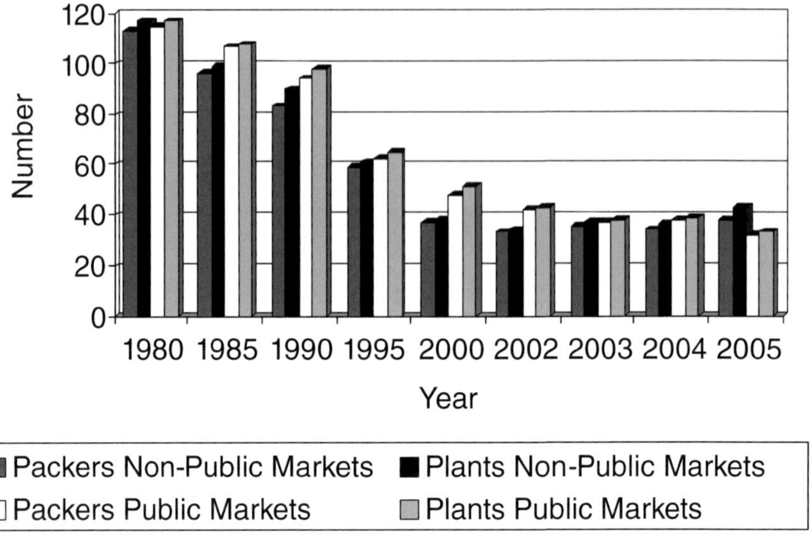

FIGURE 4-2 Number of lamb packers and plants by market type.[a]
[a]Nonpublic includes all sources except terminal markets and auctions. Public includes terminal markets and auctions.
Source: USDA (2007c).

TABLE 4-1 U.S. Federally Inspected Sheep Slaughter by Region, 2006

Region[a]	1,000 head	%
Northeast	242.2	9.5
Southeast	36.9	1.4
Midwest	770.2	30.3
Central	43.4	1.7
Mountain	1,093.5	43.0
Northwest	23.9	0.9
Other States	335.3	13.2
Total U.S.	2,545.4	100.0

[a]States in each region: Northeast = DE, MD, NJ, NY, PA, VA, WV; Southeast = AL, FL, GA, KY, MS, NC, SC; Midwest = IA, IL, IN, KS, MI, MN, MO, NE, OH, WI; Central = AR, LA, NM, OK, TX; Mountain = CO, MT, ND, SD, UT, WY; Northwest = AK, ID, OR, WA; Other States = all states not in any other region.
Source: USDA (2007e).

or boxed primals of carcass equivalents to breakers for further processing into consumer units. Secondary products include offal moving to rendering plants and pelts moving primarily to industrial leather processors. In 1990, only 38 percent of packer sales went directly to the traditional breakers, 36 percent directly to retailers (primarily supermarkets), 8 percent directly to foodservice (primarily hotels, restaurants, and institutions), and the remaining 17 percent directly to nonbreaking wholesalers marketing to both the retail and foodservice channels (Williams et al., 1991). As the demand for case-ready products has grown along with the demand for value-added cuts, packers are now doing much of this further processing themselves, although current data are not available.

Lamb Carcass Yield and Quality

Perhaps the most difficult challenge in lamb production is excess fat on lamb carcasses (Magagna, 1991; Williams et al., 1991; Tatum et al., 1992). The current market structure and pricing system reward producers and lamb feeders for weight rather than a value based on quality and yield grades. The factors that play an important role in the assessment of value of the lamb carcass include the relationship between weight (live and carcass), genotype, muscling (lean), and how they relate to fat (lipid content) of the carcass. Noticeably more lambs are being harvested when they are in the plateau of their growth curve (rather than in the positive plain of the growth curve), and as a result they are predisposed to depositing more fat than muscle. At the same time, some breeds, and individuals within a breed or population, should be harvested at lighter weights based on their individual maturity patterns to avoid excessive fat depositition whereas others (breeds, individuals within breeds, or populations) are capable of being fed to heavier weights because they have a later maturity pattern that results in heavier carcasses with relatively more muscle and less fat. The tendency is to feed all lambs (regardless of maturity pattern) to about the same average live weight, which leads to excess fat in many breeds and individuals. Without a value-based marketing system in the lamb meat industry, the tendency is to keep all lambs on feed for an extended period of time. In many cases this is done without consideration of lean gain per day. When discussing the importance of introducing a value-based marketing system, it is crucial that the discussion include an assessment of the current USDA lamb grading system. It is imperative that the lamb grading system be reviewed for its ability to accurately predict and/or assess lean versus fat.

The standards for slaughter lambs, yearlings, and sheep were revised July 6, 1992, requiring the "coupling" of quality and yield grading to identify both quality and yield when carcasses are officially graded and to require removal of most of the kidney and pelvic fat prior to grading. In addition, leg conformation score was eliminated as a yield grade factor, and the fat thickness range in each yield grade was shifted and narrowed (USDA, 2007d). The changes were made in response to requests by producers as represented by the American Sheep Industry Association (ASI) "to provide an improved communication tool to efficiently reflect consumers' preferences for lean meat products back to producers" (USDA, 2007d). The changes were overwhelmingly supported by all industry segments except lamb feeders and lamb slaughterers and processors (USDA, 2007d). Those two segments of the industry were split on the changes (USDA, 2007d).

The purpose of the grading system is to aid in the pricing and marketing of lamb. Quality grades provide an assessment of two components that influence carcass excellence: conformation and quality (fatness, maturity, and other indicators of differences in palatability of the lean flesh). Conformation is evaluated by averaging the overall thickness and shape of the carcass in the rack, loin, and leg regions. The quality grades for lamb (12 months of age or younger assessed by physiological maturity) and yearlings (12 to 24 months of age) are prime, choice, good, and utility; for sheep (older than 24 months) quality grades are choice, good, utility, and cull. Quality grades are determined based on flank streaking, leg conformation scores, and lean maturity. Sheep must exhibit a higher degree of flank streaking than lambs or yearlings to grade choice. Lamb is more easily labeled as prime with the correct conformation and flank streaking (Purdue University, 2007). Prime is the highest and choice is the second-highest quality grade demanded by consumers. Table 4-2 gives an example of lamb carcass quality grading based on flank streaking and age.

TABLE 4-2 Lamb Quality Grading

	Age of Lamb			
Flank Streaking	Young Lamb	Older Lamb	Yearling Mutton	Mutton
Abundant	Prime	Prime	Prime	Prime
Moderately abundant	Prime	Prime	Prime	Prime
Slightly abundant	Prime	Prime	Prime	Prime/Choice
Moderate	Prime	Prime	Prime/Choice	Choice
Modest	Prime	Prime/Choice	Choice	Choice
Slight	Choice	Choice/Good	Good	Good/Utility
Traces	Choice/Good	Good/Utility	Utility	Utility
Practically devoid	Good/Utility	Utility	Utility	Utility/Cull

Source: Based on data from Purdue University (2007).

Before 1992, yield grades determined the amount of proportional trimmed meat in comparison to fat and bone in the carcass. Yield grades were determined by measuring the external fat thickness between the 12th and 13th ribs of the carcass, an estimate of the kidney, pelvic, heart fat, and conformation of the leg according to the following equation:

Before 1992

$$\text{Yield Grade} = 1.66 - (0.05 \times \text{leg conformation score}) + (0.025 \times \text{percentage kidney, pelvic, and heart fat}) + (6.66 \times \text{adjusted backfat thickness})$$

The current equation that was approved for use on July 6, 1992 only uses adjusted back fat measured at the 12th rib, over the loin eye,

$$\text{Current yield grade equation} = 0.4 + (10 \times \text{adjusted backfat thickness})$$

Berg et al. (1998) reported that factors that influence the percent boneless closely trimmed retail cuts (BCTRC) are carcass muscle mass, carcass weight, internal fat, subcutaneous fat, and intermuscular (seam) fat. They also concluded that the current cutability grades (estimation of the percent BCTRC) used in the three red meat species (beef, pork, and lamb) incorporate one or more of these criteria to establish value associated with saleable product. Table 4-3 shows the differences in the current yield grading criteria for beef, pork, and lamb carcasses. Clearly, lamb yield grades are based on the least amount of information when compared to carcasses from beef and pork. Further, the lamb yield grading system is the only one for the three red meat species that does not account for muscling in the measurement equation. External fat is related to total carcass fatness. As the external fat increases, so does the numerical yield grade, resulting in a presumed

TABLE 4-3 Comparison of Yield Grade Criteria for Beef, Pork, and Lamb Carcasses

	Yield Grade Criteria		
Measurement	Beef	Pork	Lamb
Size	Carcass weight	None	None
Muscling	Ribeye area	Muscle score	None
Trimmable fat	12th rib fat thickness	Last rib backfat	12th rib fat thickness
Internal fat	Kidney, pelvic, and heart fat	None	None

Source: Berg et al. (1998) adapted from Savell (1997). Copyright 1998 by ASI. Used with permission.

reduction in carcass value (Field et al., 1963; Carpenter et al., 1964; Garret et al., 1990; Harris et al., 1990; Fritz et al., 1995). The U.S. lamb grading system is a visual appraisal of carcass fatness that leaves room for error due to the subjective nature of the assessment. The original intent of the yield grading system was to assess carcass value based on factors relative to muscle, fat, and weight, but the current system only evaluates fatness of the carcass.

When USDA graders evaluate lamb fatness, a subjective visual estimate of fat depth adjacent to the 12th rib, it is used as a predicator of percent BCTRC (Berg et al., 1998). The evaluation can be adjusted for body wall thickness and other indicators of carcass fatness. A single measurement based on a visual appraisal of fat depth can lead to the misclassification of lamb carcasses relative to retail yield (Berg et al., 1998). Heaton et al. (1993) found that visual estimation of a single fat trait (12th rib fat) was, at best, a marginal predictor of lamb carcass composition even when the most experienced lamb carcass evaluators were used. Snowder et al. (1994) reported that backfat depth only accounts for 21–22 percent of the variation in the percentage of major and total retail cuts under commercial conditions. They also found that body wall measurement for fat depth was a better indicator of total carcass fatness, explaining more than 30 percent of the variability for yield of retail cuts as compared to the use of backfat measurement over the 12th rib.

Although the yield grade calculation accounts for the fat that is trimmed prior to reaching the retail case, the calculation does not account for seam fat, which is seen by the consumer and possibly affects purchasing decisions at the retail case. If consumers make lamb-buying decisions based on the meat-to-waste ratio or on the perceived healthiness of the cut purchased, seam fat is a negative in the lamb retail case. Accurately predicting seam fat would help the lamb industry to sort and price carcasses that are desirable based on consumer acceptance and purchasing criteria.

Although the current yield grading system is not the most accurate or the most useful in determining the value of lamb carcasses, it is easily implemented during the movement of carcasses from the cooler to fabrication. Line speed in the packing plant is the largest hurdle to overcome in implementing new technologies to estimate the percent BCTRC. A number of new procedures to evaluate carcass value based on the relationship between fat and muscle have been investigated. Berg et al. (1998) reported that warm carcass weight, loin eye area, and body wall thickness, when combined with external fat depth (12th/13th rib interface), predicted percent BCTRC better than any other measures they tested with the exception of an optical grading probe measurement of a chilled carcass. They concluded, however, that taking all of these linear measurements is time-consuming and labor-intensive and would not be practical in today's lamb packing plants, particularly since carcasses are not ribbed, which allows the collection of loin eye area measurements.

Championed by producer cooperatives and direct marketers, value-based marketing systems would benefit from an accurate evaluation tool to assess BCTRC or something similar. Such a system is designed to pay producers premiums for lamb carcasses that meet particular quality and yield specifications. Although a number of other procedures for estimating the percent BCTRC from lamb carcasses have been investigated, the change in the USDA grading system to require both yield and quality grading has focused research on measures that improve accuracy over the current subjective visual assessment. Cattle producers were reluctant to sell beef in a carcass merit system as an assessment of the grading system using humans rather than calibrated technology (Savell and Cross, 1991). Cross and Belk (1994) maintained that a true value-based marketing system will not be accepted by producers unless carcass value is determined through objective mechanical instrumentation. Some of the more effective mechanical measurement techniques currently available include the following:

- *Video Image Analysis (VIA)*. Of the new technologies that have been introduced over the last decade, VIA, or more commonly referred to in the sheep industry as lamb vision system (LVS), seems to be the most promising. It is relatively accurate and has inline capabilities that do not slow line speed in a commercial packing operation. The measurements that LVS can assess include carcass length, groin to right leg length, groin to left leg length, distance from groin to end of shank, red color score for shoulder, blue color score for shoulder, red color score for loin, blue color score for loin, distance between the two legs, groin area, carcass area, total carcass width, leg area, leg width, and groin angle (Brady et al., 2003). These measurements help to assess shape and size of carcass, degree of muscularity, and relative proportions of fat and lean (Brady et al., 2003). When compared to the current yield grade equation, LVS has a more detailed inventory of factors that predict lamb carcass cutability (BCTRC). Assessment of LVS in the United States (Brady et al., 2003; Cunha et al., 2004) has validated the prediction equations and determined that both the accuracy and precision of bone-in cut yields of lamb carcasses were improved by the use of LVS compared with the current grading system. Furthermore, the authors of these reports have reported that predicted accuracy exceeded that of other methods. This equipment is in use in commercial beef plants as a method of assessing and sorting beef carcasses. Brady et al. (2003) concluded that packers would benefit from the use of LVS combined with hot carcass weight by having tighter control on inventories and producers would benefit by receiving feedback regarding lamb carcass data. The Welsh Country Food Group (2007) also reported that LVS will offer considerable benefits to both producer and processor. According a to validation study by Cunha et al. (2003) for USDA AMS, the use of LVS explained a greater proportion of the observed variation in yields of bone-in cuts from carcasses than did expert (whole number) USDA yield grades, expert (nearest tenth) USDA yield grades, or online (whole number) USDA yield grades. The equipment is being used in other countries to sort lamb carcasses and assess value.

- *Optical Grading Probes (OGP).* The basic principle behind the OGP technology is that fat that is predominantly white reflects more light than lean muscle, which would be darker in color. The OGP technology has been used in the pork industry with considerable success over the last two decades. Berg et al. (1998) reported that, in an online industrial setting, the OGP technology (1) is simple to operate, (2) is relatively inexpensive, and (3) can assess carcass composition at rapid line speeds. Hopkins et al. (1995) reported the greatest limitation of the OGP technology is its reliance on a human operator, which creates the potential for error (see also Boland et al., 1995a).
- *Ultrasound.* Although real-time ultrasound technology has been used successfully in the swine industry, the use of the technology with beef and sheep has been less successful due to the presence of hair or wool and/or variations in the thickness of pelt. The use of ultrasound is also dependent on a human operator who is highly trained and versed in interpretation of ultrasonic images. Ultrasound can be used successfully when the integrity of the image being captured is not hindered in any way prior to the image capture. Berg et al. (1998) reported that using ultrasound on carcasses after the pelt is removed introduces error in the measurements. The practice of pelt removal introduces air pockets in the subcutaneous layer of the fat, hindering ultrasonographic penetration and resulting in "noise" in the captured image. Using ultrasound on live sheep and lambs prior to harvest, while assuring an accurate reading, would require a patch of wool to be sheared to the skin at the 12th rib. Although this process would help in capturing a more accurate image, it would result in a discounted pelt credit as well as requiring additional labor to take the measurements. Ultrasound can be used as a tool for selection in a breeding flock where time and pelt credits are not at a premium.
- *Bioelectrical Impedance (BI).* BI technology has been used successfully for measurement of human body composition (Heitmann, 1994). Many of the applications of the BI technology in the livestock industry have been adopted from the human health field. With minor adjustments in programming, the same equipment is used to measure livestock carcass composition. Slanger et al. (1994) tested the BI technology to assess carcass composition in a commercial packing plant and concluded that the technology had great promise for use in the lamb industry as a way to predict kilograms of retail-ready product. Berg et al. (1998) reported that the BI

technology is simple, affordable, nondestructive, portable, and a useful tool in live animal and carcass evaluations.

- *Electromagnetic Scanning or Total Body Electrical Conductivity (TOBEC)*. As the carcass is passed through on a conveyor, muscle tissue absorbs energy whereas fat and bone do not (Berg et al., 1998). Consequently, an electromagnetic absorbance curve can be determined and used to calculate the lean versus fat or bone composition of the carcass. The TOBEC measurement process is highly accurate for determining total body composition. Researchers have determined that this method of body composition estimation is highly accurate for pork (Boland et al., 1995b) and lamb (Berg et al., 1994, 1997). The main drawback to the TOBEC measurement procedure is the amount of space required for equipment. The cost of construction and remodeling the slaughter line to allow for the equipment could be substantial for some operations. Also, the TOBEC equipment is more expensive than that required for any of the other measurement procedures previously discussed.

Without an accurate assessment of lamb carcass yield (estimation of percent BCTRC), procurement of lambs based primarily on live weight will continue to encourage the purchase of lambs that are overfinished. However, identifying a method that will accurately assess the yield of a carcass has been the challenge. Berg et al. (1998) asserted that carcass procurement based on lean yield would be a strong deterrent to marketing overfinished lambs. They concluded that building producer confidence and packer acceptance for quality grading procedures for lamb carcasses will require a carcass yield pricing system with an acceptable level of accuracy in carcass evaluation (percent BCTRC).

Live weight has traditionally been used as a measure of market readiness. Many researchers have looked at the relationship between weight (live and carcass) and carcass lean versus fat yield. Wishmeyer et al. (1996) demonstrated that the correlation between harvest weight and measures of carcass fat is positive and moderately high and that there is a negative correlation between harvest weight and measures of carcass lean yield. They reported that live weight was highly correlated with whole-body lean tissue ($r = 0.96$), ether extractable fat ($r = 0.86$), and crude protein ($r = 0.80$). Jenkins et al. (1988) found that carcass weight accounts for 91 percent of the variation in fat-free lean tissue. Slanger et al. (1994) found that carcass weight is a reliable predictor of total weight of retail cuts. Tatum et al. (1988) reported that increased carcass weight was highly associated with increased carcass fatness. Garrett et al. (1992) supported this finding and reported that yield grade 2 carcasses are significantly lighter than yield grade 3 and 4 carcasses. Berg et al. (1998) reported that live and carcass weight can explain moderate to high amounts of variation in the weight of total lean, fat-free lean, dissected carcass fat, and composition of whole body. A number of researchers (Edwards et al., 1989; Garrett et al., 1990, 1992; Berg et al., 1996, 1997) reported that the correlations between live/carcass weight and percent BCTRC, percent total dissected lean, and percent fat-free lean are small and statistically insignificant, meaning that live and carcass weight are poor predictors of retail cut yield.

Food Safety and Efficiencies

Although food safety is of concern in the meat industry, the primary public health concerns have focused on beef, pork, and poultry-based products. These three species are also the focus of the Pathogen Reduction, Hazard Analysis and Critical Control Point (HACCP) Systems regulations (USDA, 1996). These regulations require slaughter plants to test for generic *Escherichia coli* to verify that they are preventing fecal contamination. To verify that HACCP systems are effectively controlling contamination of raw products, FSIS takes samples from slaughter plants and ground-meat plants for *Salmonella* and measures the results against performance standards developed from nationwide microbial baseline surveys conducted before the regulations took effect (USDA, 1996; Morris, 2003). The requirements for developing and implementing sanitation standard operating procedures, a HACCP plan with clearly identified critical control points in the production system, and requirements for microbial testing for generic *E. coli* took effect at large meat and poultry plants (500 or more employees) in 1998, at small plants (10 to 500 employees) in 1999, and at very small plants (fewer than 10 employees or annual sales of less than $2.5 million) in 2000.

Because lamb is considered a minor species, it was not covered initially by the 1996 HACCP regulations. In 1999, however, FSIS began requiring plants slaughtering minor species such as sheep, goats, horses, mules, and other equines, as well as those that slaughter ducks, geese, and guineas, to sample and test carcasses for generic *E. coli* (USDA, 1999). As summarized in comments to the final rule, some opposition surfaced with respect to sampling sites on the lamb carcasses, as well as the frequency of sampling rate (USDA, 1999). Some packers argued that the lamb industry, with a smaller animal (frame and weight) than cattle or hogs and more animal throughput should not have to be sampled at the same frequency as required of the major species. The final rule required lamb to be tested at the same frequency (1 out every 300 carcasses harvested) as cattle or at least once a week, whichever is greater. Further sample collection occurs at three sites: flank, breast (brisket), and leg (rump).

Although foodborne illnesses have not plagued the sheep and lamb industry as has been the case for the beef, pork, and poultry industries, lamb packers appear to be responding to the changes in regulations in full force. Most have implemented their own in-house laboratories on location at the packing plant or have contracted with outside agencies to sample and evaluate their sanitation operations. One reason that there are more incidences with foodborne illness related to the major species in the United States is the variety of products from those species available to the consumer. Most bacteria are associated with the exterior portion of the product (contamination of the subcutaneous layer of fat during pelt removal) since they are aerobic in nature. With the limited ways that lamb is presented to the consumer and with few of those items in the ground or further processed form, chances for a foodborne illness outbreak are reduced. Many of the food illness outbreaks related to meat are due to contaminated and undercooked, ground, or further processed products (ground/chopped/flaked and then mixed) when bacteria can move from the surface of the product to the interior where they might be trapped in air pockets shielding the bacteria from heat during the cooking process. Foodborne illness becomes a problem when the product is not thoroughly cooked. In the case of lamb, the majority of the cuts offered are whole muscle cuts that are not further processed. Product development research leading to increased use of more of the lamb carcass will likely increase the incidence of foodborne diseases unless there is an associated research effort on pathogen control.

Meat-processing systems that focus on producing a safe, wholesome product delivered to the consumer include inverted harvest systems and technologies used in multiple hurdle applications. Although a popular meat-processing system in Australia and New Zealand, only a few processing plants in the United States have introduced inverted chain harvesting, in which the lamb is hung by its front legs after harvest and prior to evisceration. This process has been reported to have the potential to reduce labor cost using the natural weight of the pelt to pull away from the lamb carcass, reducing the amount of trim required (USITC, 1999). With less handling, there is a reduced risk of contamination (USITC, 1999). Also, with the weight of the pelt falling away from the carcass there is less opportunity for contamination from the pelt. Time of year plays a significant role in the level of potential contamination in a lamb harvest facility. Bacterial contamination from pelt/carcass contact is reduced during dry seasons of the year since reduced pelt moisture reduces bacterial activation on the pelts. The opposite is the case during wet seasons of the year. Additionally, an inverted harvest system relies on the natural musculature of the lung which confines the viscera better than the esophagus. In addition to contaminants that might be present on the pelt, leakage from the rumen through the esophagus can contaminate the breast/shoulder region, elevating bacterial counts. These types of systems have been used in Australia and New Zealand with much success over the years in producing cleaner carcasses with less microbial load upon entering the initial chill cooler. A few American packing operations have followed suit, adopting improved slaughter facilities that embrace the Australian and New Zealand design.

Many lamb slaughter plants have implemented multiple hurdle systems to minimize pathogens. The pathogens of most concern in the meat industry include *E. coli* O157:H7, *Salmonella* spp., *Listeria monocytogenes*, *Campylobacter*, *Clostridium botulinum*, *Clostridium perfringens*, *Staphlococcus aureus*, *Aremonas hydrophilia,* and *Bacillus cereus* (Huffman, 2002). These bacteria are the targets of food safety procedures of all meat-processing facilities regardless of species. Leistner and Gould (2002) described the principles of hurdle technology and concluded that if the initial microbial load is substantially reduced as a result of carcass decontamination procedures, fewer microorganisms are present and will be more easily inhibited in any subsequent processing steps. Packers also have added in-line methods to reduce microbial load on carcasses prior to reaching the chilling cooler. One company has implemented a multiple hurdle approach for beef as a comprehensive means of reducing pathogen load that includes six steps: washing, steam vacuuming, prewash/organic acid rinse, double thermal pasteurization, thermal organic rinse, and cold carcass sanitizer. In the lamb packing industry, steam pasteurization, hot water rinses, and organic acid rinses (lactic or acetic acid) have been implemented nationally. Kochevar et al. (1997) reported that a 2 percent acetic acid (vinegar) wash at 74°C is the most effective in reducing total plate count and the effects of inoculated lamb adipose (subcutaneous fat) tissue with fecal contamination. The Kochevar et al. (1997) study compared varying temperatures (16°C, 35°C, and 74°C) and four different solutions (12 percent trisodium phosphate, 2 percent acetic acid, 5 percent hydrogen peroxide, and 0.003 percent available chlorine). Other research on the preharvest reduction of bacteria on livestock includes analyses of reduced shedding and prevalence of *E. coli* O157:H7 and other pathogenic bacteria through dietary changes (roughage vs. concentrate), the use of probiotic bacteria to reinoculate the rumen working on the principle of competitive exclusion, and drinking water treatments prior to harvest (Huffman, 2002).

While the expense of these pathogen-reducing systems may be prohibitive to some smaller facilities, the payback to those able to implement these systems will be a safer product with a longer shelflife. The implementation of these systems presents problems other than the cost of the equipment as well. The systems can lead to added facilities and equipment maintenance costs because the chemical substances used are highly corrosive to metals other than stainless steel. Most lamb slaughter plants are somewhat dated and not likely to have stainless steel rails and equipment. In addition, high water temperatures, steam, and corrosive materials also create handling and human health concerns.

Processing to Enhance Value

According to a recent survey (Harris Interactive, Inc., 2007), 73 percent of American consumers have not eaten lamb within the previous 12 months. The survey also indicated the reason that many people (42 percent) do not eat lamb is that they just do not think about lamb when they are making their food purchase decisions. A study conducted by Cryovac® reached a similar conclusion (Sealed Air Corporation, 2007). The study found many people do not eat lamb because it is simply not available for purchase on a regular basis when they shop. The study found only 65 percent of the supermarkets audited carry at the most a single lamb product in the retail case and lamb accounts for, on average, 1 percent or less of the space dedicated to fresh meat sales in self-service retail cases, which was about half the space allotted to nonmeat items in those same cases. This result is consistent with the findings of Williams et al. (1991). The Cryovac study also found < 1 percent of the total lamb displayed in grocery store meat cases is preseasoned or marinated. The consequence is few options for lamb consumers in retail stores. Poultry, beef, pork, and other protein sources that account for a substantially higher share of the retail case have also seen the rapid emergence of an increasing variety of precooked, easy-to-prepare, value-added products in recent years.

Partially driven by dual-income households (65 percent of married women work outside the home) and longer work hours, almost 45 percent of Americans indicate that convenience plays a major factor in food choices because they have less time to prepare the family meal (Lyons, 2000; NPD, 2003). In response to a question regarding preparing lamb for dinner, participants in a national consumer focus groups study in 2001 responded that when they think of cooking lamb, they think of a lot of preparation time with oven cooking of three to four hours. The participants indicated that dinner needs to be prepared and on the table in 30 minutes to fit their schedules, which can be done with chicken, beef, and pork.

Although the lamb industry has lagged behind the beef, pork, and poultry industries in responding to these and other markets forces and trends affecting consumer purchasing behavior, both researchers and processors in the lamb industry are increasingly focusing on developing new technologies and new product offerings to enhance the value of lamb by doing a better job of meeting rapidly changing consumer needs. Research is focusing on enhancing tenderness, flavor, juiciness, and ease of preparation; establishing consistency within a product line; new product development, including partial and precooked convenience items; and improving product packaging and appearance. The major overall objective is to make lamb a strong contender for the center-of-the-plate, not only for meals purchased in restaurants but also for those prepared in the foodservice industry and in the home. The following are three areas where additional research and development are needed.

- *Controlled/modified-atmosphere packaging.* A reduction in bacterial counts increases product shelf life and stability particularly if the product undergoes further processing or alternative packaging systems. International processors have been using controlled-atmosphere packaging for some years by replacing oxygen with 100 percent carbon dioxide. In the United States, 100 percent carbon dioxide packaging is being used to extend the shelf life of products in the holding cooler prior to retail display, which allows both large and small retail outlets to have a constant supply of fresh lamb in their retail cases. The advent of new packaging systems has made the development and marketing of new products more feasible since the only point of contact with the product is at the packing plant, which reduces the opportunity for possible contamination or cross-contamination (with cuts from other species) and extends shelf life for up to several months. Lamb products would benefit from greater implementation of these new technologies.
- *Product development.* The lamb industry is following the lead of other meat industries in designing products that are quick to prepare, ready to eat, and shelf stable. Taking their lead from retail-level studies and "quality audits" related to beef and pork concluding that meat consumers want quality, consistency, and convenience in preparation (Purcell, 1989), researchers are exploring processing techniques used in other meat industries, such as grinding, chopping, flaking, forming, and pressing for application to lamb. For example, one major lamb processor is developing product lines geared to clientele who have minimal time to plan and prepare a meal, while at the same time ensuring an enjoyable eating experience (A. Catelli, personal communication, 2007). Fox et al. (2003) concluded that a precooked lamb curry product compared favorably to a similarly prepared beef product in terms of consumer preference and could be a viable option in the marketplace. The researchers were concerned, however, that since their results were based on blind sensory trials, many consumers would perceive a difference between an all-lamb and all-beef product if the product were labeled as such (Fox et al., 2003). Additional product development will help meet new consumer interest in lamb as a convenient and alternative protein source.
- *Value-added retail cuts.* The meat industry continues to look at new, enticing ways to present products to consumers in the retail meat case. Families are smaller today and most, aside from a desire for easy and quick-to-prepare meals, are interested in portion-appropriate food products. With the exception of a variety of chops, the majority of the lamb cuts in the retail case, primarily legs and shoulders, consist of large portions. Many consumers either do not know how to prepare large-portion cuts such as a bone-in leg or a square-cut shoulder or they simply do not have the time

to dedicate to the required preparation. Segmenting larger cuts into smaller portions that require minimal knife work and processing could more appropriately meet consumer needs and enhance their lamb-eating experience. Many lamb processors have developed products with portion size in mind. The top-selling lamb subprimals and cuts continue to be shoulder/shoulder cuts and legs, cuts that are not consistent with current consumer trends toward smaller portions (Tables 4-4 and 4-5). Segmenting cuts like the bone-in leg and/or leg roasts into smaller single- or double-serving cuts and applying postharvest applications to ensure a tender, flavorful, easy-to-prepare product could better meet consumer needs and enhance retail lamb sales.

Nutritional and Sensory Issues

Lamb meat nutrient profiles are similar to those of other protein sources with some minor differences. According to ASI, lamb compares favorably to chicken and beef and is superior to pork in terms of total fat, saturated fat, and cholesterol (Table 4-6). In contrast, Jamora and Rhee (1998) reported the total fat content of lamb (5.25 g/100 g separable lean) is about 17 percent lower than that of beef (6.33 g) and 22 percent lower than pork (6.75 g). Many of the animal proteins are reported to have a higher saturated fatty acid percentage when compared to fats of vegetable origin. However, like beef and pork, lamb fats are less than one-half saturated compared to their unsaturated fatty acid fraction. Jamora and Rhee (1998) also concluded that lamb saturated and polyunsaturated fatty acid percentages are between those of pork and beef. In addition, lamb is lower in monounsaturated fatty acids than beef or pork, but higher than veal or chicken. In terms of the protein levels, Jamora and Rhee (1998) found lamb is comparable to beef and pork. They reported similar findings for mineral levels, and in addition, they found lamb is a good source of zinc and phosphorus. Schweigert (1987) reported lamb is high in B vitamins, particularly B_{12}, and it provides more niacin than beef or pork.

Despite the comparable nutritional value of lamb and other animal protein sources, lamb consumption remains substantially lower than that of other protein sources in the United States. Part of the answer may relate to the distinct sensory properties of lamb compared to those of other animal proteins (Sink and Caporaso, 1977; Cramer, 1983; Jones et al., 1988; Jamora and Rhee, 1998). Age, sex, diet, breed, pH, and the extent and type of cooking done have all been linked to the distinctive lamb flavor profile with which American consumers are largely unfamiliar (Jamora and Rhee, 1998). In a study by conducted by Rhee and Ziprin (1996) on ground meat samples (21 percent fat content) that were pan fried to an internal temperature of 71°C, consumers as well as trained (experienced) panelists could differentiate lamb from beef and pork in blind taste tests. They concluded

TABLE 4-4 Top-Selling Lamb Subprimals, 2003–2005

Subprimals	Total U.S. Sales	California	North-east	Mid-South	Great Lakes	Plains	South Central	South-east	West	Total
	Ave. kg	----------	----------	----------	Percent of sales by cut	----------	----------	----------	----------	----------
Shoulder	700,967	15.4	38.6	8.6	9.9	2.0	2.3	16.2	7.1	100.0
Leg	700,636	11.5	39.2	11.4	11.0	2.1	4.2	13.4	7.2	100.0
Loin	253,059	9.7	36.2	11.0	11.4	2.6	4.5	16.9	7.6	100.0
Rib	129,486	11.0	51.0	6.7	5.2	0.8	4.3	14.3	6.7	100.0
Ground	68,076	10.7	31.4	7.9	11.2	3.3	7.1	14.4	14.0	100.0
Misc/Var	344,016	17.6	30.2	5.5	11.0	1.8	4.4	16.0	13.4	100.0

Source: The FreshLook Marketing Group (2007).

TABLE 4-5 Top-Selling Lamb Cuts, 2003–2005

Retail Cut	Total U.S. Sales	California	North-east	Mid - South	Great Lakes	Plains	South Central	South-east	West	Total
	Ave. kg	----------	----------	----------	Percent of sales by cut	----------	----------	----------	----------	----------
Shoulder blade steak	497,435	15.5	42.4	9.2	7.2	1.6	1.7	15.7	6.6	100.0
Leg whole	383,351	11.2	47.9	11.2	8.4	1.9	3.9	8.3	7.0	100.0
Leg roast	250,718	9.4	30.1	11.4	16.6	2.1	4.1	20.3	5.9	100.0
Loin chops	237,852	10.3	32.8	11.6	12.0	2.7	4.8	17.8	7.9	100.0
Shoulder arm steak	178,032	16.4	30.4	7.1	12.6	2.6	3.8	18.4	8.7	100.0

Source: The FreshLook Marketing Group (2007).

TABLE 4-6 Nutritional Comparison of Animal Protein Sources, 3-oz Cooked Serving Trimmed of Visible Fat

	Calories	Total Fat (g)	Saturated Fat (g)	Cholesterol (mg)
Lamb Leg	162	6.58	2.4	75.7
Pork (Fresh Ham)	179	8.02	2.8	80.2
Beef Round	164	6.59	2.4	69.0
Chicken (dark and light)	162	6.32	1.7	75.3
Turkey (dark and light)	145	4.23	1.4	64.4

American lamb nutritional composite (Percentage of U.S. Recommended Daily Allowances provided by a 3-oz serving of cooked lean lamb):
Protein: 43%; Vitamin B_{12}: 74%; Niacin: 30%; Zinc: 30%; Iron: 17%; Riboflavin: 15%; Calories: 7%.

Note: 3 ounces (oz) = 85 g.
Source: ASI (2004). Copyright 2004 by ASI. Used with permission.

that the reasons for this result were that (1) lamb inherently has the most intense flavor among the three species, (2) lamb flavor is unique or more distinct, and (3) lamb flavor is objectionable or unfamiliar to some consumers and easier to detect.

Because the flavor profile appears to be the overriding factor in consumer acceptance of lamb meat, with tenderness as a secondary contributing factor, a number of studies have explored the chemical composition of lamb as it relates to lamb meat flavor. Young et al. (1994) reported that even though lean tissue may make a minor contribution to the lamb meat flavor profile, the fatty tissue is the main source of lamb meat flavor. Kunsman and Riley (1975) indicated that lamb samples they evaluated were considerably higher in hydrogen sulfide than beef and could contribute to lamb flavor. The higher level of sulfur in lamb is associated with the production of wool and the sulfur compounds that are associated with the fiber protein (keratin). Although some sheep breeds may produce more mild-flavored lamb, research has yet to provide conclusive evidence to substantiate this claim.

Other factors found to play a role in lamb flavor intensity include chronological age (Sink and Caporaso, 1977), breed and gender (Young et al., 1994), diet (Field et al., 1983), and pH of the postmortem muscle tissue (Braggins, 1996). One factor found to affect lamb flavor over which the lamb industry has little control is consumer ability to cook lamb meat properly. The American Lamb Board (ALB) has produced educational materials to help train consumers and hotel, restaurant, and institutional trade representatives on proper cooking procedures for lamb.

LAMB MARKETING

In any market system, consumers send signals through various channels regarding the quality, quantity, and value characteristics of the products they wish to purchase. How well the signals are received through the system depends in large part on the market structure, including the number of levels through which the product must pass, the number and types of groups involved at each level, and the functions each group performs at each level. In general, the fewer the number of levels through which the product must pass and the larger the number of participants at each of those levels, the more efficient markets are at translating and transmitting signals from consumers to producers. In the lamb industry, consumer-to-producer communication is weakened by the multiple levels through which the product passes, by the fact that certain market functions have become concentrated in the hands of a few firms, and by the relatively few consumers involved in purchasing lamb. Although the stages of lamb production have not been subject to major changes over the years, efforts at greater vertical integration within the industry are evident. For example, producer-owned cooperatives not only sell feeder lambs to feedlots but also sell finished lambs to packers, carcasses to breakers, and meat products to retailers and to foodservice purveyors.

Lamb Marketing Channels

The marketing channel for lamb is anchored in the live sheep industry where a large number of geographically dispersed producers market their feeder lambs to a relatively few number of lamb feeders or directly to an even smaller number of lamb packers (see Chapter 2 for more detail). Except for the lighter-weight lambs that are marketed directly to end users, most lambs eventually are sold to packers. From packers, lamb primarily moves as carcasses or boxed primals and increasingly as boxed subprimals and further fabricated cuts into wholesale, retail, and foodservice operations. These operations involve distribution centers of major national retail chains, breakers, and many nonbreaking wholesale distributors and fabricators located closer to consumer markets. Packers also are increasingly servicing retail outlets and foodservice establishments directly with boxed primals, subprimals, or further fabricated cuts.

Breaking and Wholesaling

Breakers have traditionally played a key role in the lamb marketing system by further processing hanging carcasses or boxed primals into subprimals and individual cuts or by fabricating a more user-friendly form of lamb cuts ready for final processing by the ultimate retailer. Little specific data relating to breaker and wholesaler activities are available. According to a study by Williams et al. (1991), 31 percent of breaker lamb sales went to wholesalers, 37 percent to retailers (mostly independents and regional chains), and 25 percent to foodservice operators of all types in 1991. The study also indicates that 47 percent of breakers were concentrated in the Northeast; 20 percent were on the West Coast, mostly in California; and the remaining 33 percent spread throughout the country in 1990. Unfortunately, more recent data on both the distribution and roles of breakers are unavailable.

By locating in or near major lamb consumption areas, primarily on the East and West coasts, breakers provide service and convenience to buyers because the few lamb packers are often too far removed from these areas to service them efficiently. Breakers also serve the valuable function of distributing the various cuts across the market where individual retailers cannot generally purchase or sell lamb in carcass proportions. In addition, the low volume of sales requires the ability to purchase cuts in small lots or boxed cuts wrapped in smaller units, allowing the total purchase to be sequenced through the meat case matching the flow of demand. Even large buyers capable of purchasing directly from packers find it useful to have alternative local sources for fill-in orders as demand spikes from time to time. In addition, the local services are increasingly important to the small or startup ethnic markets where lamb is even more popular than in the general population. Breakers, of course, charge for their services and the convenience provided, with the consequence that these costs are added onto the marketing of lamb before the product reaches the consumer. Increasingly, however, packers are performing much of the initial breaking and boxing of cuts.

Wholesalers of various types are the primary suppliers for retailers and foodservice operators too small to run their own centralized purchasing operations and distribution centers. Although some specialty wholesalers may provide additional value-added processing for meat products generally or lamb specifically, such is not usually the case. Most distribute smaller quantity lots of the products they buy and function primarily to perform the services of pooling the buying power of many small customers, warehousing, and transporting pooled truckloads efficiently. In many instances, wholesalers also provide a menu of services that might include technology support, accounting, facilities planning, and specialized training. These are highly diversified operations rarely capable of being classified along product lines specifically relevant to lamb and lamb products. Although current information is lacking, data from the Williams et al. (1991) study indicate that retailers accounted for 84 percent of the sales of nonbreaking wholesalers in 1990 and foodservice operators accounted for 15 percent, with the additional 1 percent going to various other wholesalers.

Lamb Retailing and Foodservice

Lamb retailers vary widely in type and include large national chain food stores, local chain and independent food stores, local butcher shops, and foodservice groups such as hotels, restaurants, health care and similar institutions, and even the government. Little recent, publicly available data on retail lamb sales are available. Most of what is known comes from a now-dated study by Williams et al. (1991). According to that study, on average, retail food stores accounted for an estimated 36 percent of the lamb sales of packers, 84 percent of the sales of wholesalers, and 37 percent of the sales of breakers in 1990. On the other hand, the study indicated that retailers purchased three-quarters of their lamb directly from packers in 1990 with the other one-quarter coming from breakers/wholesalers. Unfortunately, no data exist to determine the share of total consumer lamb purchases accounted for by retailers.

Although government data concerning lamb at the retail level of the marketing channel are limited, useful data are available from the National Meat Case Study (NMCS), a retail meat case survey taken regularly by the Cryovac Division of the Sealed Air Corporation (2007). During the first quarter of 2007, detailed meat case data were collected from 121 major chain supermarkets in 48 large metropolitan areas across 34 states. A total of 123,204 meat packages were included, covering all of the major protein groups. The NMCS is carried out on a 2-year cycle, with 2004 and some 2002 data available for comparison with the most recent 2007 survey data.

Lamb products have increased their share of space in meat cases from 2002 to 2007, now standing at 2 percent of total linear feet, up from 1 percent in both 2004 and 2002. The average SKU (stock keeping units) count in 2007 was 4.7 per supermarket. Only veal was less among the major protein groups, at 3.2 SKUs on average. For beef, the average was 45.8, with pork at 30.9, chicken at 27.3, turkey at 8.9, and ground beef at 13.2.

For lamb, the average kilograms per package continue to decrease slightly from 0.82 in 2002 to 0.73 in 2004 to the 2007 average of 0.68. Nutritional labeling was provided on 18 percent of the lamb packages. Natural claims of all types increased from 22 percent in 2004 to 27 percent in 2007 for lamb. Cooking information was provided on 30 percent of all lamb packages, down from 36 percent in 2004.

Because of the increasing consumer demand for convenience products, the share of boneless product packages has been increasing. According to the NMCS, 59 percent of the total packages in meat cases now feature boneless products, up from 57 percent in 2004 (excluding ground meat packages). Beef had the largest share of boneless at 83 percent of all beef packages, followed by pork at 60 percent, veal at 59 percent, turkey at 56 percent, and lamb at 13 percent. Although chicken breasts are also shifting to boneless, the large portion of thighs, wings, and whole birds held the total share of boneless chicken packages to 42 percent. The lowest boneless share of packages continues to be lamb, with the share actually declining from 26 percent for all lamb in the 2004 sample.

Products carrying some form of natural or organic labeling have also been growing in prominence in meat cases according to the NMCS at 57 percent of the total, up from 44 percent in 2004 and 34 percent in 2002. In 2007, some form of natural designation was carried on 88 percent of turkey packages, 77 percent of ground beef packages, 74 percent of chicken packages, 53 percent of pork packages, 18 percent of lamb packages, 10 percent of veal packages, and 24 percent of whole-muscle beef packages. Lamb held steady from 18 percent in 2004 but increased from 10 percent in 2002.

The NMCS also indicated that from 2002 to 2007, there was a strong shift toward case-ready meat packages prepared prior to shipment to the stores. In 2007, "case ready" accounted for 64 percent of all meat case packages from all protein sources, up from 60 percent in 2004 and only 49 percent in 2002. By species, case-ready packaging was highest for turkey at 97 percent, followed by chicken at 94 percent, ground beef at 67 percent, pork at 56 percent, lamb at 60 percent, veal at 51 percent, and whole-muscle beef at 27 percent. All species either held steady or increased the case-ready share over this period. Lamb increased from a share of 47 percent in 2004 and 38 percent in 2002.

In general, the increasing penetration of lamb in retail meat cases is good news for the lamb industry, especially since the growing custom and ethnic specialty markets are not captured in surveys of this type. Meat-case merchandising is rapidly evolving as consumer tastes and preferences change. Over time, opportunities will exist to expand sales through increasing all forms of point-of-purchase consumer information and especially through on-pack consumer information such as cooking suggestions, recipes, and nutrition information. Meat products that cannot be sold before they must be discarded are extremely costly to retailers. Therefore, of increasing importance are case-ready packages allowing a portion of the shipping container to be merchandised in the case while holding the rest bagged or frozen.

Lamb Marketing Arrangements

Packers procure lambs primarily through either cash (spot) market contract transactions or various agreements that have come to be referred to as alternative marketing arrangements (AMAs) with producers (RTI, 2007). Cash or spot market transactions refer to auction barn sales; video or electronic auction sales; sales through order buyers, dealers, and brokers; and direct trades. The AMAs encompass all possible procurement alternatives to the cash or the spot market, including forward contracts, marketing arrangements, procurement or marketing contracts, packer ownership, custom feeding, and custom slaughter. In 2005, 42.2 percent of slaughter lambs were procured by packers through formula pricing arrangements, 39.4 percent through auction markets, 12.0 percent through negotiated pricing agreements, 4.9 percent as packer-owned animals, and 0.8 percent through contract production. Imports accounted for the remaining 0.7 percent of packer procurement (RTI, 2007).

According to the RTI study (2007), the marketing arrangements used by producers to sell feeder lambs to feedlots and by producers and feedlots to sell slaughter lambs to packers have two key dimensions: (1) the ownership method (e.g., sole ownership, shared ownership, or owned by another entity) and (2) the pricing method used (Figure 4-3). Different types of pricing methods are used with each type of ownership method. Formula pricing requires the specification of a formula base price. Carcasses are typically priced on a per-head basis but can be priced on a liveweight or carcass weight basis. The use of AMAs is one of the few risk-management tools available to operations because no futures market exists for lambs (Viator et al., 2007).

In selling lamb products, packers also use cash (spot) markets and various AMAs using a variety of pricing methods (Figure 4-4). Packers sell directly to buyers through cash markets and individually negotiate prices or use sealed bids or other methods to establish prices. Packers also sell their lamb products through forward contracting or other marketing arrangements using various formulae to establish prices. Whether selling through cash markets, forward contracts, or marketing arrangements of some type, pricing often involves various practices such as volume discounts and bundling. Packers also custom slaughter for a fee and transfer lamb products within a vertically integrated system using some internal transfer pricing method (Viator et al., 2007).

Lamb Market Structure and Pricing

Understanding the pricing and procurement methods used in the lamb industry at each level provides insight on how the market functions but offers little in the way of understanding how prices are actually determined in the market and who, if anyone, in the system has the ability to influence prices. The particular way in which a market is organized determines its structure[1] and heavily influences the competitive conduct of the firms in the

[1]The term "market structure" refers to the number, type, and size of firms in an industry. A market with one (or only a few) sellers has a monopolistic (or oligopolistic) structure. By the same token, a market with one (or only a few) buyers has a monopsonistic (or oligopsonistic) structure. A market with many small, homogeneous buyers and sellers is called perfect competition. A market with one or a few large sellers or buyers and many other smaller sellers or

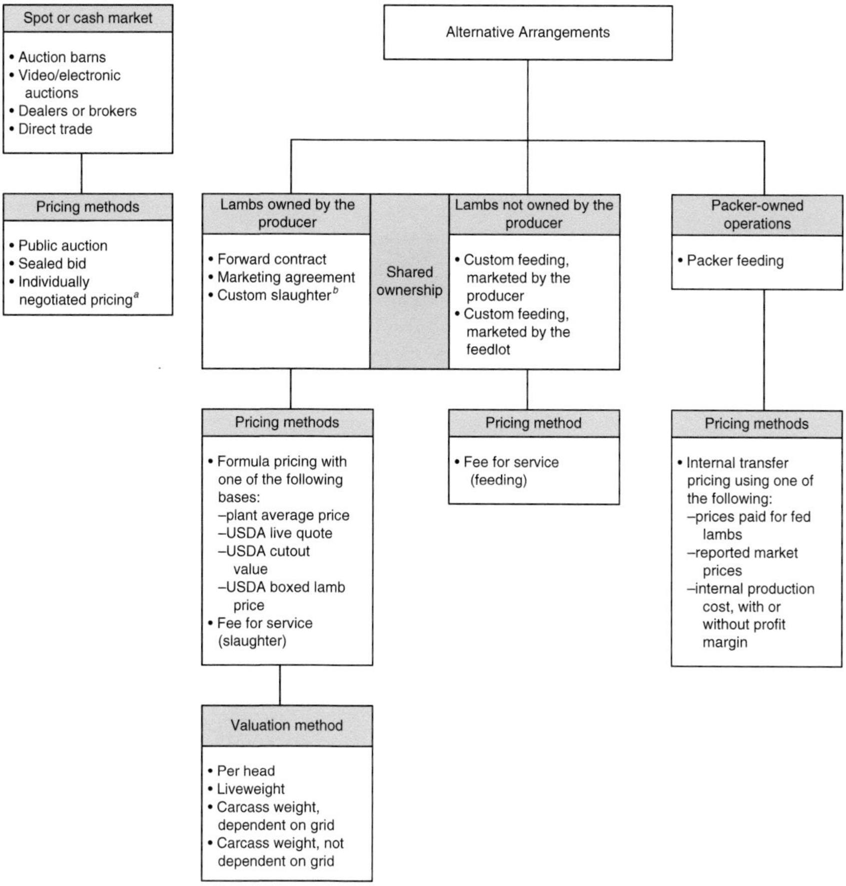

FIGURE 4-3 Marketing arrangements for sale or transfer of feeder and fed lambs by producers.

[a]Individually negotiated pricing is often benchmarked against reported prices.

[b]Custom slaughter may be coordinated by a cooperative that schedules slaughter of lambs for its producer-members.

Source: RTI International (2007).

buyers is monopolistic competition on the selling or buying side. A perfectly competitive market structure is more likely to foster strong price competition. Firms in such an industry are price takers. A concentrated market (only a few large firms) often gives rise to market power and the ability of the firm or firms in the market to control prices. In such an industry, competition is often based more on nonmarket factors than on market pressures of supply and demand. Nevertheless, concentration is a necessary but not always sufficient condition for markets to be controlled by one or more firms in the industry.

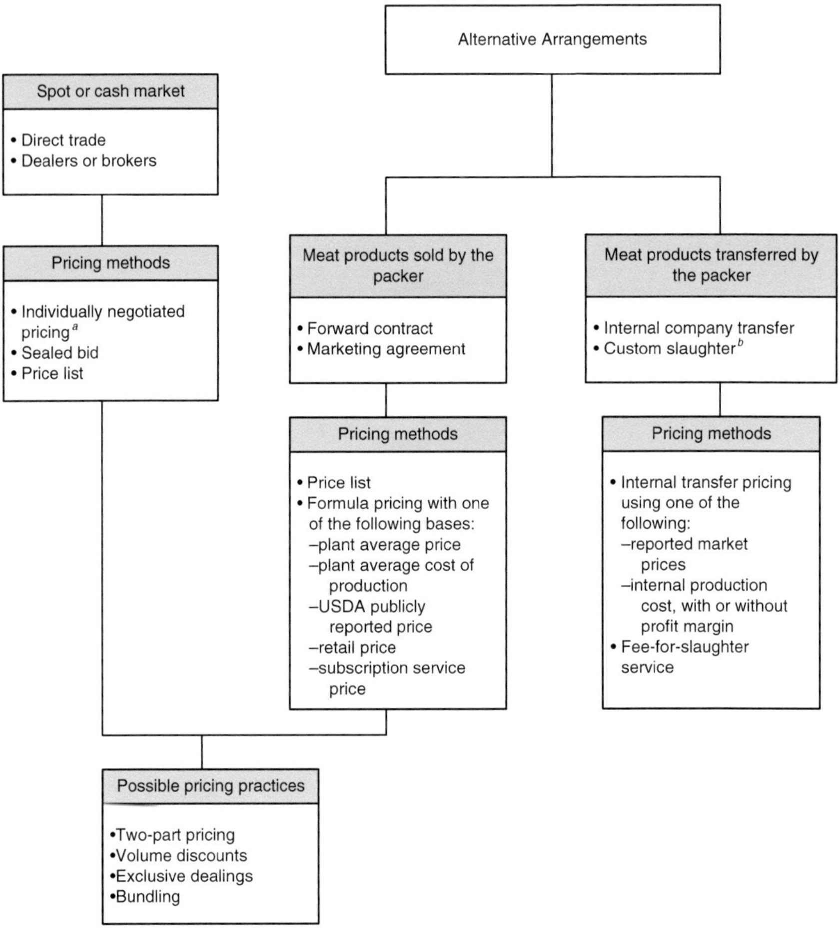

FIGURE 4-4 Marketing arrangements for sale or transfer of lamb meat products by packers.

[a]Individually negotiated pricing is often benchmarked against reported prices.

[b]Custom slaughter may be coordinated by a cooperative for its producer-members.

industry, which, in turn, dictates how prices in the industry are established and behave. The ways in which prices are determined under different types of market behavior usually differ widely. If the structure of an industry and, therefore, competitive behavior, differ markedly at different levels in the industry, prices at each industry level will likely be determined in very different ways. Consequently, to gain insight into the particular process (or processes) by which prices are determined in the lamb industry, the structural characteristics and the associated competitive behavior at each level in the lamb industry must be understood.

Producers, the original suppliers of the raw ingredients in the industry, either lamb meat or wool, closely resemble perfect competitors in economic jargon. Because of the existence of a large number of producers, the actions of individual producers typically have negligible market effects. Their collective independent actions constitute the market supply of feeder lambs and sheep. Producers sell a relatively homogeneous product (live sheep and lambs) and have some knowledge of market opportunities and prices. Because there are some differences in lamb and sheep characteristics and because they do not have perfect knowledge of all possible market conditions, producers operate within a near perfectly competitive market structure. Lamb producers are basically price takers without much market power to affect the prices at which they sell their sheep and lambs. Consequently, producers must accept the price outcomes from the interactions of entities in the lamb marketing chain.

A relatively smaller number of commercial feeders purchase lambs from producers. Commercial feeders account for a large percentage of the sheep and lambs fed. Consequently feeders operate as oligopsonists (relatively few buyers) in their actions with producers. As such, commercial feeders have some potential market power relative to lamb producers. On the selling side, commercial feeders act as oligopolists (relatively few sellers). However, as oligopolists, commercial feeders face a relatively smaller number of packers who act as oligopsonists. Consequently the interaction of commercial feeders and packers constitutes a bilateral oligopoly in economic jargon. In a bilateral oligopoly, the forces of supply and demand generally are not the primary determinants of prices.

Because there are relatively fewer packers than feeders, most feeders have few alternative buyers to whom they might sell their animals if they do not like the prices offered by their current buyer (packer). As well, once fed lambs reach slaughter weights, feeders must move them to market within a relatively short period of time. Commercial feeders subsequently contract to supply packers with a minimum number of slaughter lambs. The bargaining process with commercial feeders and packers often is one-sided, with the consequence that feeders act to a large extent as price takers in dealing with packers. Often, commercial feeders price the lambs they purchase from producers according to the prices they receive for slaughter lambs after deducting their costs and allowing for a "reasonable" profit.

Packers operate within a market structure similar to that of commercial feeders. They behave as oligopsonists in buying slaughter lambs and as oligopolists facing retailers, foodservice purveyors, or breakers in selling processed lamb. However, packers are at a distinct disadvantage in bargaining on price with large retailers and large foodservice buyers that buy in volume. Packers must operate at as close to full capacity as possible to be efficient in their operations and must move their processed lamb soon after slaughter and processing or face spoilage losses. Further, retailers and foodservice purveyors also may purchase lamb supplies directly from breakers. As a consequence, packers typically do not have much latitude in bargaining with oligopsonistic retailers.

Further, lamb is a perishable commodity and a minor item among meat products carried by retailers and foodservice agents. Hence, in this bilateral oligopoly situation, retailers and foodservice buyers realistically can make "take-it-or-leave-it" offers to packers and leave packers holding a great deal of inventory if the asking price is not to their liking. Retailers and foodservice buyers handle a myriad of items and generally are more interested in competition with rivals than in bargaining with lamb packers. Consequently, the price-determining behavior of retailers and foodservice purveyors occurs primarily as a result of interactions among rivals rather than in interactions among packers. Stated differently, whether or not a retailer has lamb in the meat case and whether or not a foodservice purveyor has lamb on the menu is much less of a problem for them than for packers for whom lamb often is either the only product or the major product of their business.

Packers are in a more advantageous bargaining position when it comes to breakers. Similar to packers, lamb tends to be the only or major product handled by breakers. The goodwill of the packer, therefore, is essential to the business of the breaker. At the same time, the breaker plays a role in carrying inventory for many small wholesalers, retailers, and foodservice purveyors. Both the breaker and the packer must move their inventory in a relatively short period of time. The level at which price is set between packers and breakers is indeterminate as in any bilateral oligopoly. But the price determination process depends more on the bargaining process than in the case of packers and retailers or packers and foodservice buyers. Over the last several years, importantly, packers have revolutionized the lamb distribution system by moving away from selling carcasses to selling boxed lamb. Much of the initial breaking and boxing of primals and subprimals now is done by the packer, diminishing both the market role and price negotiating influence of breakers.

Breakers operate as oligopolists in their interactions with lamb wholesalers and relatively small retailers and foodservice buyers. Because breakers act as oligopolists, competition among breakers is not on the basis of price but rather on the basis of service and convenience. Put another way, breakers differentiate the products they offer to buyers through the service and convenience they add. Also in this way, breakers differentiate themselves from packers.

Continuing upstream in the marketing chain, food retailers and foodservice purveyors operate as monopolistic competitors in various local areas facing a large number of consumers. Consumers act as near-perfect competitors similar to lamb producers. There are large numbers of potential consumers who are the ultimate users of the lamb product that is fed, processed, fabricated, and prepared for retail sale. Because they face the monopolistic competitive behavior of the food retailing and foodservice industries, the important element in price determination is product differentiation.

The different market structures and the price discovery processes at each level of the lamb industry create some difficulty for value preferences to migrate from consumers to producers. The oligopoly, oligopsony, bilateral oligopoly, and monopolistic competition behavior evident in the intermediate stages of the lamb marketing chain means that competition at those points is based largely on nonprice factors. Consequently, price signals from consumers reach producers only with difficulty. Enhancing the pass-through of price and quality signals from consumers to producers would require either: (1) the removal of structural obstacles in the marketing system through the creation of a more price-competitive environment or (2) the creation of more pull (and, thus, more market influence) from the consumer end.

Progress was made on the first task with the passage of the Livestock Mandatory Reporting Act in 1999 by Congress. With implementation beginning in April 2001 and ending in 2005 for lamb prices, the purposes of the mandatory price reporting (MPR) legislation were to provide market price and quantity information for cattle, hogs, lamb, and meat products that (1) could be readily understood by market participants; (2) provide information on price discovery, quantity, and quality of livestock and livestock products procured and sold under alternative marketing arrangements; (3) improve USDA price-reporting services; and (4) encourage competition. The Mandatory Reporting Act recently has been reauthorized, but implementation is not likely to occur until late 2007.

Progress towards the second task (demand creation) is being made since the creation of the ALB in July 2002. The Lamb Promotion, Research, and Information Order, better known as the American Lamb Checkoff Program, was established under the Commodity Promotion, Research, and Information Act of 1996, following calls by virtually all segments of the domestic sheep and lamb industry for the establishment of a checkoff program to enhance demand. The Lamb Checkoff Program is designed to expand market share of American lamb by: (1) getting people to ask for American lamb year-round; (2) branding American lamb as the preferred choice in the marketplace; (3) differentiating American lamb from imported lamb through advertising campaigns; (4) minimizing the volatility of seasonal product sales through targeted promotions; (5) promoting use of the whole lamb, using all cuts; and (6) leveraging and expanding ALB resources via cooperative relationships with marketing partners. To the extent that ALB is successful in enhancing the demand for lamb and increasing the number of lamb consumers, the volume of lamb moving through the marketing system will increase, which, in turn, will provide increased marketing opportunities and competition at each level in the marketing chain. With larger lamb volume flowing through the system, the firm size and share of the market needed for efficient operation as well as capital investment requirements would be lowered.

Lamb Marketing Costs and Margins

In a properly functioning market, changes in retail demand for any product subsequently lead to changes in the demand for the product at the farm level. As a product moves from the farm gate to the consumer plate, value is added at various stages. Slaughter, processing, and related activities add *form* value by transforming live sheep and lambs into the final retail cuts purchased by consumers. Inventory holding by packers, breakers, and wholesalers adds *time* value to lamb by storing when supplies are more available and distributing the product when needed. Transportation adds *place* value to lamb by moving it from the farm through the system to its final destination. Transportation between each level of the industry, from farm to slaughter to breaking/further processing to wholesaling and to retailing, adds additional costs to lamb. Finally, convenience and service value (e.g., trimming fat of heavier lamb carcasses or providing case-ready lamb cuts) is added to lamb breakers, wholesalers, retailers, and others who are geographically located near buyers.

The addition of all these values involves costs that force a wedge between the price per kilogram of lamb at retail and the price per kilogram received by producers. Increases in these costs as value is added at each level of the marketing chain tends to push farm prices down and retail prices up, as well as to reduce the farmer share of the retail dollar. The difference in price at retail and the price at the producer level in terms of retail value equivalent is called the price spread or the marketing margin. The calculated margins generally include all the costs incurred and profits realized by all firms involved in the movement of lamb from producers to consumers. An analysis of the marketing margin between prices at each level in the industry is critical to a complete understanding of the process of price determination in any industry.

Operating costs of intermediaries of the marketing chain involve labor, packaging, refrigeration, transportation, advertising, and other expenditures. In the food and fiber sector, increases in these operating costs of marketing intermediaries over time tend to exert downward pressure on farm prices. Given the aforementioned market structure of the lamb industry, packers, breakers, retailers, and foodservice purveyors typically pass the impact of increased costs back down to the producer in the form of lower live animal prices. The extent to which the costs can be passed from one level to the next in the industry depends on the market structure at each level in the industry as discussed in the previous section.

Lamb producers eventually must sell lambs where the short-run direction in prices is determined at higher levels of the marketing chain. There is virtually no opportunity for producers to influence price levels to any extent, and importantly, there are no assurances as to the security of margins above costs. In the lamb processing, wholesaling, retailing, and foodservice sectors, the market structure is different. Firms often possess some market power to influence price or other terms of trade. Consequently, in these sectors of the lamb industry, firms are better able to pass on increasing costs up or down the marketing chain and to secure margins above costs. The market power exerted in these sectors allows firms to set prices as follows: (1) cost-plus pricing or average-cost pricing, calling for the addition of some base cost as a margin to ensure profitability; (2) flexible markup pricing, calling for the markup to vary on the basis of demand considerations; and (3) price leadership, in which firms in the industry set their prices in relation to the price of a dominant firm. The dominant firm sets its price using either cost-plus pricing or flexible markup pricing.

Empirically, Capps et al. (1995) analyzed the behavior of price margins at three levels in the lamb industry: (1) slaughter to retail, (2) slaughter to wholesale, and (3) wholesale to retail. Until this study, little previous work had been done with respect to lamb margins. Problems with data availability are often cited as the explanation for the lack of effort in the analysis of lamb marketing margins. The econometric model developed by Capps et al. (1995) allowed for an analysis of the transmission of price between various marketing levels in the lamb industry. As well, their analysis helped identify and measure the impacts of several key determinants of price margins in the lamb industry, including supply and demand conditions (as represented by movements in price and output), marketing costs, seasonality, and packer concentration. They estimated the elasticities of price transmission (EPT) from slaughter-to-wholesale, from wholesale-to-retail, and from slaughter-to-retail in the lamb industry to be 0.87, 0.65, and 0.57, respectively. The EPT indicates the responsiveness of the price at one level in the industry to changes in the price at a lower level in the industry. The EPT is calculated as the ratio of the relative change in price at one level to the relative change in price at the lower level. An EPT value of 1 suggests an equal response transmission from the lower level to the higher level of the marketing chain. This type of response is consistent with perfect competition. An EPT value of zero suggests no transmission of price signals from the lower level to the higher level in the industry. This type of response is symptomatic of imperfect competition. In this case, price competition is avoided and nonprice competition is the main strategy.

The Capps et al. (1995) results indicated that price changes at the producer/feeder level are almost fully transmitted to the wholesale level, representative of a perfect competition situation. On the other hand, their results indicated that price changes are not well transmitted between the wholesale and retail levels. The potential causes for this breakdown in price transmission include (1) the low level of lamb handled by most retailers, (2) the lack of importance retailers attach to marketing lamb, and (3) the viability of non-price competition as a competitive strategy for retailers. The study also demonstrated that changes in packer concentration have relatively little effect on changes in price spreads or margins. Given that data across the 1978 to 1990 time period were used in the study and that notable structural changes in the lamb industry have occurred since 1990, this study needs to be done again with updated data to get a more current measure of price margin behavior in the lamb industry.

The impacts of market concentration on prices and margins also warrant continued investigation. Menkhaus et al. (1989) investigated the impact of market concentration on slaughter lamb prices. Brester and Musick (1995) considered the effect of market concentration on lamb marketing margins. Even though results from the literature provide limited evidence to indicate that concentration exerts a negative effect on slaughter prices, updates are needed to explore this issue more fully. The number of firms in a national market may not represent well the number of firms in a given regional market. In most regions, a handful of packers, at the most, represent the whole range of marketing opportunities for lamb producers and feeders. Consequently, research regarding the impacts of packer concentration on the lamb industry needs to account for the regional concentration trends of lamb packing.

NATIONAL LAMB DEMAND

The level of demand for lamb and the changes in that level over time are key determinants of the long-run economic viability of the lamb industry. Demand considerations assist in determining the long-range price outlook and provide the foundation for long-range investment decisions. In the livestock and poultry sectors, the demand for a product at the producer level is a *derived* demand, meaning that the demand for the live animals at the farm level is derived from the consumer demand for meat and other livestock products at the retail level. Changes in demand for meat and other livestock products at the consumer level are transmitted down the marketing channel to the producer.

Understanding and promoting demand is necessary to the expansion of any industry. By definition, demand is a schedule of the quantities that consumers are willing to buy at various prices at a given point in time in a particular market. The focus here is on lamb demand by consumers over the entire U.S. market. Analyses of national lamb demand typically attempt to quantify how sensitive consumers are to lamb prices as consumption moves from one point to another along the demand curve and to determine not only if shifts have occurred in lamb demand, but also why such shifts have occurred. The results are the bases of appropriate recommendations to improve the profitability of all segments of the lamb industry. Raising the demand for lamb is a key to the growth and expansion of the industry. Increasing demand translates into consumers purchasing more lamb at constant or higher prices, which, in turn, improves the economic viability of the entire sheep and lamb industry.

The problem facing the lamb industry, and the feature that makes the lamb industry unique among U.S. livestock industries, is the unparalleled decline in lamb production and consumption since the mid-1940s. On a retail equivalent basis, per capita lamb consumption grew from a low of about 1.8 kg in 1917 at the beginning of World War I to 3.0 kg in 1945 at the end of World War II (Figure 4-5). In the years immediately following

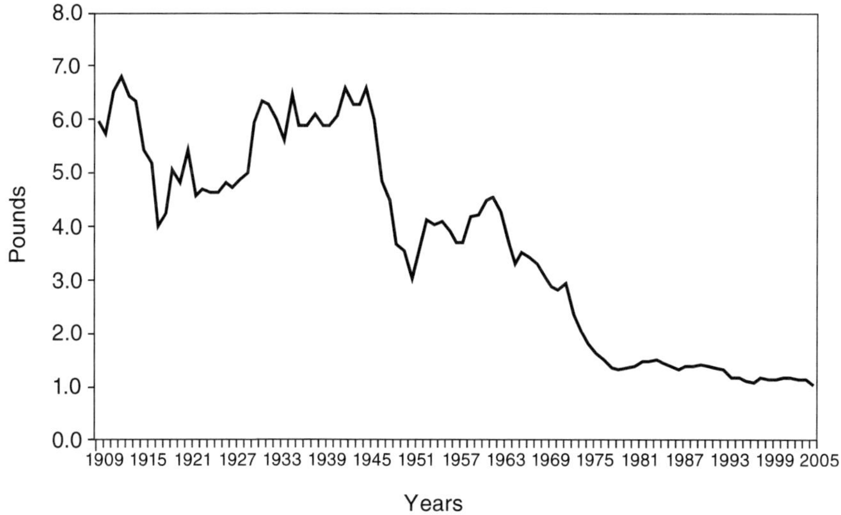

FIGURE 4-5 U.S. annual per capita consumption of lamb, 1909–2005.[a]
[a]Retail equivalent basis. (1 pound (lb) = 0.4536 kg.)
Source: USDA (2007e).

World War II, however, per capita lamb consumption dropped by more than half to only 1.4 kg in 1951. Following a slight recovery over the next decade to 2.0 kg in 1962, U.S. per capita lamb consumption began a slow, steady decline to 0.68 kg in the early 1980s, 0.59 kg in the early 1990s, and about 0.50 kg in most years since 1996.

Until recently, imports were not an important part of the U.S. lamb industry, except during the 1960s and early 1970s when a rapid liquidation in the sheep industry reduced lamb production faster than demand, opening the door for imports to fill the gap (Figure 4-6). The surge in imports during that period dissipated when the rate of decline in consumption caught up to that of production in the mid-1970s. In the early 1980s, however, the rapid decline in lamb consumption halted. Since then, the annual volume of lamb consumed has stabilized at 159 to 181 million kilograms. As a consequence, domestic production also stabilized at about the same level until the mid-1990s, when production once again began a rapid 50 percent decline between 1990 and 2005. Despite the decline in domestic production, however, consumption has remained steady, which, once again, has opened

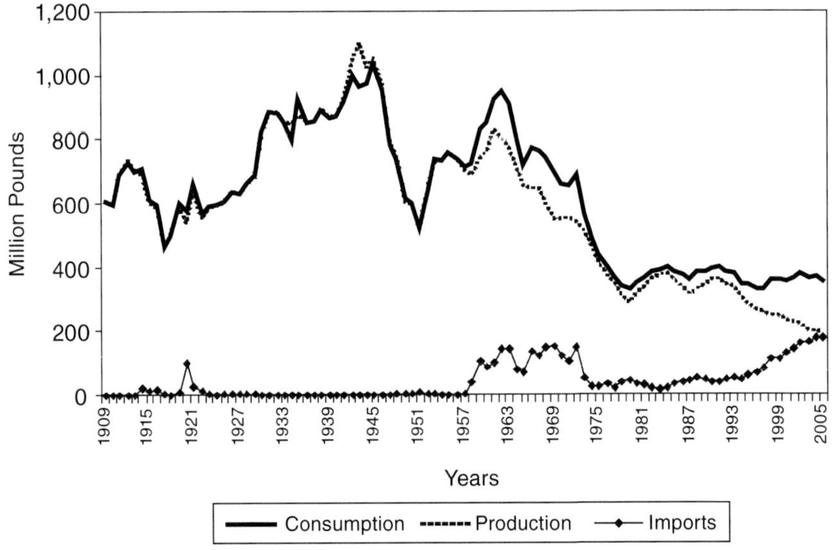

FIGURE 4-6 U.S. lamb production, consumption, and imports, 1909–2005.[a]
[a]Carcass equivalent basis. (1 pound (lb) = 0.4536 kg.)
Source: USDA (2007e).

a window for growing imports. Between 1990 and 2005, imports of lamb grew from 18.6 million kilograms, about 10 percent of domestic lamb supply, to 81.6 million kilograms, nearly equal to domestic production and half the total domestic supply.

The traditional argument that American tastes and preferences have moved away from lamb may no longer be applicable, given the steady level of consumption in recent years despite declining production. More appropriate now may be the argument that lamb is consumed fairly consistently by a small group of consumers and not at all by most consumers. Indeed, recent research shows that only 20 percent of consumers can be considered "lamb consumers," defined as those who have prepared lamb at home within the past 12 months or those who eat lamb but do not prepare it in their homes (Gross, 2007). Additionally, 35 percent of consumers have never eaten lamb, only 13 percent have prepared lamb at home during the past three months, and 16 percent no longer eat lamb.

There is some evidence that lamb demand is actually increasing. Shiflett et al. (2007) concluded that lamb demand increased by 5.7 percent during the last 10 years after having declined for many years. Much of the increased demand has been met by increased imports. The western range sheep industry produces lambs primarily for the traditional market where the lambs are weaned and then fed to nearly 63.5 kg. These lambs compete with imported lamb based primarily on price. The result has been that sheep numbers in the western states have continued to decline. Total sheep numbers in the United States have stabilized somewhat during the past four years, with growth in numbers occurring in the Northeast and upper Midwest. Much of the growth that is occurring appears to be driven by the demand for halal lambs (see Chapter 7 for more detail). Halal lambs are purchased by individuals for immediate slaughter and typically are smaller (about 27 kg) than traditional slaughter lambs. This observation suggests that the light-weight lamb market is really two markets—one where lambs are fed to heavier weight for slaughter with the meat going primarily into traditional lamb markets (i.e., retail stores and hotels, restaurants, and institutions [HRI]) and one where lighter lambs go directly to slaughter without further feeding and the meat either consumed for private use or sold through small specialty stores in large urban areas. This development is important because imported lamb appears to be continuing to gain market share from domestic lamb in traditional marketing channels, but domestic lamb is increasing in sales in specialty markets.

The decrease in domestic sheep numbers and increase in lamb imports have resulted in significant shifts in market structure in the sheep industry in the United States since 1990. Livestock feeding and processing are dominated by economies of size; that is, average per-unit production costs typically decline rapidly with volume. Consequently, most innovation in the sheep feeding and processing industries has been geared to reducing costs by maintaining volumes in fewer and fewer operations. Thus, declining domestic inventories have placed significant pressure on firms to consolidate or exit the sheep feeding and processing industries in order to remain competitive.

Lamb Demand Research

Despite its importance in potentially fostering growth and profitability in the sheep and lamb industry, research focused on understanding the economic determinants of U.S. lamb demand has been limited. Shiflett et al. (2007) provide a summary of prior lamb demand studies, including Purcell (1989), Byrne et al. (1993), Schroeder et al. (2001), and Capps and Williams (2005). The principal focus of these past investigations has been on economic and other factors affecting lamb demand. The respective demand functions are modeled using regression analysis and historical data to examine potential drivers of demand with an emphasis on measuring the elasticities of demand with respect to those demand drivers, including primarily the own-price, cross-price, and income elasticities. The factors most often found to be statistically significant in explaining changes in per capita lamb demand over the years include the real retail price of lamb, the real retail prices of beef and pork, and seasonality. Most studies have concluded that income has not been a statistically significant driver of changes on lamb demand.

The own-price elasticity measures the percentage change in consumption of a particular product due to a 1 percent change in its own price, all other factors invariant. Theory suggests that the own-price elasticity is always negative, indicating an inverse relationship between the retail level of consumption of a product and its market price. The greater the magnitude of the own-price elasticity (in absolute value), the greater the sensitivity of consumers to changes in the price. An elasticity measure greater than 1 in absolute value is considered "elastic," meaning that a given percentage change in price results in a larger percentage change in quantity demanded. In this case, demand is considered to be highly responsive to price changes. Also, if the demand for any product is elastic, then a price discount (increase) generates not only a larger percentage increase (decrease) in the quantity consumed but also an increase (decrease) in total revenue to the seller. On the other hand, an elasticity measure less than 1 in absolute value is considered "inelastic," meaning that a given percentage change in price results in a smaller percentage change in demand. Thus, demand is relatively unresponsive to price changes. For a product whose demand is inelastic, a price discount (increase) leads to not only a smaller percentage decrease (increase) in consumption, but also a decrease (increase) in total seller revenue.

Cross-price elasticities of demand refer to the percentage change in the consumption of one good due to a 1 percent change in the price of another good. If this measure is positive, then the two goods are considered substitutes. On the other hand, if this measure is negative, then the two goods are considered complements. If the cross-price elasticity is zero or not statistically different from zero, then the respective demands for the two goods are considered independent of each other. While a change in own-price moves consumption along the demand curve, a change in the price of a substitute or complement, the so-called cross-price effect, actually shifts the demand curve, all other factors invariant.

Income elasticity refers to the sensitivity of consumer purchases with respect to changes in income. With changes in income, the demand curve shifts as well, holding all other factors constant. The income elasticity of demand, by definition, is the percentage change in quantity demanded attributed to a 1 percent change in income, all other factors invariant.

The estimated own-price elasticities of per capita lamb demand across most studies have been close, ranging from –0.5 to –0.8 despite the time period analyzed (Table 4-7). In other words, research provides evidence that there is an inverse relationship between retail lamb price and the quantity of lamb purchased and that the relationship is inelastic, implying that lamb demand is not highly responsive to price changes. Those studies that included more recent data, especially data after 1999, indicated a higher degree of elasticity (more sensitivity to price) than those studies using earlier data. This finding suggests that lamb consumers may have become somewhat more sensitive to changes in price over time in terms of their willingness to buy or not buy lamb based on its price.

Most studies have found a statistically significant substitute relationship between lamb and beef, as well as between lamb and pork (Table 4-7). The estimated cross-price elasticities of lamb demand with respect to beef and pork price across most studies are also close, ranging from 0.5 to 0.6 for beef and from 0.1 to 0.4 for pork. At the same time, all but one study (RTI, 2007) concluded that lamb and chicken are independent commodities in consumption. Those studies using the most current data show greater substitutability between lamb and the other meats, although the relationships are not consistent across studies.

Indeed, a case can be made that goat meat could serve as a substitute for lamb meat, especially in the growing ethnic/religious segment of the market. But little information exists on goat meat consumption and goat prices to allow a formal determination of this supposition. At present, no research is available on cross-price elasticities between lamb and goat meats.

Also, all but one study (Shiflett et al., 2007) concluded that income is not a statistically significant driver of lamb consumption. Shiflett et al. (2007) initially found that income is statistically insignificant in explaining changes in per capita lamb demand. But they then added a trend variable

TABLE 4-7 Estimated Elasticities of U.S. Per Capita Lamb Demand[a]

			Cross-Price				Per
Study	Time Period of Analysis	Own Price	Import Lamb	Beef	Pork	Chicken	Capita Income
RTI (2007)	1970–2003	–0.523	0.293	ns	ns	0.35	ns
Shiflett et al. (2007)	1980–2005	–0.665	—	0.486	0.179	ns	0.684
Capps and Williams (2007)	1978–2006	–0.700	—	0.562	0.394	ns	ns
Schroeder et al. (2001)	1978–1999	–1.09	—	0.57	ns	ns	–0.54
Byrne et al. (1993)	1978–1990	–0.62	—	ns	0.131	ns	ns
Purcell (1989)	1970–1987	–0.51	—	ns	ns	ns	ns

[a]Dependent variable in all cases is per capita lamb consumption except for RTI study, which uses per capita consumption of only domestic lamb (excluding imported lamb).
Note: ns = not statistically significant and — = not considered in the analysis.

to their model and found a positive and statistically significant relationship between per capita lamb demand and income. This result may be spurious due to collinearity of the income and trend variables used in their analysis. The lack of broad evidence of a statistically significant relationship between income and lamb purchases may be the result of either the relatively small amount of lamb purchased or the fact that most lamb is purchased for special occasions that traditionally feature lamb.

Seasonality is another variable that all studies using at least quarterly data have found to be a statistically significant determinant of per capita lamb demand. Lamb consumption typically is highest in the first and fourth quarter of the year (e.g., see Byrne et al., 1993; Shiflett et al., 2007). Chapter 7 includes additional discussion about the effect of religious holidays on lamb demand.

Lamb Advertising and Promotion

In 1996, the Lamb Promotion, Research, and Information Order, better known as the American Lamb Checkoff Program, was established under the Commodity Promotion, Research, and Information Act of 1996, following calls by virtually all segments of the domestic sheep and lamb industry for the establishment of a checkoff program to enhance U.S. lamb demand. The 13-member ALB that administers the lamb checkoff program includes six producers, three packers or first handlers, three feeders, and one seedstock producer, all appointed by the U.S. Secretary of Agriculture. The board meets at least three times per year to establish goals and budgets for new lamb promotion programs and to evaluate the success of past promotional efforts. Board policies are implemented by a three-member staff in Denver, Colorado.

Under the order, lamb promotion programs are funded by an assessment on the sale of all feeder and market lambs and all breeding stock and cull animals. In general, the purchaser collects the assessment with a deduction from the sales proceeds of the seller. The funds are then carried forward to the point of slaughter or export, at which time the checkoff is collected and sent to the board. Those whose sales are subject to the assessment include producers, seedstock producers, exporters, feeders, direct marketers, ethnic slaughter operations, custom slaughter clients, and slaughter/packing plants (ALB, 2007). Imported sheep and lambs are assessed on weight gained in the United States. This assessment is collected from the domestic producer, seedstock producer, or first handler who takes possession of the imported animals. If sheep or lambs are imported into the United States for immediate slaughter, there is no weight assessment at the time of slaughter.

The assessment is $0.011/kg ($0.005/lb) of live lambs (ovine animals of any age) sold by producers, seedstock producers, exporters, and feeders. For lambs purchased for slaughter by first handlers, the assessment is $0.30/head. A first handler is defined as an entity that takes possession of the lambs for slaughter (including custom or ethnic slaughter) or sale directly to the consumer. First handlers are primarily packing plants but also include some producers, feeders, and direct marketers.

Marketing agencies (sale barns) are not assessed a checkoff fee but must collect assessments from the sellers and pass them on to the purchasers. Direct marketers, those who are both producers and first handlers, and those who process and market lamb or lamb products are assessed $0.011/kg ($0.005/lb) on the live weight at the time of slaughter and must pay an additional assessment of $0.30 per head. Each producer, feeder, or seedstock producer is obligated to pay its share of the assessment. The assessment is passed on to subsequent purchasers until it reaches the first handler or exporter, who then remits the total assessment. A person who is both producer and first handler is responsible for the remittance.

Initiated on July 1, 2002, the collection of assessments provides the board an annual operating budget of approximately $2 million. Administrative costs are limited to a maximum of 10 percent of collections in any fiscal year so that most of the funds are used for promotional purposes. The USDA has oversight responsibilities on the administration of the program. All activities funded with checkoff dollars must comply with the act and the order and must be approved by USDA.

Before the lamb checkoff program was approved as a mandatory program, the American Lamb Council of the ASI operated a lamb promotion program using funds made available under the Wool Incentive Program. When the Wool Incentive Program, and thus, expenditures for the promotion of lamb, were phased out in 1996–1997, an unsuccessful effort was made that year to pass a mandatory checkoff program through a producer referendum. The only funds made available for lamb promotion after the phaseout of the Wool Incentive Program in 1995–1996 and the establishment of the current lamb checkoff program in 2002–2003 was through a special grant resulting from a complaint based on Section 201 of the Trade Act of 1974. In 1999–2000, domestic petitioners alleged injury to the U.S. lamb industry from imports. The U.S. International Trade Commission ruled in favor of the domestic complainants. As a result, a lamb import tariff and a one-time assistance package for the domestic lamb industry were established to remedy the injury and facilitate industry adjustments to import competition. Through this program, $4.8 million in Section 201 relief grants for 23 lamb marketing and promotion projects were funded between 2000–2001 and 2002–2003.

American Lamb Board expenditures began in 2002–2003 and amounted to only $96,035 in real terms (adjusted for inflation) that year. Those expenditures rose to $2,433,196 in 2003–2004, dropped to $1,518,235 in 2004–2005, dropped again to $1,215,190 in 2005–2006, and dropped once more to $1,064,682 in 2006–2007. Prior to the establishment of the ALB, inflation-adjusted annual expenditures on lamb promotion by the ASI, ranged from zero to $4.2 million. Compared to the value of lamb purchases by consumers each year, the amount of funds that the lamb checkoff program collects for the promotion of lamb is extremely small. The lamb advertising-to-sales ratio (often referred to as the investment intensity ratio) over the 1978–1979 to 2005–2006 period ranged from a minimum of zero in 1999–2000 and 2000–2001 to a high of 0.23 percent in 1992–1993 and averaged 0.14 percent over the entire period. In other words, the amount of checkoff funds spent to promote lamb consumption in any given year on average has been no more than about one-quarter of 1 percent of the value of lamb sales.

To date, there has been little study of the responsiveness of lamb demand to the advertising and promotion activities that have occurred over the years. Capps and Williams (2007) developed an econometric model of lamb demand in which the effects of current and past lamb advertising and promotion efforts on U.S. lamb consumption at the retail level of the marketing channel were included using data for the 1978–1979 to the 2006–2007 time period. The analysis controls for the effects of the primary economic factors other than the lamb checkoff program that drive lamb demand, including (1) the retail price of lamb; (2) the retail prices of beef, pork, and chicken (3) personal disposable income; (4) population; and (5) inflation. In this way, the analysis isolates the specific impacts of advertising and promotion on lamb demand and allows a measurement of the change in lamb consumption (and lamb sales at fixed prices) attributable to advertising and promotion dollar expenditures, holding all other factors constant.

Capps and Williams (2005) incorporated the influence of the ALB checkoff program into their lamb demand model as a three-period moving average (current period and two lags) of inflation-adjusted ALB advertising and promotion expenditures. Using this model, they found that the ALB program had a positive but not highly significant effect on lamb demand. They reported an advertising elasticity of 0.022 between 1978–1979 and 2001–2002 (the pre-ALB period) and 0.031 between 2002–2003 and 2004–2005 (the ALB period). Though small, these advertising elasticities are consistent with those found by many other researchers across a wide variety of agricultural commodity checkoff programs (see Williams and Nichols, 1998).

In a more recent analysis of lamb demand, Capps and Williams (2007) modified their earlier model by using a polynomial distributed lag (PDL) process to capture the advertising carryover effects, as is commonly done in analyses of commodity checkoff programs (see, e.g., Kaiser et al., 2005). In addition, a square root transformation of the advertising and promotion variable was employed in the demand model to allow for both diminishing marginal returns and zero expenditures in advertising expenditures at certain time periods The results suggest that the ALB checkoff program has had a statistically significant effect on per capita lamb consumption. Their updated model explains roughly 84 percent of the variability in per capita lamb consumption over the 1978–1979 to 2006–2007 period of analysis. Besides the ALB advertising and promotion program, other statistically significant economic drivers of U.S. lamb consumption were found to be the price of lamb, and the prices of beef and pork. Again, neither income nor the price of chicken was found to have any statistically significant effect on lamb consumption. The advertising elasticity in the more recent Capps and Williams lamb demand model was estimated to be 0.0394, which is consistent with those estimated for other checkoff commodities. Using the same model with data for only the 1978–1979 through 2001–2002 period prior to the existence of the ALB, the advertising and promotion elasticity was estimated to be 0.0386, implying that the ALB promotion activities have been slightly more effective compared to past efforts at increasing U.S. lamb demand. The study concludes that the ALB advertising and promotion program generated roughly 3.43 additional kilograms of lamb purchased or $41.59 in additional lamb sales per dollar spent on advertising and promotion. This relatively high benefit-cost ratio implies that the ALB lamb promotion efforts are underfunded, a conclusion that is consistent with the experience of other commodity checkoff organizations.

Demographic Factors Affecting Lamb Demand

Williams et al. (1991) provided the first published analysis of the demographics of lamb consumers. The study used the 1987–1988 Nationwide Food Consumption Survey (NFCS) to analyze the average weekly per person expenditures on beef, pork, poultry, fish and seafood, and lamb in 1987 and 1988 by geographic location, season of the year, income quartile, race, age, and urbanization. The report concluded that the primary demographic drivers of lamb consumption included region, race, age, and income. The report concluded that the average lamb consumer lives in the Northeast or on the West Coast, comes from an ethnic background (nonwhite, nonblack, non-Hispanic), is over the age of 55, and is in the middle- to upper-income group, purchasing chops and legs most often.

The only other analysis of the demographic characteristics of U.S. lamb consumers (Williams and Capps, 2005) was based on data collected by A.C. Nielsen through its HomeScan™ Consumer Panel, a multioutlet panel that captures all consumer packaged-goods purchase information, as well as non-UPC-coded random weight perishable products like meat, on a daily basis for 7,000 to 8,000 households. Using state-of-the-art, in-home bar code scanners, participating households record daily transactions made at retail grocery stores, mass merchandiser outlets such as warehouse clubs, convenience stores, drug stores, computer stores, and by mail order or over the Internet. Purchasing households are selected for the HomeScan™ Consumer Panel to be representative of all consumers over a wide range of demographic groupings.

Because the HomeScan™ Panel is demographically balanced to represent the household population of the mainland United States, the panel data can be considered to be representative of nationwide patterns of food consumption. Because the demographic information of the purchasing households is recorded along with their purchases, the purchase information can be stratified (sliced up) and viewed by the demographic characteristics of consumers. Data on purchases of lamb for away-from-home food consumption at restaurants or elsewhere are not collected through the HomeScan™ Panels.

In their report, Williams and Capps (2005) summarized the HomeScan™ data for lamb purchases stratified by several demographic characteristics of the purchasing households, including (1) household size (number in the household), (2) household income, (3) age of the person primarily responsible for food preparation and meal planning, (4) employment status of the person primarily responsible for food preparation and meal planning, (5) education level of the person primarily responsible for food preparation and meal planning, (6) race, and (7) region where the household is located. The salient conclusions flowing from their analysis for the six years of 1998 through 2003 are the following:

Lamb Market Penetration

- An average of about 9.7 percent of all households purchased lamb each year over the period;
- Market penetration jumped from an average of 9.2 percent in the 3 years prior to the implementation of the lamb checkoff program to an average of about 10.5 percent in the two years of the dataset when the ALB began promoting lamb demand with checkoff dollars;
- Market penetration is highest among households with the following characteristics: (1) smaller household size, (2) higher income levels, (3) more mature (older) food preparers, (4) more educated food preparers, (5) unemployed (outside the home) food preparers, (6) household race classified as "black", and (7) located in the East region of the United States.

Household Lamb Purchases

- Lamb purchases per household tend to be higher for households with the following characteristics: (1) smaller household size, (2) more mature (older) food preparers, (3) unemployed (outside the home) food preparer, (4) food preparer with some college education or a high school education, (5) household race classified as "black", and (7) located in the East or West regions.
- Household income level is not strictly correlated with the quantity of lamb purchased per household.

Prices Paid by Households Purchasing Lamb

- The price per kilogram paid for lamb tends to higher among households with the following characteristics: (1) large household size, (2) higher income, (3) lower education level, (4) age of the food preparer between 25 and 65, (5) full-time employed food preparer, (6) household race classified as "white", and (7) located in the South or West regions.

Lamb Demand Index

One potential way to attempt to determine if the demand for a particular commodity like lamb is changing over time is to compute what is known as a demand index. Popularized by Purcell (1998), the demand index is a function of retail prices, per capita consumption of lamb, and the own-price elasticity of demand. While beef and pork demand indices have been calculated and available for some time (see, e.g., Kansas State University, 2007), Shiflett et al. (2007) provided the first effort to calculate a demand index for lamb. A demand index is calculated by first choosing an arbitrary base year and then calculating the percentage change in per capita consumption from the base period to the current period using the real retail price of lamb and a measure of the own-price elasticity for lamb. Then, using the assumed own-price elasticity and the calculated percentage change in per capita consumption, the corresponding percentage change in price is calculated. This percentage change in price then is used to calculate the change in price from the base period that would have occurred if the demand curve did not change between the base period and the period of interest. This price is called the demand constant price (DCP). Finally, the observed price for the period of interest is compared to the DCP. If the observed price is greater than the DCP, then demand is concluded to have increased by the percentage difference in the prices between the base period and the period of interest. If the observed price is less than the DCP, then demand is assumed to have decreased. The demand index for the period of interest, therefore, is taken to be the observed price in the period divided by the DCP.

The lamb demand index with a 1980 base year as calculated by Shiflett et al. (2007) for 1980 through 2005 under alternative lamb demand own-price elasticities is presented in Table 4-8. Although the lamb demand index is not invariant with respect to the own-price elasticity assumed, the differences in the calculated indices for the different assumed elasticities are not large. As calculated by Shiflett et al. (2007), the lamb demand indices indicate that lamb demand trended downward by almost 40 percentage points between 1980 and 1996 and then stabilized before recovering about 6–7 percentage points through 2005.

Although the common understanding is that lamb demand has dropped precipitously over the years, a comparison of the lamb demand index from Shiflett et al. (2007) and the beef demand index published by Mintert (2007) using the same 1980 base year actually shows that beef demand has dropped even more precipitously than lamb demand, nearly 50 percentage points, between 1980 and 1998 (Figure 4-7). Even though the Mintert beef demand index indicates that beef demand has since recovered by some 10–14 percentage points, beef demand is still lower compared to the 1980 base period level than is the case for lamb.

Several structural changes in the early 1990s likely gave lamb demand a boost. First, the lamb processing industry updated its technology and improved its product offerings, as discussed earlier. Technology enhancements included spray washing of carcasses and gas-flush packaging, which has resulted in longer shelf life and improved meat color. At the same time, the United States, as well as Australia and New Zealand, began producing more consumer-ready, fresh products to fit better into modern consumer diets and schedules. Also, the sharp contraction of the U.S. sheep industry in the mid-1990s due to the repeal of the National Wool Act in November 1993 had bottomed out by the early part of the current decade, allowing

TABLE 4-8 Lamb Demand Index with Alternative Elasticities, 1980 = 100 (Base)

	Assumed Own-Price Elasticity		
Years	–0.66	–0.56	–0.76
1980	100.0	100.0	100.0
1981	96.5	97.3	95.9
1982	95.7	98.1	94.0
1983	90.3	92.5	88.7
1984	90.8	93.7	88.8
1985	82.6	83.9	81.6
1986	78.8	79.3	78.4
1987	75.6	75.1	75.9
1988	82.5	83.2	81.9
1989	80.3	80.8	79.9
1990	81.9	82.8	81.3
1991	78.8	79.2	78.4
1992	73.5	73.2	73.8
1993	71.5	70.6	72.1
1994	62.9	60.9	64.4
1995	62.9	60.8	64.5
1996	61.8	59.4	63.7
1997	61.8	59.3	63.8
1998	66.5	64.4	68.1
1999	62.7	60.6	64.4
2000	62.0	59.7	63.8
2001	63.6	61.4	65.3
2002	69.8	67.7	71.5
2003	68.7	66.2	70.6
2004	68.9	66.4	70.8
2005	65.8	63.0	68.1

Source: Shiflett et al. (2007). Copyright 2007 by American Lamb Board. Used with permission.

greater availability of lamb for consumption and lessened upward pressure on retail price. Finally, as mentioned earlier, the ALB began promoting lamb consumption in mid-2002.

The various demand indices do not describe why lamb demand is changing; they are merely measurement tools. Also, the demand index is often calculated using a single estimate of the own-price elasticity for a given time period. Estimates of own-price elasticities, however, can range rather widely and can be quite different for different time periods. No research has been

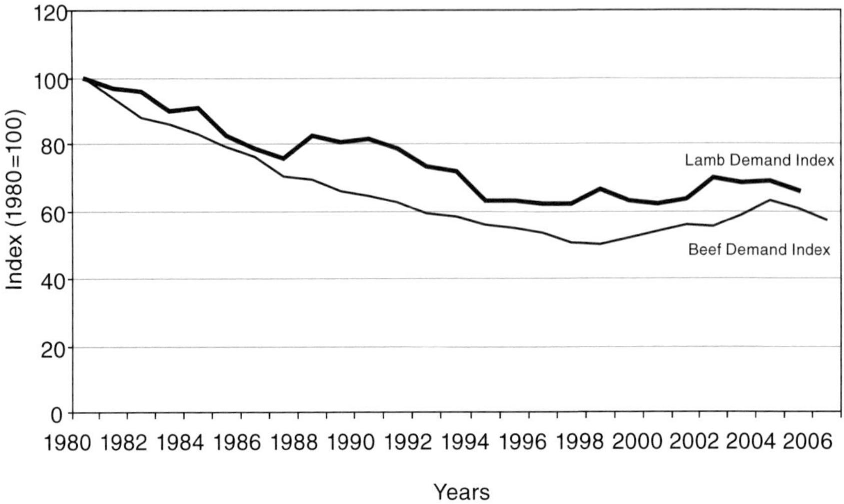

FIGURE 4-7 Comparing beef and lamb demand indices, 1980–2006.
Sources: Lamb Demand Index, Shiflett et al. (2007); Beef Demand Index, Mintert (2007). Copyright 2007 by American Lamb Board and AgManager.Info (Dr. James Mintert, Department of Agricultural Economics, Kansas State University). Used with permission.

done to investigate the sensitivity of the various meat demand indices to changes in the own-price elasticity measures used.

LAMB AND MUTTON TRADE

International trade issues have and continue to be a fundamental component of the overall health of the U.S. sheep industry. The competitiveness of the U.S. sheep industry in a global market is important because relatively few trade restrictions are associated with sheep products imported into the United States. Also, at least on a global scale, the United States is also not a major sheep-producing country and is not among the top 25 countries in terms of sheep numbers. In particular, the comparative advantage in production affords Australian and New Zealand lamb producers a competitive advantage over domestic producers. As a consequence, foreign competitors from Australia and New Zealand pose significant risks to the U.S. sheep and lamb industry. Despite the decline in U.S. lamb production since the 1940s, however, only recently have imports become a notable force in the industry. In the 1960s and early 1970s, imports grew to 20 percent to 30 percent of the level of domestic production. During this period, the decline in domestic production outpaced the decline in consumption (see Figure 4-6). The rate of decline in domestic consumption eventually caught up to that of production, however, so that imports again disappeared. The elimination of the Wool Act in the mid-1990s precipitated another sharp decline in U.S. sheep inventories and lamb production that has not yet abated. Consumption, however, has not followed suit, leading to growing imports. Between 1990 and 2005, imports increased from roughly 18 million kilograms in 1990 on a carcass weight basis to approximately 81.6 million kilograms (see Figure 4-6). By 2005, lamb imports were almost equal to domestic production, making up half of the total domestic consumption of lamb.

Over the period from January 2005 to January 2007, lamb imports varied from roughly 2.7 million kilograms per month to 8.2 million kilograms per month (Figure 4-8). Australia and New Zealand supply the majority of lamb imported to the United States. Relative to 2005, lamb imports were up 2 percent in 2006 to 66.8 million kilograms. Imports from Australia were up 4 percent in 2006 and imports from New Zealand were down 1 percent in 2006 (Figure 4-9). Over the quarterly period from January 2003 to January 2007, imports from Australia ranged from 1.81 million kilograms to nearly 6.35 million kilograms. Imports from New Zealand ranged from 0.91 million kilograms to 3.63 million kilograms over the same period. The import share of the U.S. lamb supply increased from about 35 percent in January 2002 to about 50 percent in January 2007 (Figure 4-10).

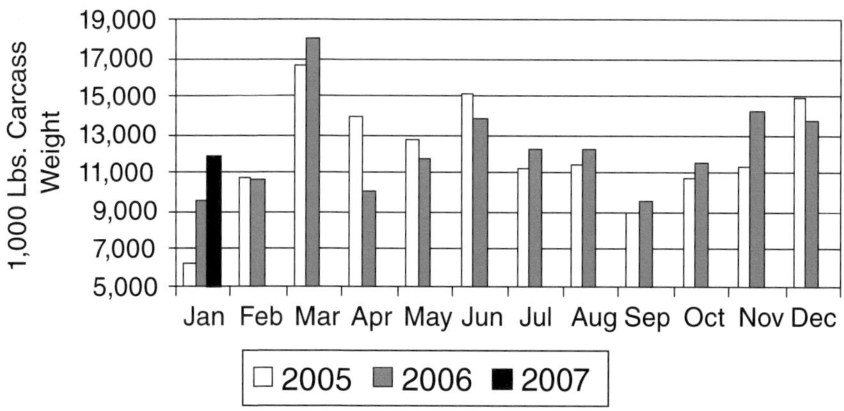

FIGURE 4-8 Monthly lamb imports, January 2005–January 2007. (1 pound (lb) = 0.4536 kg.)
Source: USDA (2007f).

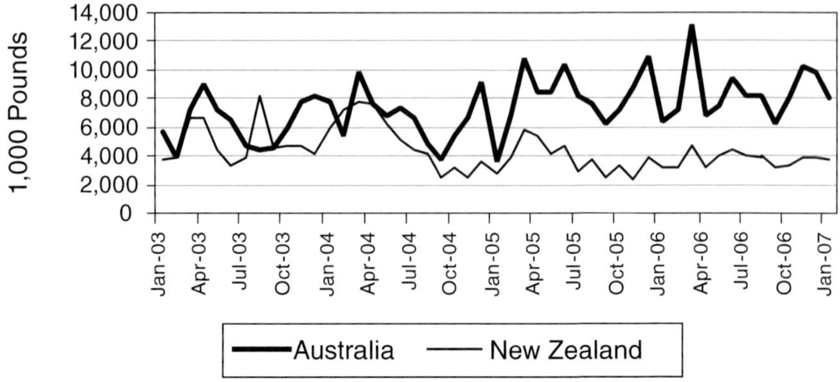

FIGURE 4-9 Lamb imports from Australia and New Zealand, January 2003–January 2007. (1 pound (lb) = 0.4536 kg.)
Source: USDA (2007f).

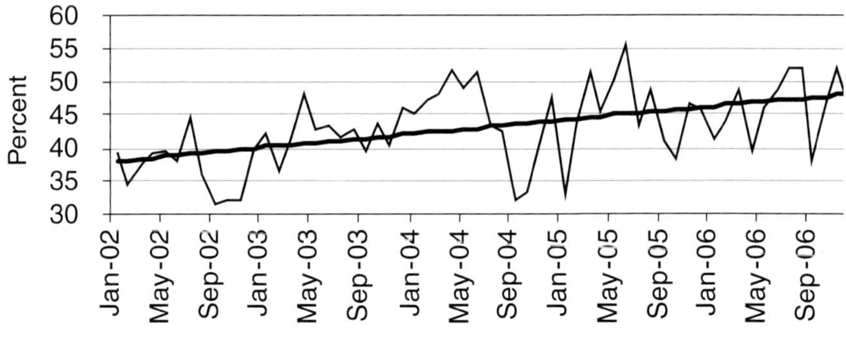

FIGURE 4-10 Imported lamb as a share of total lamb availability, January 2002–January 2007 (data line and trend line are shown).
Source: USDA (2007f).

Countervailing Duty on Lamb Imports

During the 1985 to 1990 period, the U.S. Department of Commerce imposed a countervailing duty on imports of New Zealand lamb meat. The U.S. government determined that New Zealand lamb industry subsidies were at least partially responsible for increasing import market shares. Babula (1997) investigated the effects of this countervailing duty on U.S. lamb supply, demand, and price at the meat packing-wholesale level. This econometric investigation used monthly data from January 1981 to May 1994. The econometric results indicated that the countervailing duty increased domestic wholesale lamb prices by 10 percent; reduced domestic quantity demanded for lamb by 3.5 percent; decreased imports from New Zealand by 11 percent; and increased imports from Australia by 92 percent.

Testimony by foreign producer interests, as reported by Babula (1996), maintained imported lamb is sufficiently different from U.S. domestic lamb so as not to displace U.S. quantities or to suppress the price of domestic lamb. On the other hand, testimony by domestic producers, as reported by Babula (1996), suggested domestic and imported lamb are close substitutes so that imports displace U.S. production and suppress prices of domestically produced lamb. The questions that are begged, then, are (1) whether and to what degree lamb imports suppress price and displace quantities of U.S.-produced lamb and (2) whether and to what degree U.S. consumers consider fresh domestic and frozen imported lamb as substitutes.

Using a six-equation vector autoregression model based on annual data from 1961 to 1994, Babula (1996) found lamb imports had mild adverse effects of less than 1 percent on U.S. lamb meat output, price, and revenue. The implication is that imports during that period had little effect on the U.S. sheep and lamb industry. Consequently, import limits were an ineffective means of bolstering domestic lamb production and price. Babula (1996) also found increases in domestic lamb production displace imports to a far greater extent than increased imports displace domestically produced lamb. Given this finding, it follows then that successful promotion efforts of the ALB could be effective in deterring imports. This situation could be the case if U.S. consumers prefer primarily fresh, larger-cut, and primarily grain-fed lamb over the primarily frozen, smaller-cut, and primarily range-fed imported product. Babula (1996) also found domestic and imported products are substitutes to a degree but are neither perfect substitutes nor independent products.

A correlation analysis of lamb price and imports conducted by Babula (1996) provided some support for the conclusion that imports have a small negative effect on lamb price. Based on monthly data for January 2005 through January 2007 (Figure 4-11), the correlation coefficient between lamb imports and gross carcass value was found to be –0.07, implying a slight negative effect of imports on the U.S. wholesale price of lamb.

Factors Affecting Recent Import Growth

U.S. imports of lamb and mutton have increased rather substantially since the mid-1980s, with very sharp increases after 1994. Imports, which currently account for nearly half of U.S. lamb consumption, are primarily from Australia and New Zealand. Imports from Oceania account for more

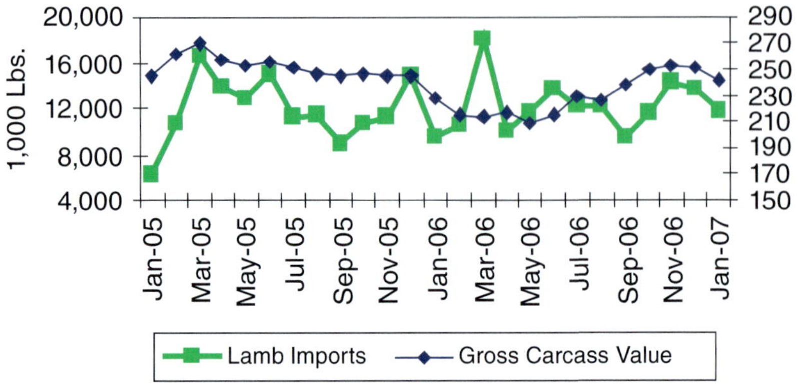

FIGURE 4-11 Relationship between imports and U.S. wholesale values, January 2005–January 2007. (1 pound (lb) = 0.4536 kg; 1 hundredweight [cwt] = 45.36 kg.)
Source: USDA (2007f).

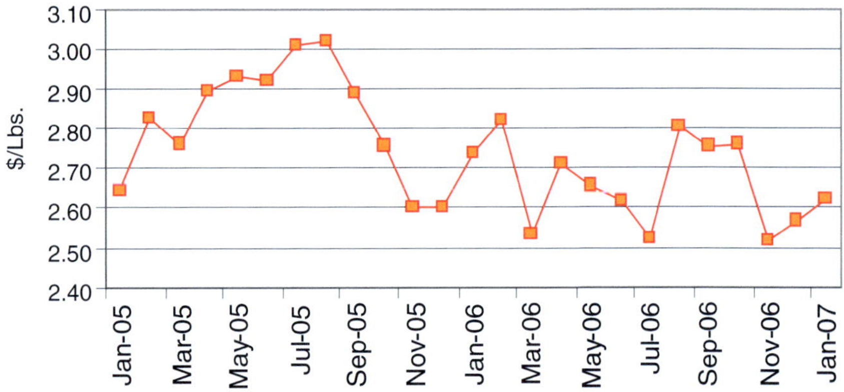

FIGURE 4-12 Quarterly nominal unit values of lamb imports, January 2005–January 2007. (1 pound (lb) = 0.4536 kg.)
Source: USDA (2007f).

than 98 percent of all U.S. imports (Muhammad et al., 2007). A number of factors likely are behind the recent surge in imports, including the continuing, rather stable level of consumption in the face of declining production. Also important are prices and exchange rates. The unit values of lamb imports have declined since early 2005 (Figure 4-12), providing some stimulus

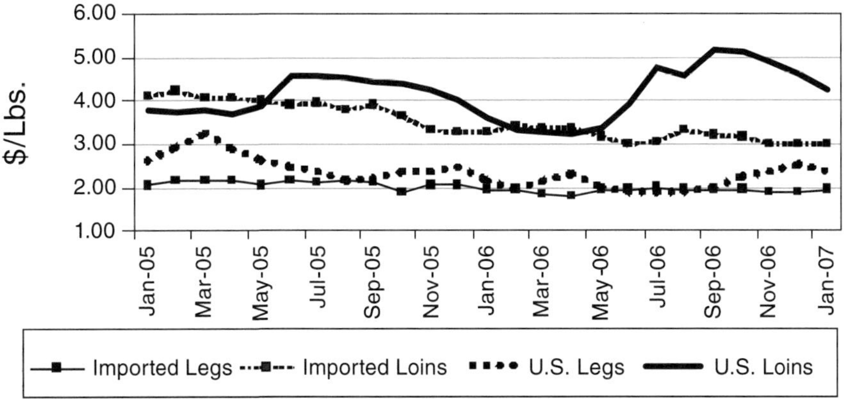

FIGURE 4-13 Wholesale prices of domestic and imported loins and legs, January 2005–January 2007. (1 pound (lb) = 0.4536 kg.)
Source: USDA (2007f).

to imports. U.S. wholesale prices of domestic loins and legs typically move together with the prices of imported legs and loins (Figure 4-13).

Imported lamb is differentiated by source of country of production (Australia or New Zealand) and by quality (frozen or chilled). In the early 1990s, chilled lamb imports accounted for about 25 percent of total lamb imports from Australia and New Zealand. Since that time, the share of chilled imports has been on the rise; currently, chilled lamb imports account for roughly 45 percent of total imports. To be sure, the growth in chilled imports can be attributed in part to the improvement in distribution infrastructures (Boal, 2001). Own-price elasticities for New Zealand frozen, New Zealand chilled, Australian frozen, and Australian chilled lamb have been recently estimated to be in the neighborhood of –0.24, –1.21, –1.01, and –1.24, respectively (Muhammad et al., 2007). Given these recent estimates, the demand for chilled imports tends to be relatively more elastic than the demand for frozen imports. Own-price elasticities associated with chilled imports suggest that Australia and New Zealand exporters of chilled products have the ability to increase their revenues with price reductions, all other factors invariant. The cross-price elasticities among Australia and New Zealand frozen and chilled imports are relatively small. As well, the signs of the respective cross-price elasticities suggest that frozen and chilled lamb imports from both countries can be either substitutes or complements. Consequently, this recent evidence on cross-price sensitivities suggests that increasing the price of one type/source of imported lamb has relatively little impact on the demand for other types/sources of lamb. However, when relative prices change, frozen lamb is more likely to be replaced with chilled lamb than the other way around (Muhammad et al., 2007). This finding underscores the preference of domestic consumers for the chilled product.

The exchange rate—the number of units of a foreign currency that can be exchanged for one unit of domestic currency—is one of the most important factors affecting the level and the destination of agricultural exports. For those involved in making the decision of where to buy and where to sell in the global marketplace, the process is one of converting one currency to another at the prevailing rate of exchange and comparing the ensuing prices. In this light, trade economists unequivocally suggest the exchange rate between the U.S., Australian, and New Zealand currencies is a potentially key determinant of the level of imports between the United States and Oceania. On the basis of Section 201 of the Trade Act of 1974, a tariff rate quota (TRQ) was imposed on lamb imported from Australia and New Zealand between 1999 and 2001. But despite the TRQ, currency exchange rates still made the U.S. market profitable for foreign exporters (US ITC, 1999). Over the period 1999 to 2001, the U.S. dollar appreciated against Australian and New Zealand currencies by roughly 20 and 25 percent, respectively. In situations where the U.S. dollar appreciates in value, Australian and New Zealand currencies decline in value. The cost of foreign exchange to U.S. importers subsequently decreases, thereby lowering imported lamb prices and increasing the quantity demanded of imported lamb in the domestic market. The appreciation of the U.S. dollar allowed Australia and New Zealand to effectively manage the TRQ even at tariff rates of 40 percent in 1999 and 32 percent in 2000 (Muhammad et al., 2007).

Recently, the exchange rate between the U.S. dollar and Australian and New Zealand currencies has fallen. The exchange rate between the U.S. dollar and the Australian dollar (USD/AUD) fell from 0.762 USD/AUD in 2005 to 0.753 USD/AUD in 2006, a drop of about 1 percent over that period (Figure 4-14). Over the same period, the exchange rate between the U.S. dollar and the New Zealand dollar (USD/NZD) fell from 0.704 USD/NZD in 2005 to 0.648 USD/NZD in 2006, a drop of about 8 percent. In essence then, Australian and New Zealand currencies have risen in value relative to the U.S. dollar. Hence the foreign currency cost to U.S. importers also has increased, resulting in higher imported prices and a subsequent decline in quantity demanded of imported lamb, all other factors invariant. Relative to 2005, imports from Australia were up 4 percent in 2006 and imports from New Zealand were down 1 percent in 2006.

Given that a sizable share of domestic lamb consumption comes from foreign sources, it is important to consider the exchange rate between the U.S. dollar and Australian and New Zealand currencies. Without question, changes in the aforementioned exchange rates can affect the quantity of

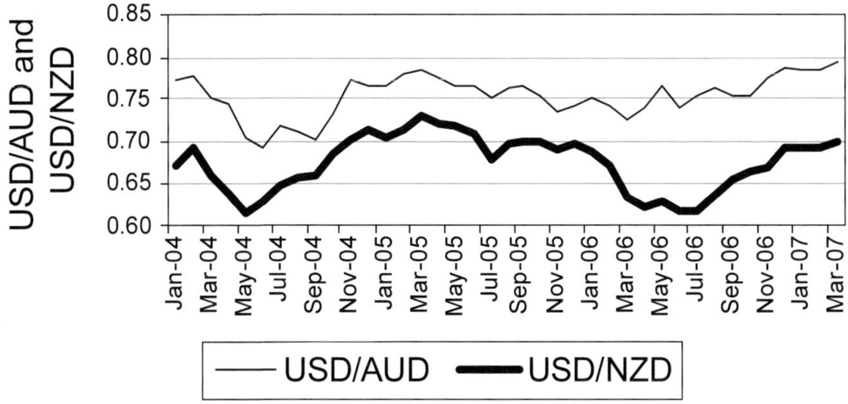

FIGURE 4-14 Exchanges rates between the United States and Oceania over the period January 2004–March 2007.
Source: Antweiler (2007). Copyright 2007 by Werner Antweiler. Used with permission.

imported lamb from Australia and New Zealand. Currently, the exchange rate is favorable to domestic lamb producers, given the decline in the value of the U.S. dollar relative to Australian and New Zealand currencies.

Lamb and Mutton Exports

The volume of U.S. lamb exports over the period 1990 to 2004 pales in comparison to the volume of imports. Exports typically consist of mutton or lower-valued cuts that are not desired by domestic consumers. The majority of exports (about 75 percent) historically have gone to Mexico. Japan is the other main importer of U.S. lamb. Lamb and mutton exports increased 97 percent in 2006 to 8.3 million kilograms (Figure 4-15).

Competitive Advantage in Global Sheep and Lamb Markets

Models of competitive advantage suggest that factor conditions or endowments, demand conditions, firm strategy and rivalry, related industries, government, and chance or uncertainty determine the competitive advantage a firm or country possesses in the global marketplace. In reference to international trade, the "principle (or law) of comparative advantage" suggests that countries gain by producing those commodities in which they have the greatest comparative advantage or those in which they have the least comparative disadvantage.

The following is a brief discussion of various points as they relate to the competitive position of U.S. sheep products. Factor conditions are related

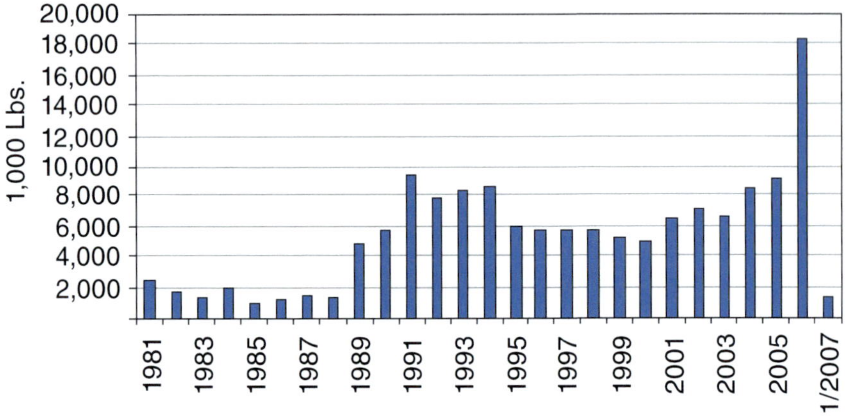

FIGURE 4-15 U.S. lamb and mutton exports from 1981 to January 2007. (1 pound (lb) = 0.4536 kg.)
Source: ASI (2007). Copyright 2007 by ASI. Used with permission.

to costs of production related to factors such as human resources, physical resources, knowledge resources, capital resources, and infrastructure. The United States has a large land base with rangeland located primarily in the West that is well suited for sheep production. Given that Australia and New Zealand are the primary global competitors with the U.S. sheep industry, then clearly those large competitors are also well suited to sheep production. This is especially true for New Zealand with its temperate climate and access to 12-month grazing. Australia has temperate as well as dry locations with large populations of sheep. Specialized labor related to sheep production is more abundant in New Zealand and Australia than in the United States because of the size and importance of the sheep industry in these two countries relative to the United States. The capital structure associated with sheep production is larger in Australia and New Zealand than in the United States with approximately 95 million and 40 million head of sheep existing in Australia and New Zealand, respectively, compared to just over 6 million head of sheep in the United States.

Infrastructure related to logistics favors the United States in the domestic American market because of the vast distances lamb carcasses must travel from New Zealand and Australia to reach the American market. However, New Zealand is an island nation with all locations relatively close to the availability of relatively cheap ocean transportation. Australia's logistics are complicated by the sometimes long distances that sheep and/or meat must travel to reach ocean ports. Ocean freight is a relatively cheap method for shipping commodities and the proximity of New Zealand and Australia to Asian markets have made them dominate in those markets. All three countries have a relatively highly educated workforce that can support the sheep industry.

Although domestic demand for sheep products is strong in New Zealand and Australia, compared to the United States, both countries have relatively small populations (about 20 million in Australia and more than 4 million in New Zealand, compared to slightly more than 300 million in the United States). Domestic annual per capita consumption of sheep and goat meat in the United States is about 0.51 kg compared to 14.05 kg and 22.14 kg in Australia and New Zealand, respectively (FAO, 2007). This observation suggests a much stronger domestic market for sheep meat in Australia and New Zealand. This continuing strong domestic market as well as the importance of exports translate into a greater influence of the sheep and lamb industry on the overall agricultural economies and agricultural policy making of these countries relative to the United States. The presence of related industries is an important component of competitive advantage. Related domestic firms allow for easier communication and cooperation in developing products and services that can support industry. Although all three countries have large capital investments in livestock systems (e.g., beef, dairy, and poultry), the U.S. system is by far the largest of the three. Other supporting industries include input providers such as veterinary services, livestock feed companies, and public and private sector research. Consequently, all three countries appear to have a strong set of supporting industries. However, the United States is the home base for many important technology and input providers. Along with the relative size of the U.S. livestock sector, the supporting infrastructure provides some advantage to the U.S. sheep industry over its global rivals.

Issues related to firm strategy and rivalries appear to favor the Australian and New Zealand sheep industries over the United States. Both Australia and New Zealand have well-organized and dedicated export infrastructures. They are also both export dependent in terms of their sheep industries and are aggressive exporters. Although recent growth in the dairy industry in New Zealand has reduced sheep numbers somewhat, New Zealand has a well-organized marketing plan for its livestock products (see Meat New Zealand, 2007). Both Australia and New Zealand are actively marketing their products as being natural from animals that are allowed to graze freely throughout their lives. They are also large producers of Halal products with large amounts of live animals and carcasses being transported to the Middle East. The U.S. lamb industry historically has had a much less well-funded campaign to promote its products, although the ALB has made progress in this regard in recent years, as discussed elsewhere in this chapter.

Thus, the American sheep industry has some advantages in producing sheep in terms of a natural resource base and a strong cadre of supporting industries. However, it faces strong international competition from countries that are at least as well suited for producing sheep, have much stronger domestic market bases for their sheep industries, and have well-organized marketing efforts. Consequently, successfully competing with New Zealand and Australia will require the U.S. lamb industry to focus on differentiating American lamb from imported lamb, either by quality or some other characteristic in the minds of American consumers.

LAMB BYPRODUCT MARKETS

In addition to lamb carcasses and cuts, packers sell a number of byproducts that are used in many commercial applications (Table 4-9). Most lambs dress from 48 percent to 52 percent, meaning that 48 percent to 52 percent of the live weight is the hanging carcass after removal of the pelt, viscera, organs, and other parts. Williams et al. (1991) indicated margins tend to be thin in the lamb packing industry so that the profit in lamb packing is often in the sale of the byproducts. Such tends to be the case across all livestock meat-packing operations. Consequently, the meat industry and specifically the lamb industry actively work to develop ways to market byproducts and rendered products.

The primary byproducts of lamb slaughtering include pelts and/or wool, fats and fatty acids, viscera (especially the intestines), manure, and bones, horns, and hooves. Wool as a byproduct is a relatively frequent phenomenon. As discussed in more detail in Chapter 5, the declining fortunes of the U.S. sheep industry over the years resulted in a corresponding decline in the relative return to wool production so that lamb and mutton production eventually became the primary product of sheep production. Besides wool, which is discussed extensively in Chapter 5, some of the major byproducts are the following: (1) edible byproducts, (2) rendered byproducts, (2) pet-food products, (3) pelts, and (4) lanolin.

Edible Byproducts

A number of lamb byproducts are considered to be edible, depending on the ethnic background of buyers, including blood, blood plasma, bone, intestines (large and small), cheek trimmings, fat, hearts, liver, kidney, spleen, sweetbreads, tail, testicles, tongue, and tripe (stomach). This list may differ depending on the region of the United States or the ethnicity of buyers.

TABLE 4-9 Uses for Sheep and Lamb Byproducts

Pelt and Wool		
• Lanolin	• Rouge base	• Felt, carpet
• Clothing	• Insulation	• Footwear
• Drum heads	• Rug pads	• Woolen goods
• Yarns	• Asphalt binder	• Baseballs
• Artists' brushes	• Textiles	• Upholstery
• Sports equipment	• Ointment base	• Pelt glue
• Fabrics	• Tennis balls	• Paint and plaster binder
• Pelt products	• Worsted fabric	
Fats and Fatty Acids		
• Explosives	• Rennet for cheese	• Dog food
• Solvents	• Industrial oils	• Protein dog food
• Chewing gum	• Industrial lubricants	• Mink oil
• Paints	• Stearic acid	• Oleo
• Makeup	• Cosmetics	• Margarine
• Ceramics	• Antifreeze	• Oleo shortening
• Medicines	• Crayons	• Herbicides
• Shoe crème	• Floor wax	• Shaving cream
• Dish soap	• Tallow for tanning	• Protein hair conditioner
• Tires	• Chemicals	• Protein hair shampoo
• Paraffin	• Rubber product	• Creams and lotion
• Chicken feed	• Insecticides	
• Biodegradable detergents	• Candles	
Intestines		
• Sausage casings	• Surgical sutures	• Tennis racquet strings
• Instrument strings		
Manure		
• Nitrogen fertilizer	• Phosphorus	• Minor minerals
• Potash		
Bones, Horns, and Hooves		
• Syringes	• Bone jewelry	• Bone charcoal for high-grade steel
• Gelatin desserts	• Bone meal	• Plywood and paneling
• Rose food	• Emery boards and cloth	• Shampoo and conditioner
• Piano keys	• Ice cream	• Dice
• Marshmallows	• Laminated wood products	• Collagen cold cream
• Potted meats	• Horn and bone handles	• Crochet needles
• Pet food ingredients	• Collagen/bone for plastic surgery	• Cellophane wrap and tape
• Bandage strips	• Bone china	• Glycerin
• Bone charcoal pencils	• Wallpaper/wallpaper paste	• Photographic film
• Gelatin capsules	• Dog biscuits	• Fertilizer
• Adhesives/adhesive tape	• Steel ball bearings	• Neatsfoot oil
• Phonograph records	• Malts and shakes	• Abrasives
• Combs and toothbrushes		
• Buttons		

Rendered Byproducts

Over time, the rendering industry has developed to dispose of nonedible portions of the slaughtered animal. One-third to one-half of each animal produced for meat, milk, eggs, and fiber is not consumed by humans (Meeker and Hamilton, 2006). The most valuable use of most animal byproducts is as feed ingredients for livestock, poultry, aquaculture, and companion animals (Meeker and Hamilton, 2006). The Food and Drug Administration (FDA) regulates the use of animal wastes in animal feeds. Currently, the FDA bans the use of ruminant-derived MBM in feeds intended for ruminants as a precaution against the spread of transmissible spongiform encephalopathy (TSE). The restriction has limited the opportunities for the use of beef and lamb byproducts in the ruminant feeds. Taylor et al. (1995) reported that while rendering lowers the infectivity of the prion protein associated with TSEs, the infectivity is not totally inactivated.

Rendering is the process of cooking that inactivates bacteria, viruses, protozoa, and parasites, usually by the introduction of steam at temperatures of 115.5° to 143.3°C (Meeker and Hamilton, 2006). In general, rendering separates the fat from the protein and solid materials associated with bone. With further processing, a large portion of the moisture is released in the rendered product. Meeker and Hamilton (2006) also reported that the rendering industry in the United States produces approximately 5.08 billion kilograms of animal-derived proteins and 4.94 billion kilograms of rendered fats yearly. Further, they reported that about 85 percent of this production is utilized as animal feed ingredients and that the use of rendered animal fats in the chemical, rubber, and oleochemical industries make up the second-largest market. Not all of the rendered lamb byproduct used in the United States is produced domestically. In fact, a fair amount of rendered products are imported from Australia and New Zealand to meet the domestic demand for lamb- and sheep-derived rendered byproducts.

Pet Food

The global pet food and products industry is growing rapidly and expected to continue growing. Aldrich (2006) estimated pet food is a $53 billion industry globally. Dog and cat food sales in the United States have reached a combined total of $14.5 billion with exports of nearly $1 billion. Aldrich (2006) also indicated lamb meal has been a popular ingredient for the better part of the last 15 years. The popularity of lamb meal as a pet food ingredient escalated over the last decade with increasing concerns about animal nutrition, health, and well-being. Lamb meal is thought to be easier for most animals to digest and results in a lower level of hypersensitivity (food-related allergies). The demand for lamb-derived byproducts initially outstripped the supply due to the novelty of designer pet foods.

"Lamb meal analogs," made of other protein meals, were rumored to have entered the market but tight controls due to BSE and scrapie issues and new DNA-typing technology (see Kremar and Rencova, 2003) have eliminated this potential competition (Aldrich, 2006).

There is limited information about lamb meal and the analytical composition that it offers to pet diets. The protein quality of lamb meal is reported to be comparable to MBM and about 75 percent of chicken byproduct meal (Johnson and Parsons, 1997; Johnson et al., 1998). Johnson et al. (1998) also reported the digestibility of the essential amino acid lysine and threonine and the nonessential sulfur amino acid cystine were low in lamb meal diets, possibly as a result of a high concentration of wool contamination in the lamb meal. Cystine, a sulfur-based amino acid, is elevated in keratin-based products, such as wool, and is not highly digestible. Aldrich (2006) reported that the effects of lamb meal in dog or cat diets on palatability, shelf life, aroma, or appearance are lacking in the literature. Nevertheless, the pet food aisle of most supermarkets suggests that lamb meal is a popular ingredient in pet foods.

Pelts

Sheep pelts are one of the more common, and most lucrative, byproducts of the lamb packing industry. Pelt prices have been positively associated with slaughter lambs prices because pelt sales represent the largest component of byproduct income for lamb packing operations (Greer and Ward, 2000). Variations in average pelt prices across the grading scale have been extreme over the last 15 years (Figure 4-16). With the introduction of hair sheep into the United States and the increase in small producers on the eastern seaboard that are favoring these type of sheep (considered to be easy-care sheep because there is no need to shear), high-quality wool pelts are harder to secure in the packing industry. Using hair-type sheep in a crossbreeding program with wool-type sheep may result in wool pelts, but with a lower fiber quality and in some cases with kemp fibers (hair type). The differences in pelt prices are not as great as they used to be in the industry. However, they do play a role in the price determination of harvest-ready lambs. Much of the tanning industry is outside the United States. Many of the pelts are initially processed and shipped to tanneries outside the United States and then returned as finished products. According to ASI (2006), there is only one raw skin processor in the United States that has a tannery processing approximately 5,500 lambskins a day (4,500 to 6,000) from domestic lamb production. In addition, this facility processes approximately 1 million Australian skins and 200,000 Irish/English skins that move through the plant annually.

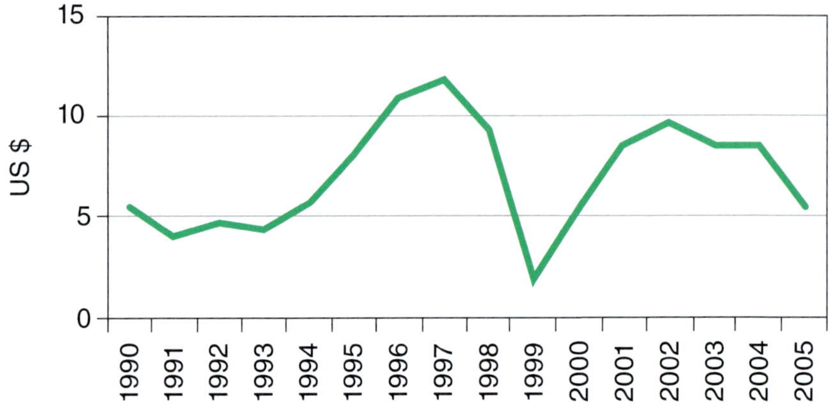

FIGURE 4-16 Pelt prices, average of fall clips (No. 1, No. 2, No. 3, and No. 4). Source: USDA (2007g).

Lanolin

Lanolin, sometimes referred to as wool wax, wool fat, or grease, is a substance that is secreted from the sebaceous gland that is associated with the fiber follicle under the surface of the skin. This type of greasy substance can be used as a skin ointment, a waterproofing agent, and an ingredient of shoe polish. The benefits that lanolin gives to the sheep, as viewed by humans, are seen as potential benefits to humans. For example, the ability of greased (lanolin) wool to wick moisture allows its use as a waterproofing agent. During the scouring process, raw wool is stripped of the natural grease that is associated with the fleece. This waxy substance, known commercially as lanolin, is most commonly used in the cosmetic and pharmaceutical industries. Lanolin must be refined and purified to be used for these purposes. Further, lanolin can be used in some industrial applications in the lubrication industry (such as motor oil, auto lubrication, ink, and adhesives).

Pharmaceutical and Research Uses

According to the USDA, more than 24,000 sheep are used each year in research in the United States and even more worldwide as research institutions and government agencies look for ways to improve various aspects of the human existence. For example, Protherics is an Australian company using 4,500 Merino × Border Leicester wethers to produce antibodies used in medical research (Adelaide Bio News, 2005). Also, sheep blood, which is used in some cases in the production of biomedical supplies, traditionally has been collected at slaughter facilities. Transgenic animals are being developed that have a gene replacement or alteration in their genetic code for a specific purpose. Dr. Esmail Zanjani, for example, has been working with transgenic sheep in the area of stem cell research in the hope that sheep will be able to incubate organs that one day will be transplanted into human recipients (Reno Gazette Journal, 2005). The use of sheep and lambs in research related to orthopedics (BBC News, 2001) and artificial organ technology (MC3, 2006) has also been reported.

Miscellaneous Uses

Sheep have been shown to be able to make contributions to waste management both as consumers and producers (Glenn, 1994). Wool has been identified as a sorbent for oil spills. With its high tensile durability and ability to retard flame, wool has been identified as an excellent choice for oil spill contamination cleanup (Millsaps Sorbent and Environmental Laboratory, 1993). Glenn (1994) reported that wool mats, for use in the landscaping industry as an alternative to mulch, have become commercially available. In addition to other properties that mulch provides, the wool mats biodegrade in 2 to 5 years and provide nitrogen, potassium, sulfur, and other trace minerals, aiding in both plant and environmental health. The wool mats provide an outlet for low-grade, pigmented wools and serve as an environmentally friendly alternative to herbicides and plastic sheeting often used in the landscape industry. Pilot studies over 11-week periods have been conducted using sheep manure to treat and successfully reduce petroleum contaminates (Kamnikar, 1992). The Department of Energy has investigated the use of water-soluble coal compounds in conjunction with rumen fluid to produce industrial and liquid fuels. Organisms isolated from the sheep rumen have been found to biodegrade pyrolizidine alkaloid toxins that have been found in tansy ragwort plants (Glenn, 1994). A probiotic is being developed as a result of this finding to protect cattle, saving the industry close to $20 million annually in Oregon, northern California, and Washington (Wachenheim et al., 1992a,b).

POLICIES AND REGULATIONS RELATED TO THE LAMB INDUSTRY

A number of policies directly impacting the sheep industry were discussed in Chapter 2. In this section of Chapter 4, policies and regulations related to the lamb industry are discussed. In general, the U.S. livestock meat industry is faced with a wide array of governmental regulations. These governmental regulations typically affect either directly or indirectly various aspects of the marketing system (McCoy, 1981). The justification for such regulations is the enhancement of the general welfare, a broad term construed to include health concerns, economic well-being, food safety, social conditions, and other factors.

Meat Inspection

Meat inspection is concerned with the wholesomeness, cleanliness, and truthfulness in labeling of meat and meat products. The purpose of meat inspection is to safeguard health by (1) eliminating diseased and otherwise unwholesome meat from human consumption, (2) maintaining sanitary conditions during slaughtering and processing, (3) preventing the addition or use of harmful ingredients, and (4) preventing false or misleading labeling of meat and meat products. Historically, prominent pieces of legislation in this regard include the Meat Inspection Act of 1906 and the Wholesome Meat Act of 1967. The cost of federal meat inspection is about $0.55/kg of red meat (McCoy, 1981).

Packers and Stockyards Act

The Packers and Stockyards Act was enacted in 1921 to provide relief for livestock producers from anticompetitive practices of packers and market agencies. The act originally was administered by the Packers and Stockyards Administration but now is administered by the Agricultural Marketing Service of the USDA. In general, the act is designed to regulate the business practices of those who engage in the buying and selling of livestock and meat that enter interstate and international trade. Given the oligopsonistic/oligopolistic nature particularly of packers in the marketing channel for lamb, this act is important to the lamb industry.

Barriers to Trade—Import Quotas and Tariffs

Congress enacted the Meat Import Law (P.L. 88-482) in August 1964, which allows import quotas for fresh, chilled, and frozen beef, veal, mutton, and goat meat. The purpose of the law was to limit annual imports of specified meats, including carcass and boneless meat. Importantly, this bill does not pertain to pork or lamb products. However, as discussed previously, during the 1985–1990 period, the U.S. Department of Commerce imposed a countervailing duty on imports of New Zealand meat. Additionally, on the basis of Section 201 of the Trade Act of 1974, a tariff rate quota (TRQ) was imposed on lamb imported from Australia and New Zealand between 1999 and 2001. Currently, no U.S. import quotas or tariffs on lamb are in effect. Most countries with exportable surpluses are actively engaged in efforts to stimulate trade. With regard to the lamb industry, key promotional agencies are the Australian Meat Board and the New Zealand Meat Board.

Mandatory Price Reporting

Before 2001, information on livestock and meat market transactions reported by USDA was based on data voluntarily submitted by market participants. However, an increasing number of livestock transactions were being made under long-term marketing arrangements (LMAs) that set sales terms well before delivery of the animals for slaughter (Perry et al., 2005). Because the terms of LMAs were not reported during that period, USDA livestock price and volume data were increasingly based on a declining number of transactions. Concerns emerged that the cash market prices reported by USDA did not reflect an increasing share of livestock sales. Along with growing concentration in the meat packing industry at the time, the increasing lack of transparency in livestock transactions fueled concerns of packer manipulation of markets.

The Livestock Mandatory Reporting (LMR) Act of 1999 (P.L. 106-78) was the legislative response to these concerns. The LMR Act required major meatpackers to report all transactions covering hog, cattle, and lamb purchases and commitments to USDA. The act also required packers to report the details of fresh wholesale beef and lamb transactions. In implementing LMR, the intent was to facilitate price discovery in the industry. Initial producer response to the LMR Act was negative primarily because of implementation problems that severely reduced the amount of price and volume data reported. At the same time, the data did not show that contract prices were higher than cash prices, as many in the industry expected. A recent study by RTI (2007) found that the primary effect of the LMR Act for the lamb industry has been to reduce price risk rather than to influence the level of the price paid for slaughter lambs. The study found that implementation of the LMR Act in 2001 increased the slaughter lamb price by only 0.129 percent. Further, the study concluded that LMAs have had only a small effect on slaughter lamb prices. The study found that a 10 percent increase in slaughter lamb purchases through formula pricing increases the slaughter lamb price by an estimated 2.54 percent. In contrast, a 10 percent increase in slaughter lamb purchases through cash markets increases slaughter prices by an estimated 2.68 percent. On the other hand, a 10 percent increase in packer ownership was found to reduce slaughter lamb prices by only an estimated 0.23 percent.

Country-of-Origin Labeling (COOL)

In 2002, Congress mandated COOL for beef, lamb, pork, poultry, and fish. COOL has been an issue strongly supported by livestock producers but generally opposed by meat packers, processors, and retailers. Congress amended the Agricultural Marketing Act of 1946 and enacted this legislative act as part of the Farm Security and Rural Investment Act of 2002. The economic impacts of COOL for the beef industry have been examined (Anderson and Capps, 2004), but virtually no information exists regarding the economic impacts of COOL for the lamb industry.

Livestock Risk Protection (LRP)-Lamb Insurance Policy

The Livestock Risk Protection (LRP)-Lamb Insurance Policy is a new price risk management tool that provides producers and feeders of lambs with the opportunity to insure lambs they own against unexpected price declines. Implemented in September 2007, lamb producers may select coverage prices for 13-, 26-, or 39-week insurance periods. The LRP policy provides coverage to lamb producers in 27 states (Arizona, California, Colorado, Idaho, Illinois, Indiana, Iowa, Kansas, Michigan, Minnesota, Missouri, Montana, Nebraska, New Mexico, Nevada, North Dakota, Ohio, Oklahoma, Oregon, Pennsylvania, South Dakota, Texas, Utah, Virginia, West Virginia, Wisconsin, and Wyoming). To date, about 300,000 lambs have been covered by this insurance policy. For additional details, see the USDA Risk Management Agency (RMA) website, http://www2.rma.usda.gov/livestock. There is currently no other type of price insurance, exchanges offering futures contracts, or derivative contracts on lamb prices for lamb producers.

MAJOR ACCOMPLISHMENTS, OPPORTUNITIES, AND CHALLENGES OF THE U.S. LAMB INDUSTRY

This chapter examined the U.S. lamb industry with particular interest in the current status of the industry and changes taking place in lamb production, marketing, consumption, and trade, as well as the market forces and government policies that influence their pattern of change. To close this chapter, the major accomplishments, opportunities, and challenges of the U.S. lamb industry are discussed.

Major Accomplishments of the U.S. Lamb Industry

Major accomplishments of the U.S. lamb industry may be delineated primarily into two broad categories: (1) the development of technological innovations designed to more accurately assess the value of carcasses, to improve food safety, to extend shelf life in the retail meat case, and to create more convenient products; and (2) the development of ways to stimulate demand such as centering attention on new uses of lamb products, product promotion and advertising done through the creation of the ALB, direct marketing of lamb, and niche marketing of lamb. A third major accomplishment deals with the aforementioned LRP-Lamb Insurance Policy.

- *New technologies to accurately assess the value of carcasses.* These developments will be instrumental in the adoption of some sort of value-based marketing plan. If the industry can accurately assess the true value of a lamb carcass, producers and feeders may have the opportunity to be paid for lambs that are superior in carcass conformation. The current pricing mechanism generally rewards producers primarily on the basis of weight. The level of superiority will be set forth by the minimum criteria of the grid, which may change depending on the market requirements and which, in turn, are likely to change over time. While there are methods for sorting carcasses in the commercial lamb industry, there are concerns about consistency and accuracy. Both of these criteria need to be met in order to maximize production of truly superior products on a consistent basis from production run to production run, as well as plant to plant. With the adoption of these technologies and the facilitation of value-based marketing systems used by some cooperative organizations in the lamb industry, producers and feeders can be compensated for superior lamb carcasses.
- *Improved safety of lamb products.* Modernization and inverted kill systems along with intervention programs for bacterial contamination have improved the safety of lamb products in the retail sector. Also, the industry has achieved a marked increase in product longevity in storage. Food safety is one of the major concerns of consumers about meat products. Implementing these aforementioned intervention programs helps to assure consumers that lamb is a safe alternative in the retail meat case. The industry will need to continue educating consumers about food safety issues in regard to lamb.
- *Improvements in packaging and shelf life.* New technologies in the area of packaging have improved shelf life of lamb products. The meat industry has seen a change in the retail case related to packaging and display of red meat products. With the advent of new packaging technologies, meat products in general and lamb in particular are experiencing extended shelf life, allowing an increase in retail shelf display time. These new package systems currently are being used in packing plants, resulting in reduced chances for bacterial (pathogenic or spoilage) contamination. Consequently, the lamb industry stands to benefit from improvements in packaging and extended shelf life.
- *Improved convenience of lamb products at retail.* Meat processing companies are producing products that are more convenient, reducing time

spent in meal planning and preparation. Partially or fully prepared retail meats are on the rise. Catelli Brothers, for example, is striving to develop preseasoned products with cooking instructions included to help consumers when purchasing and preparing lamb for at-home consumption (A. Catelli, personal communication, 2007). Others in the industry are looking at ways to assure consumers that they can purchase lamb, successfully prepare it at home, and have a positive eating experience. Consumer tastes and preferences for convenient meat products are not to be ignored. Keeping up with consumer signals to the meat industry is both a challenge and an opportunity. Segments of the lamb industry have shown a desire to meet consumer tastes and preferences regarding product convenience dimensions.

- *Additional uses of lamb and lamb byproducts.* The sheep and lamb industry has focused on developing additional alternative uses of lamb and lamb products. The lamb industry continues to look at ways to utilize the post-production and processing of waste material in the byproducts industry. Products from the lamb industry currently are being used in health and beauty supplies, waste management, environmental management, medical research, and other places. The continued development of ways to use the entire lamb pre- and post-harvest increases the value of the sheep to producers as well as consumers.
- *Creation of the American Lamb Board.* The establishment of the ALB to promote lamb is one of the most important self-help efforts ever undertaken by the lamb industry in an attempt to turn around the long-run decline of lamb consumption in the United States. While the funds made available for promotion are modest compared to most major checkoff commodity organizations, research has shown that the funds are being effectively invested to maximize the impact of each dollar spent.
- *Growth of direct marketing.* Many lamb producers are taking proactive approaches to selling their own products directly to customers and foodservice operators and bypassing traditional marketing channels. The number of individual sheep and lamb producer websites is rapidly increasing. Freezer market lambs are another growing direct marketing method. Lambs are sold live to customers and arrangements are made with a custom slaughter facility for processing and packaging. The growth and importance of this market is considered in more detail in Chapter 7.
- *Growth of the organic lamb market.* A number of sheep and lamb producers are converting to organic production systems. There was rapid growth of this part of the industry during the period 1997 to 2002, although numbers today have been growing less rapidly. Organic lambs are raised in production systems that promote and enhance biodiversity and biological cycles and minimize the use of off-farm inputs. Lambs are raised without the use of antibiotics and growth hormone stimulants. Lambs intended for meat markets must be raised organically from the last third of gestation.

Although lamb may be well suited for organic production, limited capacity and availability of processors who handle organic lamb products may play a role in limiting the growth of the market for certified organic lamb. A large portion of the western range lambs would qualify as organic or natural lamb products, but no major lamb packer is certified for organic or natural production and processing. The growth and importance of the organic lamb market is considered in more detail in Chapter 7.

Major Opportunities and Challenges Facing the U.S. Lamb Industry

Opportunities and challenges are two sides of the same coin. The key opportunities and challenges facing the U.S. lamb industry pertain to international trade issues, emerging ethnic markets, understanding of the competition, improvement of the competitive position of domestic producers, adoption of technology, and product promotion.

The primary opportunities for growth, development, and enhanced competitiveness of the U.S. lamb industry appear to be the following:

- *Recent depreciation of the U.S. dollar and use of alternative marketing arrangements.* Imported lamb is differentiated by source country of production (Australia or New Zealand) and by quality (frozen or chilled). As the consumption of lamb in the United States becomes increasingly dependent on sources from Oceania, the responsiveness of importers to domestic and foreign prices gives insight into the behavior of importers in the presence of a declining domestic industry. Without question, lamb quality is a key issue for the U.S. market. Chilled imports currently are preferred to frozen imports. U.S. producers are in a prime position, from a transportation cost standpoint, to capitalize on consumer preferences for chilled and fresh lamb. As well, from the standpoint of current exchange rates, Australian and New Zealand currencies are appreciating against the U.S. dollar, thus making imported lamb relatively more expensive relative to domestically produced lamb. Further, the use of AMA, such as forward contracts or marketing agreements, as opposed to the use of cash transactions, may enable U.S. operations to more effectively compete with increasing foreign imports. Besides the LRP-Lamb Insurance Policy instituted in fall 2007, the use of these AMA is one of the few risk management tools available to operations given that no futures market exists for lambs.
- *Promotion of lamb as a tasty, healthy, convenient, and safe meat protein source.* As the industry continues to look at ways to develop products for consumers to enjoy a palatable eating experience, concerns for a tasty, healthy, convenient, and safe product are a priority. Although lamb is not much different from other protein sources, relatively speaking, with regard to nutritional attributes, finding ways to differentiate lamb and elevate it

in the eyes of the consumer indeed presents opportunities to increase the demand for lamb.

- *Better positioning of lamb in the U.S. meat market.* Knowledge of cross-price elasticities allows the identification of competitors to lamb, namely beef and pork in that order, in the retail meat case. With this information, opportunities exist to better position lamb vis-à-vis chief competitors. Knowledge of consumer sensitivity to changes in retail prices also permits the opportunity to implement pricing strategies designed to increase revenue to retailers and packers in the lamb marketing chain.
- *Large potential market expansion.* Opportunities also are evident through the ALB to persuade consumers who have not consumed lamb previously to consume lamb for the first time. Nearly one-third of all U.S. households have never eaten lamb at all. Developing profiles of households who have not yet eaten lamb could be useful for targeting these households in promotion campaigns with the hopes of improving market penetration of lamb.
- *Emergence of new lamb markets.* The emergence of new markets for lamb products presents arguably the best opportunity for growth of the lamb industry. The growth in the number of Muslims who reside in the United States is one example. According to a recent study conducted by JWT, an advertising agency, these Muslims are, on average, wealthier and better educated than the general population (The Economist, 2007). At the same time, nontraditional markets for lamb serving several ethnic groups appear to be growing rapidly as discussed in more detail in Chapter 7.

Several major challenges face the lamb industry from the production side, including the following:

- *Improving the competitive position of domestic producers.* The elastic demand for chilled lamb imports suggests that Australia and New Zealand exporters of chilled products have the ability to increase their revenues with price reductions, all other factors held constant, which will create additional pressures on domestic producers of chilled/fresh lamb. Given the current open U.S. borders to lamb imports, U.S. producers will be challenged to increase their production efficiency and lower their costs in order to improve their competitive position in the domestic market.
- *Adoption of a value-based grading system that accurately sorts carcasses based on quality and yield.* Developing a system that accurately assesses value on which packers and producers/feeders can agree and trust will be a major challenge. Whatever system is developed will likely be automated and have the capability to uniformly assess carcass value from processor to processor and from day to day within a processing plant. Such

an automated system will have to fit into current plant designs and must be in keeping with current processing plant line speeds.

A number of challenges also face the U.S. lamb industry from the consumer side that affect not only production and profitability, but also the ability of researchers to conduct needed analyses of lamb demand to enhance decision-making in the sheep and lamb industry, including the following:

- *Lack of a long-standing retail price series.* A major problem for both research on lamb demand and decision-making in the lamb industry is the absence of a long-standing retail price series. As a consequence, research on lamb demand has considered only limited time periods or has used proxy data series for retail price, such as wholesale prices (Purcell, 1989) or imputed retail prices (Byrne et al., 1993; Schroeder et al., 2001). The USDA collected monthly lamb retail prices from 1950 to 1981. The American Sheep Industry Association then continued the collection of retail prices on a bimonthly basis from January 1986 to June 1992 and again from September 1993 to December 1995, leaving holes in the price series between 1981 and 1986 and between June 1992 and December 1993. The USDA commenced collecting monthly prices again from January 2001 to August 2005 under the umbrella of the MPR program, leaving another hole in the retail price data from December 1993 to January 2001. The USDA has once again discontinued reporting a retail lamb price, so that the latest data on the retail price of lamb are from August 2005. The U.S. Bureau of Labor Statistics (U.S. BLS, 2007) reports monthly price indices for lamb and organ meats, as well as for lamb and mutton. The former series runs from December 1977 to the present while the latter series only runs from December 1997 to December 2005. Efforts in dealing with gaps in retail prices are a key need for conducting effective analyses of lamb consumer behavior.
- *Measurement and reporting of per capita consumption of lamb.* Per capita consumption often is thought to be synonymous with demand. Per capita consumption, however, is calculated by USDA as cold-storage lamb stocks at the beginning of the year plus production plus imports minus ending stocks divided by the population of the United States. Consequently, this measure is more akin to disappearance than to consumption. As Shiflett et al. (2007) noted, there is no precision in measurement of the per capita lamb demand series. The USDA publishes per capita lamb demand with only one significant digit. The result often is a series that shows very little variability. Such lack of measurement precision (or variability) complicates any efforts to estimate demand models. Shiflett et al. (2007) and Capps and Williams (2005, 2007) used quarterly measurements of per capita lamb consumption posted on the Livestock Marketing Information Center (LMIC) website, which carries a number of significant digits.

Another problem with the reported data on per capita lamb demand is that the data do not differentiate between American lamb and lamb from either Australia or New Zealand. The imported share of lamb in the U.S. market has risen steadily in recent years, up to 45 percent currently. Breaking out and reporting lamb consumption by country of origin is necessary to understand changes occurring in consumer lamb-purchasing behavior. For example, ALB promotional activities are intended to enhance the demand for domestically produced lamb. Without separate data on the consumption of domestic and imported lamb, however, research cannot determine whether the ALB lamb promotion program has a generic impact on lamb consumption or primarily impacts the consumption of domestically produced lamb as intended. These and other issues related to imported lamb, such as the extent to which consumers consider domestic and imported lamb to be substitutes, cannot be reliably addressed unless separate series on lamb demand by country of origin are available.

• *Consideration of other factors potentially influencing lamb demand.* Most research on lamb demand has considered the demand effects of retail lamb prices, competing retail meat prices (specifically, beef, pork, and chicken), income, seasonality, and advertising. A case can be made that goat meat could serve as a substitute for lamb meat, especially in the growing ethnic/religious market segment, but not much information exists on goat meat consumption and goat meat prices. At present, no research is available on cross-price elasticities between lamb and goat meats.

The effects of diet, health, and nutrition information on lamb consumer purchasing behavior also have not been explored to any degree. Additionally, neither away-from-home consumer lamb purchasing behavior nor the demographic characteristics of lamb consumers has been explored adequately.

• *Increasing the presence of lamb in the foodservice/HRI sector.* Shiflett et al. (2007) suggested that the foodservice sector accounts for an estimated 37 percent of domestic lamb volume and is growing. Potential opportunities may exist for increased lamb demand in the foodservice sector, or hotels, restaurants, and other institutions. In the United States, the share of the food dollar spent away from home is nearly 50 percent (Jensen, 2006). Increased training of chefs and overall increased awareness of American lamb could increase lamb offerings in the foodservice sector (Shiflett et al., 2007). Consequently, targeted marketing efforts aimed at the foodservice sector would likely prove effective in increasing the demand for U.S. lamb.

• *Improved understanding of the demographic characteristics of lamb consumers.* There exists a pressing need to extend beyond the traditional price and per capita consumption series to provide improved and more detailed socioeconomic profiles of consumers in different market areas so that product offerings can be tailored to meet the desires of consumers. The only published study documenting demographic characteristics of lamb consumers used the 1987–1988 Nationwide Food Consumption Survey (NFCS) data to build a profile of lamb consumers (Williams et al., 1991). The demographic characteristics considered included geographic location, season of the year, income quartile, race, age, and urbanization. The study provided information on average weekly per-person expenditures of households by those demographic characteristics for beef, pork, poultry, fish and seafood, and lamb in 1987 and 1988. As well, the study provided a definitive picture of the profile of a lamb consumer in the United States. Region, race, age, and income were the major demographic factors found to influence the probability of consuming lamb in the United States. The Williams et al. (1991) analysis, however, needs to be updated with more current information. Importantly, ethnic markets need to be considered as well in order to better understand the demand for lamb.

- *Research on the demand for specific cuts of lamb.* Most research on lamb demand has considered lamb in the aggregate. Little research has been done regarding the retail demand for specific cuts of lamb, such as legs, chops, shoulder cuts, racks, shanks, ground lamb, and stew meat. Williams et al. (1991) examined the demand for selected individual lamb products in Houston using weekly retail scanner data. The cuts included various types of lamb chops, leg of lamb, lamb shank, and ground lamb. In most cases, the demand for each individual cut was found to be quite responsive to changes in price with own-price elasticities ranging from –1.66 to –3.17. This finding is consistent with demand theory in which the demand for specific components of a product (e.g., lamb versus all food and lamb cuts versus lamb in the aggregate) is expected to be more price responsive than demand for the product itself. The more price elastic response of lamb cuts than lamb demand in the aggregate to price changes largely reflects the greater degree of substitutability among cuts of lamb than among lamb and other types of meat and foods. This type of information is needed to assist retailers and foodservice purveyors in pricing and price-based promotion of lamb cuts so as to maximize lamb sales revenues. Retail scanner data will be highly useful in conducting these types of analyses in support of retail lamb marketing efforts.

REFERENCES

Adelaide Bio News. 2005. Harvesting the antibodies. Online at: www.bioinnovationsa.com.au. Accessed April 30, 2008.

ALB (American Lamb Board). 2007. How the checkoff works. Online at: http://americanlambboard.org. Accessed June 7, 2007.

Aldrich, G. 2006. Rendered products in pet food. Pp. 159–178 in Essential Rendering: All About the Animal By-products Industry, D. Meeker, ed. Alexandria, VA: National Renderers Association.

Anderson, D. P., and O. Capps, Jr. 2004. Country-of-origin labeling and the beef industry. Choices 19(4):1–2.

Antweiler, W. 2007. Pacific Exchange Rate Service. Saunder School of Business. The University of British Columbia, Vancouver, British Columbia, Canada. Online at: http://fx.sauder.ubc.ca/. Accessed April 10, 2008.

ASI (American Sheep Industry Association). 2004. Fast facts about American lamb. Online at: http://www.sheepusa.org/index.phtml?page=site/text&nav_id=06b30b9c925056e3635d644c8ae9a29b. Accessed April 30, 2008.

ASI. 2006. Sheep industry news: ASI board tours industry businesses in Greely. Online at: http://www.sheepusa.org/index.phtml?page=site/newsandevents&nav_id=82ee74b2354b223f831548af9ef21f5a. Accessed April 30, 2008.

ASI. 2007. First-quarter 2007 sheep industry in review. Report prepared for the American Lamb Board.

Babula, R. A. 1996. An empirical examination of U.S. lamb-related import and domestic market relationships near the farmgate. J. Int. Food Agri. Marketing 8(2):65–82.

Babula, R. A. 1997. Economic effects of a countervailing duty order on the U.S. lamb meat industry. Agr. Resource Econ. Rev. 26(1):82–93.

BBC News. 2001. Sheep yield osteoporosis clues. Online at: http://news.bbc.co.uk/1/hi/sci/tech/1480925.stm. Accessed August 8, 2007.

Berg, E. P., J. C. Forrest, D. L. Thomas, N. Nusbaum, and R. G. Kauffman. 1994. Electromagnetic scanning to predict lamb carcass composition. J. Anim. Sci. 72:1728–1736.

Berg, E. P., M. K. Neary, J. C. Forrest, D. L. Thomas, and R. G. Kaufman. 1996. Assessment of lamb carcass composition from live animal measurement of bioelectrical impedance or ultrasonic tissue depths. J. Anim. Sci. 74:2672–2678.

Berg, E. P., M. K. Neary, J. C. Forrest, D. L. Thomas, and R. G. Kauffman. 1997. Evaluation of electronic technology to assess lamb carcass composition. J. Anim. Sci. 75:2433–2444.

Berg, E. P., M. K. Neary, and J. C. Forrest. 1998. Methodology for identification of lamb carcass composition. Sheep Goat Res. J. 14(1):65–75.

Boal, F. 2001. Sheepmeat: a niche meat product or acceptable world animal protein source? Agribusiness Consulting and Research Services, Rabobank, Wellington, New Zealand.

Boland, M. A., E. P. Berg, J. T. Akridge, and J. C. Forrest. 1995a. The economic impact of operator error using optical probes to predict pork carcass value. Rev. Agric. Econ. 17:193.

Boland, M. A., K. A. Foster, A. P. Schinckel, J. R. Wagner, W. Chen, E. P. Berg, and J. C. Forrest. 1995b. Alternative pork carcass evaluation techniques. I. Differences in prediction of value. J. Anim. Sci. 73:637–644.

Brady, A. S., K. E. Belk, S. B. Levalley, N. L. Dalsted, J. A. Scanga, J. D. Tatum, and G. C. Smith. 2003. An evaluation of the lamb vision system as a predictor of lamb carcass red meat yield percentage. J. Anim. Sci. 81:1488–1498.

Braggins, T. J. 1996. Effects of stress-related changes in sheepmeat ultimate pH on cooked odor and flavor. J. Agric. Food Chem. 44:2352.

Brester, G. W., and D. C. Musick. 1995. The effect of market concentration on lamb marketing margins. J. Agric. Appl. Econ. 22:145–156.

Byrne, P., O. Capps, Jr., and G. W. Williams. 1993. U.S. demand for lamb: The other red meat. J. Food Distrib. Res. 24(1):69–86.

Capps, O., Jr., and G. W. Williams. 2005. Measuring the effectiveness of lamb advertising and promotion: An updated analysis. Commodity Market Research Report No. CM-01-05. Texas Agribusiness Market Research Center, Texas A&M University, College Station. September.

Capps, O., Jr., and G. W. Williams. 2007. Is lamb promotion working? Commodity Market Research Report No. CM-01-07. Texas Agribusiness Market Research Center, Texas A&M University, College Station. November.

Capps, O., Jr., P. J. Byrne, and G. W. Williams. 1995. Analysis of marketing margins in the U.S. lamb industry. Agr. Res. Econ. Rev. 24(2):233–240.

Carpenter, Z. L., G. T. King, F. A. Orts, and N. L. Cunningham. 1964. Factors influencing retail carcass value of lambs. J. Anim. Sci. 23:741–745.

Cramer, D. A. 1983. Chemical compounds implicated in lamb flavor. Food Technol. 37:249.

Cross, H. R., and K. E. Belk. 1994. Objective measurements of carcass and meat quality. Meat Sci. 36:191.

Cunha, B. C., K. E. Belk, J. A. Scanga, S. B. Levalley, J. D. Tatum, and G. C. Smith. 2003. Validation of regression equations that utilize lamb vision system (LVS) output to predict lamb carcass fabrication yields. USDA-AMS. Online at: http://www.ams.usda.gov/AMSv1.0/getfile?dDocName=STELPRD3319271. Accessed May 1, 2008.

Cunha, B. C., K. E. Belk, J. A. Scanga, S. B. Levalley, J. D. Tatum, and G. C. Smith. 2004. Development of validation of equations utilizing lamb vision system output to predict lamb carcass fabrication. J. Anim. Sci. 82:2069–2076.

Edwards, J. W., R. C. Cannell, R. P. Garrett, J. W. Savell, H. R. Cross, and M. T. Longnecker. 1989. Using ultrasound, linear measurements and live fat thickness estimates to determine the carcass composition of market lambs. J. Anim. Sci. 67:3322–3330.

Field, R. A., J. D. Kemp, and W. Y. Varney. 1963. Indices for lamb carcass composition. J. Anim. Sci. 22:3322–3330.

Field, R. A., J. C. Williams, and G. J. Miller. 1983. The effect of diet on lamb flavor. Food Technol. 37(5):258.

FAO (Food and Agriculture Organization of the United Nations). 2007. FAOSTAT—consumption. Online at: http://faostat.fao.org/site/345/default.aspx. Accessed April 30, 2008.

Fox, J. A., L. S. Vander Wal, P. Udomvarapant, D. H. Kropf, E. A. E. Boyle, and C. L. Kastner. 2003. Consumer evaluation of pre-cooked lamb. Sheep Goat Res. J. 18:65–68.

Fritz, K. D., M. D. Menzies, S. L. Boleman, J. W. Savell, and C. L. Skaggs. 1995. Effect of breed type, sex class, and antemortem evaluation of yield grade on carcass components of market lambs. Proc. Int. Congr. Meat Sci. Techn. 41:206–209.

Garrett, R. P., J. W. Savell, S. G. May, H. K. Johnson, and H. R. Cross. 1990. Role of yield grade and carcass weight on the composition of lamb carcasses. J. Anim. Sci. 68:1299–1310.

Garrett, R. P., J. W. Savell, H. R. Cross, and H. K. Johnson. 1992. Yield grade and carcass weight effects on the cutability of lamb carcasses fabricated into innovative style subprimals. J. Anim. Sci. 70:1829–1839.

Glenn, J. S. 1994. The role of sheep and sheep products in waste management. Sheep Goat Res. J. Special Issue: The Role of Sheep Grazing in Natural Resource Management. Online at: http://www.sheepusa.org/index.phtml?page=site/newsandevents&nav_id=601e0a31bbf6a0ef56f1f0591aa0dc78&volume=Special%20Issue:%201994%20--%20The%20Role%20of%20Sheep%20Grazing%20in%20Natural%20Resource%20Management. Accessed May 2, 2008.

Greer, H. C., and C. E. Ward. 2000. Regional differences in slaughter lamb marketing and prices. Sheep Goat Res. J. 16(2):52–57.

Gross, J. B. 2007. American lamb attitude and usage study 2006. Presentation made at the American Lamb Board Meeting, San Antonio, TX, January 25.

Harris Interactive, Inc. 2007. Trends survey—shopper. Online at: http://www.harrisinteractive.com. Accessed May 2, 2008.

Harris, J. J., J. W. Savell, R. K. Miller, D. S. Hale, D. B. Griffin, L. C. Beasley, and H. R. Cross. 1990. A national market basket survey for lamb. J. Food Qual. 13:453–465.

Heaton, K. L., J. B. Morgan, J. D. Tatum, J. W. Wise, R. P. Garrett, H. G. Dolezal, H. D. Loveday, and G. C. Smith. 1993. Field studies to document efficacy of visual assignments of lamb carcasses to appropriate yield grades. Sheep Res. J. 9:7–15.

Heitmann, B. L. 1994. Impedance: A valid method in assessment of body composition? Euro. J. Clin. Nutr. 48:228.

Hopkins, D. L., M. A. Anderson, J. B. Morgan, and D. G. Hall. 1995. A probe to measure GR in lamb carcasses at chain speed. Meat Sci. 39:159–165.

Huffman, R. D. 2002. Current and future technologies for the decontamination of carcasses and fresh meat. Meat Sci. 62:285–294.

Jamora, J. J., and K. S. Rhee. 1998. The uniqueness of lamb: Nutritional and sensory properties. Sheep Goat Res. J. 14(1):53–64.

Jenkins, T. G., K. A. Leymaster, and L. M. Turlington. 1988. Estimation of fat-free soft tissue in lamb carcasses by use of carcass and resistive impedance measurement. J. Anim. Sci. 66:2174–2179.

Jensen, H. 2006. Consumer issues and demand. Choices 21(3):165–169.

Johnson, M. L., and C. M. Parsons. 1997. Effects of raw material source, ash content, and assay length on protein efficiency ratio and net protein ratio values for animal protein meals. Poult. Sci. 76:1722–1727.

Johnson, M. L., C. M. Parsons, G. C. Fahey, Jr., N. R. Merechen, and C. G. Aldrich. 1998. Effects of species raw material source, ash content, and processing temperature on amino acid digestibility of animal by-products meals by cecectomized roosters and ileally cannulated dogs. J. Anim. Sci. 76:1112–1122.

Jones, D. K., R. Leu, J. T. Kemp, J. W. Savell, and H. R. Cross. 1988. Consumer evaluation of sodium reduced, restructured lamb roasts. J. Food Qual. 11:235.

Kaiser, H. M., J. M. Alston, J. M. Crespi, and R. J. Sexton (eds.). 2005. The Economics of Commodity Promotion Programs: Lessons from California. New York: Peter Lang Publishing.

Kamnikar, B. 1992. Bioremediation of contaminated soil. Poll. Eng. (November):50–52.

Kansas State University. 2007. Livestock and Meat: Marketing. AgManager.info. Online at: http://www.agmanager.info/livestock/marketing/graphs/. Accessed May 1, 2008. Last updated, January 22, 2007.

Kochevar, S. L., J. N. Sofos, S. B. LeValley, and G. C. Smith. 1997. Effect of water temperature, pressure and chemical solution on removal of fecal material and bacteria from lamb adipose tissue by spray-washing. Meat Sci. 45:377–388.

Kremar, P., and E. Renova. 2003. Identification of species-specific DNA in feedstuffs. J. Agric. Food Chem. 51:7655–7658.

Kunsman, J. E., and M. L. Riley. 1975. A comparison of hydrogen sulfide evolution from cooked lamb and other meats. J. Food Sci. 40:500.

Leistner, L., and G. W. Gould. 2002. Hurdle Technologies: Combination Treatment for Food Stability, Safety and Quality. New York: Plenum Publishers.

Lyons, K. 2000. How America's eating has changed since the beginning of the 20th Century. EFood Rap Vol. 10, No. 15, Purdue University School of Consumer and Family Science, West Lafayette, IN.

Magagna, J. 1991. Value-based marketing—A panel discussion. Proc. Rec. Meat Conf. 44:133.

MC3. 2006. Product pipeline: MC3 artificial lung (Biolung®). Online at: http://www.mc3corp.com/case_studies/artificial_lung_bio/. Accessed May 1, 2008.

McCoy, J. H. 1981. Livestock and meat marketing, 2nd Ed. Westport, CT: Avi Publishing Company, Inc.

Meat New Zealand. 2007. Production information. Online at: http://www. meatandwoolnz.com/guide/en/product_info.htm. Accessed May 1, 2008.

Meeker, D. L., and C. R. Hamilton. 2006. An overview of the rendering industry. Pp. 1–16 in Essential Rendering: All About the Animal By-products Industry. Alexandria, VA: National Renderers Association.

Menkhaus, D. J., G. D. Whipple, and C. E. Ward. 1989. Concentration in the lamb slaughtering industry: Impact on lamb prices. SID Sheep Res. J. 6:25–29.

Millsaps Sorbent and Environmental Laboratory. 1993. Final report on sorbency evaluation for western textile products.

Mintert, J. 2007. Annual choice retail beef demand index. Department of Agricultural Economics. Kansas State University. Online at: http://www.agmanager.info/ livestock/marketing/graphs/Meat%20Demand/Beef%20Demand/Annual%20Beef%20Demand/RetBfDmnd-IndexCPI.htm. Accessed May 2, 2008.

Morris, C. E. 2003. Multiple hurdles minimize pathogens. Food Engineering. Online at: http://www.foodengineeringmag.com/CDA/Archives/3ffe92e5a22f8010VgnVCM100000f932a8c0____. Accessed March 28, 2007.

Muhammad, A., K. G. Jones, and W. F. Hahn. 2007. The impact of domestic and import prices on U.S. lamb imports: a production system approach. Agr. Res. Econ. Rev. 36 (2):293–303.

NPD. 2003. Eating Habits in America. NPD Group. Online at: www.npd.com.

Perry, J., J. MacDonald, K. Nelson, W. Hahn, C. Arnade, and G. Plato. 2005. Did the mandatory requirement aid the market? Impact of the Livestock Mandatory Reporting Act. Outlook Report No. LDPM13501. Economic Research Service, U.S. Department of Agriculture, Washington, DC. September. Online at: http://www.ers.usda.gov/publications/ldp/sep05/ldpm13501/. Accessed May 2, 2008.

Purcell, W. D. 1989. Analysis of demand for beef, pork, lamb, and broilers: Implications for the future. Research Bulletin 1089. Research Institute on Livestock Pricing, Virginia Polytechnic Institute and State University, Blacksburg.

Purcell, W. D. 1998. Measures of changes in demand for beef, pork, and chicken, 1975–1998. Research Bulletin 3-98. Research Institute on Livestock Pricing, Virginia Polytechnic Institute and State University, Blacksburg.

Purcell, W. D., J. Reaves, and W. Preston. 1991. Economics of past, current, and pending change in the U.S. sheep industry with an emphasis on supply response. Research Bulletin 6-91. Research Institute on Livestock Pricing. Virginia Polytechnic Institute and State University, Blacksburg.

Purdue University. 2007. Lamb, from producer to consumer: Lamb meat product marketing. Online at: http://ag.ansc.purdue.edu/sheep/ansc442/Semprojs/marketing/LambMeat ProductMarketing.htm. Accessed May 1, 2008.

Reno Gazette Journal. 2005. Stem cell work raises hope for organ transplants. Online at: http://corporate.recruitingnevada.com/newsarticle.php?id=1099. Accessed May 1, 2008.

RTI (Research Triangle Institute International). 2007. Livestock and Meat Marketing Study. Volume 5: Lamb and Lamb Meat Industries. RTI Project No. 0209230. Final report prepared for the Grain Inspection, Packers and Stockyards Administration, U.S. Department of Agriculture, Washington, DC, February.

Reynolds, R. G., and B. Gardiner. 1980. Supply response in the Australian sheep industry: A case of disaggregation and dynamics. Austr. J. Agr. Econ. 24:196–209.

Rhee, K. S., and Y. A. Ziprin. 1996. Identification and acceptance of lamb versus beef and pork by consumers and experienced sensory panelists. J. Muscle Foods 7:243.

Richie, M. M. 1979. New Zealand beef and sheep supply relationships. Austr. J. Agri. Econ. 23:102–115.

Savell, J. W. 1997. Animal Science 307 lecture notes. Texas A&M University, College Station.

Savell, J. W., and H. R. Cross. 1991. Value-based marketing: Current status. Proc. Recip. Meat Conf. 44:117–120.

Schroeder, T. C., R. J. Jernick, R. Jones, and C. Spaeth. 2001. U.S. lamb demand. Sheep Goat Res. J. 17:14–19.

Schweigert, B. S. 1987. The nutritional content and value of meat products. Pp. 127–165 in The Science of Meat and Meat Products, J. F. Price and B. S. Schweigert, eds. Westport, CT: Food and Nutrition Press, Inc.

Sealed Air Corporation. 2007. National meat case study. Cryovac Food Packaging Division. Online at: http://www.sealedair.com/products/food/nmcs.html. Accessed May 2, 2008.

Shiflett, J. S., W. D. Purcell, D. Marsh, and P. Rodgers. 2007. Analysis of lamb demand in the United States. Report to the American Lamb Board, January.

Sink, J. D., and F. Carporaso. 1977. Lamb and mutton flavor: Contributing factors and chemical aspects. Meat Sci. 1:119–125.

Slanger, W. D., M. J. Marchello, J. R. Busboom, H. H. Meyer, L. A. Mitchell, W. F. Hendrix, R. R. Mills, and W. R. Warnock. 1994. Predicting total weight of lamb retail-ready lamb cuts from bioelectrical impedance measurements taken at the processing plant. J. Anim. Sci. 72:1467–1474.

Snowder, G. D., R. A. Field, and J. R. Busboom. 1994. Efficacy of body wall thickness and backfat depth for estimating percentage yield of retail cuts of lamb. Sheep Goat Res. J. 10(3):153–159.

Tatum, J. D., J. W. Savell, H. R. Cross, and J. G. Butler. 1988. A national survey of lamb carcass cutability traits. SID Res. J. 5:23.

Tatum, J. D., M. S. DeWalt, S. B. LeValley, J. W. Savell, R. P. Garrett, F. L. Williams, Jr., and J. W. Wise. 1992. Development of lamb classification and production systems to facilitate marketing based on carcass cutability. 1992 Sheep Research Highlights. Fort Collins: Colorado State University.

Taylor, D. M., S. L. Woodgate, and M. J. Atkinson. 1995. Inactivation of the bovine spongiform encephalopathy agent by rendering procedures. Vet. Rec. 137:605–610.

The Economist. 2007. Marketing to Muslims: food, fashion, and faith. August 2.

The Freshlook Marketing Group. 2007. Data for lamb for 2003–2005. Online at: http://www.freshlookmarketing.com/. Accessed May 1, 2008.

U.S. BLS (U.S. Bureau of Labor Statistics). 2007. Lamb price indices. Online at: www.bls.gov. Accessed May 1, 2008.

USDA (U.S. Department of Agriculture). 1996. Pathogen reduction; hazard analysis and critical control point (HACCP) systems; final rule. Fed. Reg. 61(144):38805–38989. July 25. Online at: http://www.fsis.usda.gov/OPPDE/rdad/FRPubs/93-016F.pdf. Accessed July 25, 2007.

USDA. 1999. Generic E. coli testing for sheep, goats, equines, ducks, geese, and guineas. Fed. Reg. 64(228):66547–66553. Online at: http://frwebgate1.access.gpo.gov/cgi-bin/waisgate.cgi?WAISdocID=179567267383+0+0+0&WAISaction=retrieve. Accessed November 29, 2007.

USDA. 2006. Red Meat Yearbook. Economic Research Service, Washington, DC. Last updated January 2006. Online at: http://usda.mannlib.cornell.edu/Mann Usda/viewDocumentInfo.do?documentID=1354. Accessed May 2, 2008.

USDA. 2007a. Agricultural Outlook: Statistical indicators. Economic Research Service, Washington, DC. August. Online at: http://www.ers.usda.gov/Publications/AgOutlook/AOTables/. Accessed May 2, 2008.

USDA. 2007b. Quick Stats: National Agricultural Statistics Service. Online at: http://www.nass.usda.gov/Data_and_Statistics/Quick_Stats/index.asp. Accessed May 2, 2008.

USDA. 2007c. Packers and Stockyards Statistical Report: 2005 reporting year. GIPSA Report No. SR-07-1. Grain Inspection, Packers and Stockyards Administration, Washington, DC, February. Online at: http://archive.gipsa.usda.gov/pubs /2005_stat_report.pdf. Accessed May 2, 2008.

USDA. 2007d. United States Standards for Grades of Slaughter Lambs, Yearlings, and Sheep. Effective date July 6, 1992. Agricultural Marketing Service, Washington, DC. Online at: http://www.ams.usda.gov/AMSv1.0/getfile?dDocName=STELDEV3062573. Accessed May 2, 2008.

USDA. 2007e. Food availability data system. Economic Research Service, Washington, DC. Last updated February 15, 2007. Online at: http://www.ers.usda.gov/data/foodconsumption/FoodAvailIndex.htm. Accessed May 2, 2008.

USDA. 2007f. Market and trade data. Foreign Agriculture Service. Online at: http://www.fas.usda.gov/markettradedata.asp. Accessed May 2, 2008.

USDA. 2007g. Summary livestock and grain market news reports. Agricultural Marketing Service. Online at: http://amsdev.ams.usda.gov/lsgtest/NewMN/NatSum.htm. Accessed May 2, 2008.

USITC (U.S. International Trade Commission). 1999. Lamb meat industry adjustment plan. Investigation No. TA-201-68. USITC Publication 3176. Washington, DC.

Van Tassell, L. W., and G. D. Whipple. 1994. The cyclical nature of the U.S. sheep industry. J. Agr. Resour. Econ. 19:267–279.

Vere, D., G. Griffiths, and R. Jones. 2000. The specification, estimation, and validation of a quarterly structural econometric model of the Australian grazing industries. CRC for Weed Management Systems Technical Series No. 5, University of Adelaide, Glen Osmond, Australia.

Viator, C. L., S. C. Cates, M. K. Ruth, S. A. Karns, and G. Brester. 2007. Cash versus contract marketing in the U.S. lamb industry. Sheep Goat Res. J. 22:32–41.

Wachenheim, D. E., L. L. Blythe, and A. M. Craig. 1992a. Effects of antibacterial agents on in vitro ovine ruminal biotransformation of the hepatoxic pyrrol-izidine alkaloid jacobine. Appl. Environ. Microbiol. 58(8):2559–2564..

Wachenheim, D. E., L. L. Blythe, and A. M. Craig. 1992b. Characterization of rumen bacterial pyrroli-zidine alkaloid biotransformation in ruminants of various species. Vet. Hum. Toxic. 34(6):513–517.

Welsh Country Food Group. 2007. An evaluation of the use of video image analysis to predict the classification and meat yield of sheep carcasses. http://www.hybucigcymru.org/uploads/MediaRoot/416.pdf. Accessed May 1, 2008.

Whipple, G. D., and D. J. Menkhaus. 1989. Supply response in the U.S. sheep industry. Amer. J. Agr. Econ. 71:126–135.

Whipple, G. D., and D. J. Menkhaus. 1990. Welfare implications of the Wool Act. Western J. Agr. Econ. 15:126–135.

Williams, G. W., and J. P. Nichols. 1998. Effectiveness of commodity promotion. Consumer and Product Research Report No. CP-01-98, Texas Agribusiness Market Research Center, Texas A&M University, College Station. May.

Williams, G. W., and O. Capps, Jr. 2005. Household level lamb consumption patterns. Commodity Market Research Report No. CM-02-05, Texas Agribusiness Market Research Center, Texas A&M University, College Station, September.

Williams, G. W., O. Capps, Jr., R. Dietrich, J. W. Edwards, R. A. Field, H. L. Goodwin, Jr., D. B. Griffin, K. Litzenberg, J. P. Nichols, J. W. Savell, G. Smith, D. Tatum, and J. B. Ward. 1991. Assessment of marketing strategies to enhance returns to lamb producers. Commodity Research Report No. CM-1-91, Texas Agribusiness Market Research Center, Texas A&M University, College Station.

Wishmeyer, D. L., G. D. Snowder, D. H. Clark, and N. E. Cockett. 1996. Prediction of live lamb chemical composition utilizing electromagnetic scanning (TOBEC). J. Anim. Sci. 74:1864–1872.

Young, O. A., G. J. Cruickshank, K. S. Maclean, and P. D. Muir. 1994. Quality of meat from lambs grazed on 7 pastures in Hawkes Bay. N. Z. J. Agric. Res. 37:177–186.

5

The U.S. Wool Industry

The United States was the world's fifth largest wool-producing nation in the 1940s. At the time, wool was considered to be the primary product of sheep production with lamb and mutton as byproducts of wool production. As the fortunes of the U.S. sheep industry declined over the years, so did the relative return to wool production. As a consequence, sheep producers and researchers have turned their attention to improving lamb and mutton production. Today, the United States accounts for < 1 percent of the world's wool production (Anderson et al., 2007). This chapter reviews the current status of the U.S. wool industry, with a particular focus on the challenges and future opportunities facing the industry. After reviewing the production, marketing, use, pricing, trade, and government policies related to wool, the chapter concludes by offering some insight on the future course of the industry.

WOOL PRODUCTION

The production of wool is a continuous, year-round process influenced by a large number of factors, including genetics, nutritional status, lactation, and other stress factors. The majority of U.S. sheep are shorn in the late winter and early spring months each year, although some producers shear in the fall. In the western range states, producers normally shear prior to the onset of the lambing season. Exceptions include many producers that lamb on the range in the fall or winter where climate risks result in shearing at or close to lamb weaning time in the spring. Some range producers that lamb in the winter in sheds shear prior to lambing. Another source of wool is from feeder lambs. Lamb feeders in higher rainfall climates shear lambs to be feedlot finished, and also do so in regions with hot summer periods when the lambs enter the feedlots. An estimated 70 percent of the lambs finished in feedlots are shorn (McDonnell, personal communication, 2007).

In 2006, the 14 western range and intermountain states accounted for 72 percent of sheep and lamb inventories but produced 77 percent of the U.S. wool clip and received 88 percent of the income from wool sales (USDA, 2007). The production cost and return budgets presented in Chapter 2 (Tables 2-3 and 2-4), for example, show that wool provided 14.5 percent of the income in Nevada public land range sheep production systems in 2006 and < 2 percent of the income in a 50-ewe farm flock operation, barely covering shearing costs. A frequent debate in the U.S. sheep industry is whether wool is a liability or an important economic component of sheep production that is often neglected. Wool may be a liability to sheep producers in one of three situations:

- *High-rainfall production areas.* Wool breeds of sheep are not well adapted to high rainfall conditions. For this reason, there is growing interest in the hair breeds of sheep in the subtropical regions of southeastern states.
- *Farm flock production systems that emphasize lamb production.* These operations generally utilize medium wool breeds or crossbreds that produce lower-quality fleeces that often do not generate enough income to pay for shearing costs.
- *Remoteness of wool markets.* In this case, producers often have difficulty finding qualified shearers and market outlets for wool, providing some incentive to shift to hair breeds.

Wool is an important component of sheep production systems in other situations:

- *Range sheep production systems where arid rangelands limit the potential for increased lamb production.* The best adapted range maternal ewe breeds all include some Merino genetics in their origin, from 100 percent in the Rambouillet and Merino to 75 percent in the Targhee and 50 percent in the Columbia. These breeds are all good wool producers.
- *Marketing of wool versus marketing of meat and milk products.* Because wool is an easily stored commodity, producers can delay the marketing of wool, but must sell their meat and milk products more quickly. Some producers store their wool for a year or more for financial or market considerations.
- *Remoteness of lamb markets.* The distance of their continent from major world lamb markets together with the storability of wool are pri-

mary reasons that Australian sheep producers historically have emphasized wool production and have become the dominant suppliers to world wool markets.

- *Availability of niche markets.* Some producers, particularly those with smaller flocks, have become specialty producers of wools, merchandising to hand spinners and weavers. These may be longer and coarser wools for easier spinning or unique colored wools. With aggressive marketing strategies, these producers may earn from $100 to $150 in wool income per sheep in the flock.

WOOL MARKETING

Wool producers use one of three primary market preparation methods at shearing (original bag, bellies out and the fleeces not tied individually, and table skirted and classed), and one of four primary sales mechanisms (wool warehouses, wool cooperatives, wool pools, and private treaties).

Worsted and Woolen Systems

Worsted and woolen are the two basic systems used to process wool from the clean, scoured state through to yarn. The wools flowing through each system are, thus, determined by their characteristics. The worsted system utilizes longer staple and finer grades of wool for higher-quality cloth and finished fabrics. Worsted yarn is spun from fibers that have been carded and combed, resulting in relatively parallel fibers and a smooth yarn. The woolen system utilizes coarser grades of wool for yarn production that ranges in end uses from heavier fabrics for outerwear to drapes, upholstery, and carpets. Woolen yarns are spun from fibers that have been carded but not combed. As a result, the fibers are randomly arranged, yielding a relatively rough yarn. Other wools may be used for felt, insulation, batting, and as adsorbents and filters (ASI, 2002).

The physical characteristics of wool as sold by producers are determined by several objective measurements determined on full-length staples removed from random positions in the bale. These characteristics include fiber diameter, variability of fiber diameter, clean wool yield, staple length, staple strength, quantity and type of vegetable matter, and number of black fibers present. Objective measurements are obtained from core samples of bales/sale-lots of wool and determined by internationally approved machine and laboratory analyses. Subjective measurements may include color or stains, condition of staple tips, crimp, and style or handle.

Fiber Diameter

All international wool markets describe fiber diameter in microns, with superfine wool as low as 14–17 microns up to coarser wool from 27–35 microns. Staple wools < 26 microns are generally used in the worsted system, with quality of fabric increasing in the lower micron ranges. Variability of fiber diameter is important both within a fiber and within a fleece. As the fiber diameter decreases, market value generally increases. Premiums often are paid for wool with low average fiber diameter and low variability of fiber diameter.

Yield

Yield is the percentage of clean wool in a grease wool sample, usually adjusted to 12 percent moisture, 1.5 percent alcohol extractives (oil), and 0.5 percent mineral content. The wool trade often refers to this situation as "clean wool fibers present" (CWFP). Yield in the U.S. wool clip may vary from 40 to 70 percent. Fleece density, or number of fibers per area of skin, and staple length are the principal genetic factors influencing yield. Yield also can be affected by management practices, such as where the animals are managed (pasture, range, croplands, drylot), moisture content of fleece at shearing, and cleanliness of shearing area. Clean wool yield can have a major influence on grease wool market price. A wool clip with a 60 percent yield has 50 percent more clean wool than a clip with a 40 percent yield.

Staple Length

Length is important to both yield and the system in which the fiber will be used. For example, finer wools < 22 microns must be ≥ 70-mm staple for combing and spinning to yarn in the worsted system. Coarser wools from 23 to 26 microns must have a minimum 76-mm staple length. Because wool grows in length every day, disease and/or nutritional stress may cause weaker (tender) fibers or breaks in the staple length and decrease its use and value. Position of the break in the staple is also critical to determining staple length. Another concern regarding staple length is second cuts in the shearing process. Shearers may miss cutting part of the staple and then cut a second time to present a smooth appearance when finished. Second cuts, if left in the staple length fleece, are considered a contaminant. Staple strength is critical in processing wools to finished yarns and fabrics, and measurement of strength is becoming a standard objective measurement for international wool marketing. Staple strength is inherited and can also be affected by nutrition, animal health, and other management practices.

Vegetable Matter

Contaminants such as dirt and oils can be washed out of raw wool in the scouring process. However, vegetable matter, such as burrs and stems, generally cling to the wool fibers and must be removed during mechanical processing (carding and combing). Increasing levels of vegetable matter reduce clean wool yield and increase the cost of processing. Other contaminants of concern include polypropylene twine, usually from hay baler twine that is very difficult to remove from the fleece; use of nonscourable branding paint (scourable paints are readily available); and black fibers and kemp or medullated fibers in a white fleece.

Wool Market Preparation Methods Used by Growers

Wool is generally bagged for market in one of three different methods: (1) original bag (OB) method in which bellies, stained wool, and other inferior fleece portions are not removed from the fleece and no table skirting occurs; (2) bellies out and fleeces not tied individually (BOU) method in which some producers may do some fleece skirting to remove short and more contaminated fleeces but no table skirting occurs; and (3) table skirted and classed (TSC) method, which is the recommended method for international and premium domestic wool markets. In the TSC method, bellies typically are removed from the fleece by the shearer and the wool-handling crew removes most of the manure tags and urine-stained wool as it is shorn. The fleece is then thrown open on a slatted table to allow skirting to remove any vegetable matter, stained wool, second cuts, and shorter or off-grade parts of the fleece—normally around the head and lower parts of the leg area of the fleece. A qualified classer then estimates fiber diameter, staple length, and strength of the fleece to determine the grade of the fleece to sort into separate bins. Short (less than staple) fleeces and tender fleeces generally are packaged separately.

When a sheep is properly shorn, the fleece can be laid out on a table or floor and be seen as one piece. Skirting is the process of removing from fleeces the stained or inferior wool that grows on the belly and legs of the sheep (Lupton et al., 1992). Table skirting is simply placing the fleece on a table and finishing the skirting process. Classing is the preliminary sorting of the fleece according to its estimated physical properties.

Australia is the largest producer of wool and sets the international standard in the marketing of wools through preparation and class. In Australia, most wool is skirted and then subjectively classed by spinning quality number (fiber diameter or fineness range), staple length, color, condition, style, and soundness. Classers produce as few lines as possible from a wool clip while maintaining uniformity within a line and eliminating contamination

TABLE 5-1 Percentage of Wool Prepared by Growers for Sale to Major Wool Warehouses Using One of the Three Major Preparation Methods, 2007 Wool Clip

	Market Preparation Method (%)		
Market	OB	BOU	TSC
Roswell Wool Co., Roswell, NM	5–7	60–65*	30–35
Sonora Wool & Mohair, Co., Sonora, TX	10–15	75*	10–15
Eden Wool & Mohair Co., Eden, TX	15–20	75*	5–10
Producers Marketing Cooperative, Inc. (PMCI), Mertzon, TX	10	60*	30
Utah Wool Marketing Cooperative, Tooele, UT	10	40	50
Center of the Nation Wool, Inc., Belle Fourche, SD	15	45	40

*Several warehouses reported that, even though the BOU wool is not TSC, an increasing number of producers are doing some sorting of their BOU wool on the basis of grade, staple length, level of contaminants, and/or color.
Source: Survey of major warehouses by authors.

of the clip with stained, pigmented fibers and all foreign material (Lupton et al., 1996). Subsequently, most lots are objectively measured (prior to sale) for clean yield, vegetable matter content, average fiber diameter (and variability), staple length, staple strength, and color (Lupton et al., 1989).

Most wool in the United States continues to be sold as OB wool (bagged without any further processing) (Table 5-1). A recent telephone survey of major wool-marketing warehouses in the western states by the authors indicates some trend toward further preparation of wools by producers. The six warehouses[1] surveyed merchandise approximately 50 percent of the wool produced in the western states (Table 5-1). The survey suggests that growers increasingly recognize the importance of wool preparation for market. Producers Marketing Cooperative, Inc. (PMCI), a grower-owned marketing cooperative, handles approximately 20 percent of the Texas wool clip, with approximately one third of their wool from growers that are owner/members of the co-op. All owner/member wool clips are TSC by co-op staff and normally sell at a 15 percent to 25 percent premium. Warehouse managers in the survey agree that the highest-selling wool clips are those that have been using TSC for several years, establishing predictable reputation clips. With the 2007 wool clip selling at near-record prices, TSC has been an important method of preparation for growers of better wool clips.

The warehouse managers also indicate that 20–30 percent of the OB

[1]Roswell Wool Co., Roswell, NM; Sonora Wool & Mohair, Co., Sonora, TX; Eden Wool & Mohair Co., Eden, TX; Producers Marketing Cooperative, Inc. (PMCI), Mertzon, TX; Utah Wool Marketing Cooperative, Tooele, UT; Center of the Nation Wool, Inc., Belle Fourche, SD.

and BOU wool sold through their warehouses is purchased by a major U.S. wool buyer and moved to its private warehouse for TSC and then merchandised to major U.S. and international wool market outlets. This observation suggests that producers not preparing their wool clip using TSC are selling at a substantial discount, since buyers clearly expect a cost-plus-profit return on investment for the further processing.

Wool Marketing Mechanisms

In the United States, wool is generally marketed through one of four different marketing mechanisms: (1) warehouse system, (2) marketing cooperatives, (3) pools, and (4) private treaties. Sales options within each mechanism have become fewer with the decline in sheep numbers. The primary objective of all four marketing mechanisms is to provide wool buyers adequate volumes of wool that can be purchased with confidence that their uniformity is accurately represented. All TSC and most BOU wool sold in the United States is core tested with the results available to buyers. Most buyers know the specifications of their orders for wool and are not interested in purchasing wools that do not meet their requirements. The importance of meeting international standards to realize potential market values has increased as exports of U.S. wool have grown.

Wool Warehouse System

Major U.S. wool-producing regions have access to one or more commercial wool warehouses. Growers consign their wool to the warehouse that has the major lines or classes within a clip core tested for the previously described objective measurements. Warehouses may then sell their wool consignments by either (1) sealed bids from and/or direct negotiations with buyers or (2) public auctions with adequate volume to attract major national and international buyers. Reputation wool clips may be sold separately within TSC grades. Smaller wool clips may be combined with other clips to achieve adequate volume within grades to attract buyers. Growers may instruct warehouse managers to obtain predetermined minimum bids as their agents, or they may trust the manager to get the best available price for their wool.

Wool Marketing Cooperatives

Many areas of the United States are too remote from wool warehouses and have inadequate volume to attract a commercial warehouse. The largest wool-marketing cooperative is Mid States Wool Growers Cooperative with assembly warehouses at its headquarters in Canal Winchester, Ohio, and in Kansas City, Missouri. Mid States also has assembly stations in several locations throughout the Midwest and the East for grower delivery to minimize freight costs for smaller clips. Wool delivered from each grower to their central warehouses is then sorted into grades and wool types and pooled with other clips to obtain adequate volumes in the various grades and types to attract buyers. Mid States provides an important service to farm flock producers remote from traditional wool markets, and provides an assembly service for buyers. Although California Wool Growers Marketing Cooperative is the primary wool market outlet for California growers, the cooperative has contracted with Roswell Wool to sell its wool to obtain greater market access. Producers' Marketing Coop., Inc. (PMCI) in Mertzon, Texas, may be a model for future cooperatives of growers that produce higher-quality wool in the major wool producing areas.

Wool Pools

Growers in several regions, particularly in the Northwest and North Central states, have used a variation of the cooperative system called wool pools. Montana, for example, continues to operate a number of such pools. A group of growers that produce similar sheep and wool will combine their clips and offer them for sale at a central shipping point. The more progressive wool pools provide objective measurement information on the wool offered for sale. The pools provide an opportunity for smaller growers to pool with other growers with similar wool in 11,340- to 13,608-kg truckload lots. Wool pool members may jointly own a wool baler and may use a common shearing crew.

Private Treaty

Direct sales were a traditional method growers utilized to sell their wool clip. Growers with larger reputation clips would contact buyers and negotiate a sale price. Although some private treaty sales continue in the major wool-producing states, this sale method has declined for at least three reasons: (1) the use of objective measurements, (2) recognition by growers that buyers were generally more knowledgeable of the wool market than growers, and (3) the desire of major buyers to bid on larger quantities assembled at a central location with known specifications and objective measurements. There continue to be reputable buyers in the North Central and eastern states that provide a market outlet for farm flock producers and lamb feeders. Producers with small flocks that produce lower-quality wool may give the wool to the shearer as partial payment for shearing costs. Most niche market wools are merchandised by growers as individual fleeces or smaller amounts or as natural hand-spun yarns.

WOOL PROCESSING

Once raw wool is purchased, several steps are required before it is processed into end products. As stated previously, most fabrics for clothing are produced from finer wools (< 26 microns) in the worsted system. Carpets, rugs, heavier blankets, and some upholstery products are generally produced from longer staple, coarser wools primarily using the woolen system; however, an increasing percentage of carpet and blanket yarns are produced on a modified or semiworsted system. With the exception of a limited number of producers of blankets, the woolen system is almost nonexistent in the United States. China, India, Turkey, and, to a lesser extent, some Eastern European nations, now dominate carpet, rug, and heavier blanket production from the woolen system. Since the worsted system is the primary use of wool in the U.S. textile industry, brief description of the steps from raw wool to retail products is appropriate. A more detailed description is presented in the SID *Sheep Production Handbook* (ASI, 2002).

- *Scouring.* Scouring is the removal of impurities from grease wool using water, detergent, and sometimes a mild alkali. Lanolin is the primary grease base in wool, and it is removed as a merchantable byproduct. Water and other materials, such as dirt and vegetable matter, that are removed in the scouring process can be treated in conventional sewage processes.
- *Carbonizing.* Wool clips contaminated with excessive and harder-to-remove vegetable matter, such as burrs and thorny branch segments, are carbonized using an aqueous sulfuric acid treatment to carbonize the cellulosic vegetable matter and followed by heating to convert the defective material into carbon. The carbon is then crushed and shaken from the wool. The acid base is then neutralized and the wool rescoured to complete the process. Carbonizing is expensive, often resulting in shorter and weaker wool fibers. In addition, the wastewater must be treated separately to meet environmental quality requirements. No specific data are available, but only a minor portion of the U.S. wool clip requires carbonizing. The TSC process should remove these types of vegetable matter contaminants from the fleece.
- *Drying.* The wet scoured wool is then dried. The wool is first mechanically squeezed, then deposited on conveyers at uniform depths to pass through continuous flow heated air dryers.
- *Carding.* Carding disentangles and orients the fibers uniformly for further processing, and is achieved by passing the wool through cylinders and rollers covered with short wires varying from 500 to 2000 wires per square centimeter, depending on the type of wool and end-use objectives.
- *Combing.* Combing removes vegetable matter and short and tangled fibers and orients the longer fibers in a more or less linear configuration. The fibers are combined to form a continuous combed rope or sliver that is then

fed into rollers for further combing, and then coiled into cans or bundles for delivery.

- *Top-finishing.* Further blending of the combed fibers into a uniform weight and thickness per unit of length and winding into a ball known as top is accomplished by combining (drafting) several combed slivers that are drawn (drafted) together to provide more uniform material. For some yarns destined to be woven into multicolored fabric patterns, dyeing may occur at the top-finishing stage.
- *Roving, spinning, winding, and twisting.* Roving, a process similar to spinning, reduces the top to a more uniform size or thickness prior to spinning. Spinning is the final drawing to the desired yarn fiber count or thickness and includes a predetermined amount of twist and delivery of the yarn to an appropriate package. Wrapping two or more yarns together to form a multiply yarn is called twisting or plying.
- *Weaving.* Weaving interlaces two sets of yarn to form fabric, with warp yarns running lengthwise and weft yarns crosswise. A modern loom is computerized to control multiple functions to including speed, specific weave, and color functions. The woven cloth moves to a roller synchronized with warp speed in a continuous operation and holds the cloth to a specified width.
- *Knitting.* Knitting interlaces yarn in a series of connected loops by needles to control a fabric. Specialty yarns may be produced for hand knitting, whereas mechanized commercial knitting machines are very rapid and require stronger yarns. Warp knitting machines produce flat fabric similar to woven cloth, while specialized machines may knit socks, gloves, and fully fashioned garment patterns.
- *Finishing.* Technically, everything that happens to wool fabric after leaving the loom until it is ready for the cutter is regarded as finishing. This may include inspecting for defects, scouring, dyeing, and other chemical treatments to increase flame resistance, reduce shrinkage, improve breathability, and increase comfort. The finished cloth is then ready for textile manufacturers to prepare finished products for consumers.
- *Wool felts.* Not covered in the above process is the production of wool felt textiles. Felt is characterized by the densely matted fibers of the wool used. Felts are produced by the combined actions of mechanical work, chemical reactions, moisture, and heat. Although current felt production records are not available, U.S. consumption of wool fibers in felts in 1986 was nearly 4 million kilograms in terms of clean wool.

Current Status of U.S. Wool Processing Industry

The wool processing industry has shrunk significantly in the United States in the last 20 years and is discussed in depth in the following section on wool demand. The majority of wool processing in the United States is located in North Carolina and South Carolina and other southeastern U.S. states along the eastern seaboard, and in the Northeast. The global shift in the textile industry is largely attributed to low wages in China, India, and other developing nations, but this shift is also due to rapid modernization of wool processing plants, particularly in China and India (WOOLNEWS.net, 2007). The surviving textile mills in the United States have largely modernized to utilize new technologies and tailor their products to specific consumer needs, such as the U.S. military (ASI, 2007a). Many of the steps in wool processing require chemicals and create byproducts that result in substantial expense to comply with environmental regulations, while environmental regulations are considerably less in China, India, and other developing nations. Other issues of concern to the U.S. wool processing industry include the removal of textile tariffs and quotas, and the claim that China has artificially kept its currency at a low value relative to the U.S. dollar. Another concern often mentioned is the decline in U.S. sheep and wool production. However, demand for wool apparel goods has increased in the United States in the last 10 years, and export of the U.S. wool clip has increased from traditional levels of approximately 33 to 70 percent of the wool clip in fiscal year 2007. In fact, China now consumes more U.S. raw wool than the domestic wool processing industry (ASI, 2007b).

WOOL DEMAND

Clean, graded, scoured wool is used primarily for processing into apparel and carpets. Wool used for carpets is mostly imported and is shorter, coarser, and less uniform than wool used for apparel. While some apparel wool is produced domestically, the majority (about 88 percent) is imported (ASI, 2007a). Apparel wool is used primarily for clothing such as tweeds, flannels, and knits for blankets. At the mill level, wool competes with a large number of both natural fibers, including cotton, flax, and silk, and synthetic fibers, such as nylon, rayon, acrylic, and polyester. However, wool can function in a complementary role with other fibers in the production of fiber-blend textile goods. The same is the case at the retail level. While wool textiles compete with other fiber textiles for the consumer dollar, wool complements the demand for other fiber textiles in many blended fiber textiles.

Domestic Mill Demand

The dominant fibers in textile processing are cotton and synthetic fibers. From 1995 through 2005, synthetic fibers accounted for an average of 69.9 percent of all fibers used by U.S. mills, with cotton accounting for

28.7 percent (USDA, 2006). Wool accounted for an average of only 0.6 percent, with flax and silk accounting for the remaining 0.8 percent.

Since 2000, the U.S. domestic mill use of all fibers, including wool, has been in a general downward trend, reflecting the growing concentration of the textile industry in developing countries, where wages are relatively low compared to the United States (Figure 5-1). According to the National Council of Textile Organizations, more than 350 U.S. textile plants have closed since 1997 and the industry has lost nearly 200,000 jobs in the past five years (NCTO, 2007). Wool has experienced the largest percentage decline in mill use, falling by an annual average of nearly 11 percent between 1999 and 2005, compared to the 8 percent average annual decline in the mill use of cotton, the 6 percent decline in the use of flax and silk, and the 2 percent decline in the use of synthetic fibers over the same period (Figure 5-2). The decline in domestic mill use of wool began after World War II due to the reduction in use by military service personnel. Mill use of wool has also suffered from the continuing decline in sheep numbers, the marketing and promotion programs of the cotton industry designed to increase cotton demand at the mill level and at the retail level, and a shift in consumer tastes and preferences toward lighter-weight casual clothing.

In addition, wool has competed with a growing number of other fibers

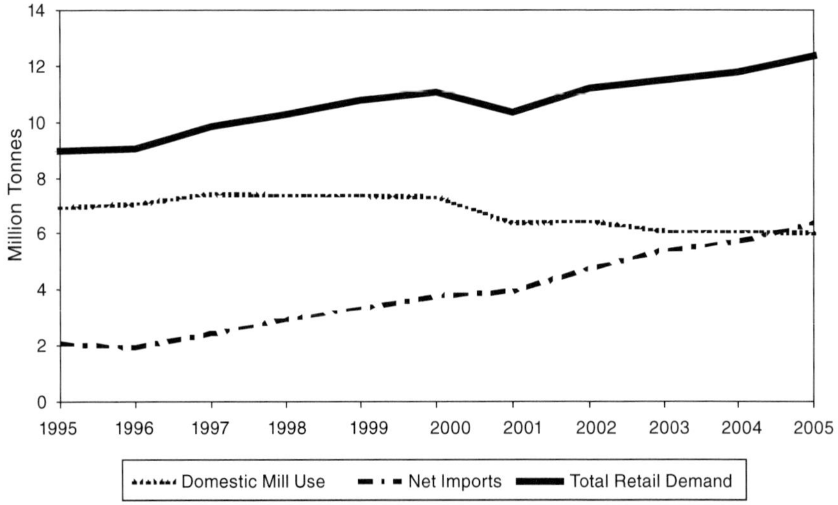

FIGURE 5-1 U.S. fiber demand at the mill and retail levels, 1995 to 2005.[a]
[a]All fibers including cotton, wool, synthetic fibers, flax, and silk.
Source: Based on data in USDA (2006).

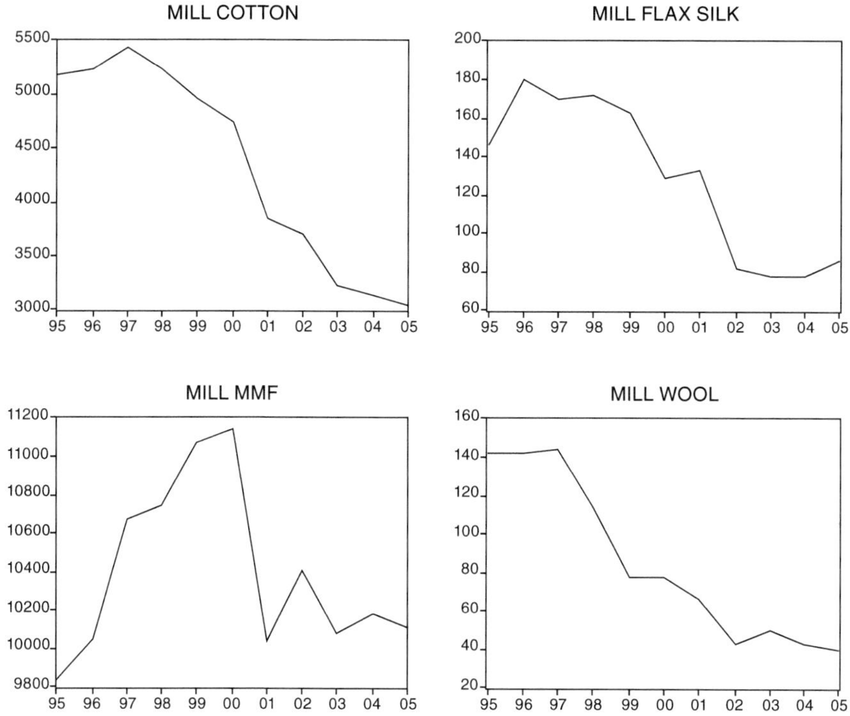

FIGURE 5-2 Mill use of cotton, flax and silk, synthetic fibers, and wool, 1995–2005. Units are 1,000 lb (1 lb = 0.4536 kg).
Source: USDA (2006).

in the production of textile goods over time. As of 2003, the most recent information reported by the USDA, noncellulosic fibers and cotton accounted for almost 92 percent of world textile fiber production (Capps and Williams, 2006). The share accounted for by noncellulosic fibers has risen since 1980 while the shares accounted for by wool, cotton, flax, and hemp have declined. The share of global textile production on a raw-fiber-equivalent basis accounted for by wool fell from a high of 5.7 percent in 1982 to a low of 2.3 percent in 2003.

Perhaps more important than any other factor in the sharp decline of textile manufacturing in the United States, however, was the replacement of the Multi-Fiber Arrangement (MFA) with the World Trade Organization (WTO) Agreement on Textiles and Clothing (ATC) in 1995. Under the MFA, a large portion of textiles and clothing exports from developing countries to the United States and other industrial countries was subject to quotas under a special regime outside normal GATT rules. On January 1, 1995, the international textiles and clothing trade began a process of fundamental change under the ATC, as the industrial countries began eliminating their import quotas over a 10-year period that ended on January 1, 2005. As a consequence, U.S. imports of textile products have grown rapidly as raw fiber textile milling has increasingly shifted to China, India, Eastern Europe, and other countries where labor costs are lower. According to Wilcox (2007), labor costs in textile manufacturing in these countries are as low as \$1–\$2 per hour compared to \$15–\$20 per hour in the major industrial nations. Fewer or less restrictive environmental regulations in these countries may also be a factor. Other traditional wool-processing regions, including Western Europe, Great Britain, and Australia, are grappling with similar declines in raw wool processing and increasing imports of wool textiles.

There is growing concern in the U.S. textile industry, however, that the opening of the U.S. market to textile imports has facilitated extensive government intervention by China and other developing country governments in their textile and apparel export sectors to set prices at artificially low levels (ATMI, 2003). Claims of anticompetitive actions by the Chinese include currency manipulation to subsidize exports, subsidization of nonprofitable state-owned textile and apparel manufacturers, export tax rebates, and loans by China's central banks to achieve a competitive advantage against foreign competition. The concern is that over the past 3 years these and other subsidies have led to an average 58 percent drop in the prices of Chinese textile exports where quotas have been removed. These concerns have motivated talks between the U.S. and Chinese governments with a resulting agreement in August 2007 to increase import limits for certain cotton, wool, synthetic fiber, silk blend, and other vegetable fiber textiles and textile products produced or manufactured in the People's Republic of China (USDC, 2007).

The Retail Demand for Wool Textiles

The total U.S. use of wool and all other fibers at the retail level is measured by USDA as domestic mill use of the fibers plus the raw-fiber-equivalent of imported textiles. As at the mill level, the dominant fibers at the retail level are cotton and synthetic fibers. While the use of wool by U.S. mills has been declining, imports of wool textiles have been increasing (Figure 5-3). The growing imports of wool textiles have more than compensated for the decline in the domestic mill use, leading to an increase in the total retail consumption of wool in the United States. The growth in retail demand for wool textiles has been high enough in recent years to boost per capita demand for wool from 0.50 kg in 1999 to 0.63 kg in 2005 on a raw-fiber-

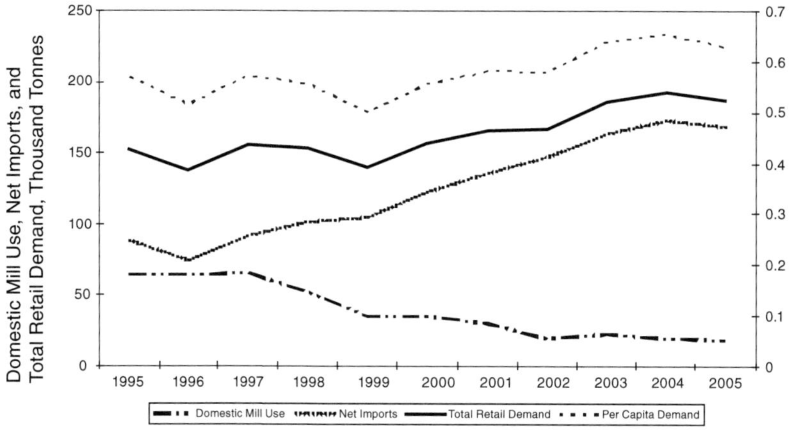

FIGURE 5-3 U.S. wool demand at the mill and retail levels, 1995–2005
Source: USDA (2006).

equivalent basis (USDA, 2006). The trend has been the same for other fibers, with growing imports more than compensating for the declining domestic production of textiles, leading to increased total and per capita demand (see Figure 5-1). Between 1995 and 2005, the per capita usage of synthetic fiber textiles increased from 18.5 kg to 22.7 kg, while per capita usage of cotton textiles increased from 13.4 kg to 17.1 kg (USDA, 2006).

Traditional wool apparel goods were heavier, more durable products worn primarily in colder climates and winter months and were not responsive to changes in styles and fashion trends. They included men's and women's suits, topcoats, and sweaters. The first major challenge to tradition came from the advent of synthetic fibers that were lower cost and more responsive to changing fashion trends and consumer tastes, including "wash and wear" convenience. Cotton and other natural fibers responded to this challenge more rapidly than wool. Similar shifts to synthetic fibers occurred within nonapparel products such as carpets, blankets, and upholstery due primarily to lower costs. As a result, wool consumption has become a small portion of the textile industry.

Recent trends in the textile industry indicate that wool could develop a stronger market position in certain areas. Significantly lower production costs in China and India are resulting in retail apparel products including wool, wool blends with other natural fibers such as alpaca and cashmere, and even wool blends with synthetic fibers that are more cost competitive. Lighter-weight, finer wool apparel goods in men's and women's wear are increasing slightly in market share (Wilcox, 2007). Wilcox (2007) reported that in 2006, China produced 39 percent of the world supply of apparel goods, 80 percent from imported wool. It consumed 60 percent of this production internally, indicating that its emerging economy now consumes approximately 24 percent of the world's apparel goods production. According to ASI (2007c), the U.S. military is the largest consumer of U.S. wool, due to congressional requirements that domestic products be used when available. The military is also a major contributor to wool research.

Raw Wool Trade

In general, the United States exports raw fibers, both natural and synthetic, and imports textiles, as well as some raw fibers, including wool. Germany was the leading destination for U.S. raw wool exports in 2002, a distinction achieved by China beginning in 2004 (Table 5-2). According to the American Wool Trust report to the U.S. Congress in January 2007, exports of U.S. greasy wool to China have increased from 1,054 metric tons (tonnes) in 2003–2004 to 1,446 tonnes in 2004–2005 to 2,713 tonnes in 2005–2006, a trend that is expected to continue (ASI, 2007c). German imports, on the other hand, dropped by nearly 30 percent over the same period. In 2005, the two countries together accounted for over 70 percent of U.S. wool exports. Other major U.S. wool export destinations include India, Belgium, and Mexico.

China now purchases almost 50 percent of the world market supply of raw wool for finishing to textile products, plus utilizing all of its internal wool production. China produces 39 percent of the world's wool apparel products, up from approximately 20 percent in 2000, and is the dominant producer of machine-manufactured wool carpet and rug products (Wilcox, 2007). Internal consumption of wool apparel products sold at retail in China is now 60 percent of their total production. As a consequence, the emerging Chinese economy is consuming 24 percent of the world wool apparel goods, almost triple the internal wool use duing 2000 (Wilcox, 2007). China is now also the largest sheep-producing country in the world. A recent Australian Wool Innovation (AWI) raw wool market research report estimated China's wool production at 100,000 tonnes and raw wool imports at 400,000 tonnes (WOOLNEWS.net, 2007). The AWI also reported that major Australian wool buyers and processors are investing in joint ventures with Chinese manufacturing companies and that China is investing heavily in state-of-the-art new technology to modernize its textile industries. Wilcox (2007) suggested that while these transitions may be painful, the lower-cost, higher-quality products produced in China, India, and other developing economies will likely result in wool apparel goods becoming more cost competitive with other fibers.

TABLE 5-2 U.S. Raw Wool Exports (Thousands of Pounds) by Country of Destination, Clean Yield, 2002–2005

	Shorn Wool				Unshorn Wool				Carbonized Wool			
Country	2002	2003	2004	2005	2002	2003	2004	2005	2002	2003	2004	2005
Belgium	891.3	433.5	54.0	456.1	—	—	—	—	—	—	—	—
Canada	11.4	60.8	62.0	94.5	582.4	213.3	61.2	27.3	—	—	—	12.2
China, Mainland	1,136.7	2,417.3	2,855.7	2,209.0	—	246.9	1,101.6	2,864.6	—	—	186.9	—
France	137.2	—	22.4	42.0	27.5	40.8	—	236.2	—	1.8	—	—
Germany	2,714.6	3,868.8	1,847.7	1,961.6	—	137.2	65.1	713.9	—	—	—	—
Guatemala	6.3	23.1	16.0	14.9	—	—	2.7	—	—	—	—	—
Hong Kong	4.8	86.0	—	—	—	—	26.7	1.7	—	17.3	3.6	7.3
India	394.8	1,333.1	629.0	414.4	—	67.7	157.7	290.1	11.1	130.4	—	—
Italy	313.4	233.9	238.9	41.3	35.4	—	408.2	1,247.5	—	—	—	—
Japan	0.3	29.5	8.1	4.5	—	22.8	—	—	—	—	—	—
Korea	—	36.4	2.8	—	—	5.1	44.1	9.6	—	—	—	—
Mexico	936.2	301.7	156.2	458.4	31.5	—	6.5	—	130.7	2.4	3.0	137.8
Poland	140.8	185.7	215.5	85.5	22.1	27.4	—	—	—	—	—	—
Portugal	—	—	—	—	—	—	—	—	—	—	1.6	13.2
Spain	166.9	—	1,665.3	—	—	—	—	—	—	—	—	—
Turkey	24.9	—	—	—	—	—	—	—	—	—	—	—
United Kingdom	72.3	42.2	28.1	4.5	300.0	428.1	259.5	461.8	—	—	38.8	—
Other	322.4	523.7	652.6	120.9	46.5	80.0	221.7	405.5	—	70.1	124.7	85.9
Total	7,274.3	9,575.7	8,454.3	5,907.4	1,045.4	1,269.3	2,355.0	6,258.2	141.8	221.9	358.7	256.4

Units are 1,000 lb (1 lb = 0.4536 kg).
Note: — = No exports.
Source: USDA (2006).

Raw wool imports to the United States have traditionally been larger than U.S. exports, making the country a net raw wool importer. Nearly 80 percent of coarse U.S. raw wool imports (not-finer-than 46's) came from New Zealand in 2005 (Table 5-3). Australia, on the other hand, accounted for 72 percent of the finer wool (48's-and-finer) in 2005. Other major sources of U.S. coarse wool imports in 2005 included the United Kingdom (8 percent), Argentina (6 percent), and Australia (4 percent). Other major sources of U.S. fine raw wool imports in 2005 included Canada (10 percent), New Zealand (7 percent), and South Africa (5 percent).

Research on the Demand for Wool and Competing Fibers

Research on the demand for fibers has focused predominantly on cotton and synthetic fibers particularly at the mill level (e.g., Donald et al., 1963; Dudley, 1974; Stennis et al., 1983; Dickerson, 1999; Capps and Williams, 2006). A recent analysis of global cotton and fiber markets used a modified version of a multi-equation, econometric simulation model developed by the Cotton Economics Research Institute (CERI) at Texas Tech University (Pan and Mohanty, 2005) to study the own-price elasticities or price sensitivities of cotton and other fiber demands at the mill and retail levels across various countries (Capps and Williams, 2006). They found that the mill demand for cotton ranged from –0.14 in the United States and India to –0.74 in Egypt over the study period of 1976 to 2003. That is to say, given a 10 percent change in the price of cotton the quantity demanded of cotton for mill use ranges from 1.4 percent to 7.4 percent in the opposite direction. The U.S. cotton mill demand price elasticity was close to the elasticity estimate of –0.17 previously reported by Capps et al. (1997) but smaller than the –0.40 mill demand price elasticity reported by Murray et al. (2001) and lower than the –0.30 elasticity reported by Lowenstein (1952). Shui et al. (1993) reported a much higher U.S. cotton mill demand price elasticity of –0.60.

In addition, the Capps and Williams (2006) study found that for all countries except the United States, the estimated cross-price elasticities for polyester in the cotton mill demand equations (i.e., the responsiveness of cotton demand to a change in the price of polyester) were positive and smaller in magnitude than the corresponding own-price elasticities for cotton. The implication is that polyester and cotton are substitutes in foreign cotton mill use. With respect to the U.S. cotton mill demand, however, the polyester cross-price elasticity was estimated to be negative and larger in magnitude than the own-price elasticity, implying that cotton and polyester are complements in U.S. cotton mill use. The finding of complementarity between cotton and polyester in U.S. cotton textile milling is consistent with the conclusions of a number of other studies, including Capps et al. (1997) and Murray et al. (2001). In the latter study, the elasticity of cotton mill

TABLE 5-3 U.S. Raw Wool Imports (Thousands of Pounds) by Country of Origin, Clean Yield, 2002–2005

	Not-finer-than 46's				48's-and-finer			
Country	2002	2003	2004	2005	2002	2003	2004	2005
Argentina	363.2	632.1	454.5	679.8	19.2	0.5	—	61.0
Australia	439.3	155.3	694.7	430.6	8,126.0	3,601.4	4,248.0	4,499.7
Belgium	43.9	—	16.7	—	—	—	—	33.3
Brazil	—	—	—	—	52.8	74.5	33.1	—
Canada	117.1	72.1	43.0	82.0	840.6	350.7	638.8	593.8
Chile	—	—	—	—	—	—	—	—
France	—	—	—	—	—	—	—	—
Italy	0.1	—	—	—	20.9	—	25.0	1.4
Mexico	34.8	279.1	117.0	16.6	97.6	69.7	76.8	69.6
New Zealand	10,277.5	11,757.1	12,137.6	9,668.6	498.7	377.8	403.9	415.3
South Africa	54.0	62.2	202.8	224.8	528.5	377.8	506.7	285.0
Spain	11.8	—	—	—	39.6	—	—	—
Switzerland	—	2.4	—	—	—	0.2	2.3	—
United Kingdom	2,702.0	2,528.0	2,732.5	966.2	15.5	1.6	56.4	106.4
Uruguay	—	41.9	—	—	272.4	132.1	212.5	128.9
Other	115.4	219.1	55.7	111.4	13.8	0.1	0.6	25.4
Total	14,159.1	15,749.4	16,454.7	12,180.0	10,525.5	4,986.2	6,204.1	6,219.8

Units are 1,000 lb (1 lb = 0.4536 kg).
Note: — = Not available
Source: USDA (2006).

demand with respect to polyester price was found to be –0.13, somewhat lower than the estimate of –0.21 found by Capps and Williams (2006). Capps et al. (1997) estimated the polyester cross-price elasticity to be –0.5479. The intuition behind the notion of complementarity rests on the development of blended fabrics, at least in the United States, incorporating elements of synthetic fibers and cotton.

For U.S. synthetic fiber mill demand, Capps and Williams (2006) found that the estimated own-price elasticity was –0.04 and the estimated cross-price elasticity with regard to cotton was –0.10. Because this study was the first to report estimates for U.S. synthetic fiber mill demand, there are no results from other studies with which to compare. However, the results are consistent with those found for U.S. cotton mill demand, providing some confidence that cotton and synthetic fibers are complements in mill use in this country. In other words, when prices for either or both cotton or synthetic fibers rise (fall) in the United States, less (more) of both cotton and synthetic fiber is demanded by U.S. mills.

At the retail level of textile fiber markets, Capps and Williams (2006) found that demand for textiles across all countries, including the United States, is inelastic with respect to both the prices of textiles and income. In the United States, the own-price elasticities of cotton fiber textile demand and synthetic fiber textile demand were estimated to be –0.41 and –0.61, respectively. In foreign countries, the estimated own-price elasticities for all textiles ranged from as low as –0.07 in the EU-15 (the original 15 European Union countries) to as high as –0.52 in Pakistan. The estimated income elasticities ranged in magnitude from 0.02 in Taiwan to over 0.90 in Mexico and the United States. For the United States, the Capps and Williams study concluded that the demand for synthetic fiber textiles is slightly more income inelastic (0.60) than is the case for cotton fiber textiles (0.92). Given that the estimated income elasticities of textiles are positive and less than unity in magnitude across all countries, the implication is that consumers consider textile goods to be necessities rather than luxury goods regardless of where they live.

Clements and Lan (2001) analyzed the pattern of world mill level demand for cotton, wool, and chemical (synthetic) fibers. A distinctive feature of their analysis was the use of the systemwide approach to model jointly the demand for the three fibers. Much of the previous research had rested on the use of single-equation demand functions. The benefit of the systems-wide approach is the ability to capture the demand interrelationships among the fibers in a theoretically satisfactory way. Using data for various OECD countries for the periods 1974 and 1992, Clements and Lan (2001) estimated the (compensated) own-price elasticities to be –0.136 for cotton, –0.020 for wool, and –0.155 for chemical fibers. They found the world demand for each of the fibers is quite inelastic at the mill level. They also found the income elasticities of the mill demand for cotton to be 0.803, 0.532 for wool, and 1.304 for chemical fibers. The implication is that consumers consider wool and cotton to be necessities, whereas chemical or synthetic fibers are viewed as luxuries. Finally, they estimated the cross-price elasticities to be in the range of 0.001 and 0.006 between cotton and wool, in the range of 0.135 and 0.152 between cotton and synthetic fibers, and in the range of 0.003 and 0.014 between wool and synthetic fibers. Hence, all three fibers are viewed as weak substitutes at the mill level.

Except for the Clements and Lan (2001) study, little other research has focused on estimating the own-price, cross-price, and income elasticities for wool, either domestically or globally. This information is critical to ensure a proper picture of the demand situation for wool in the United States. Demand analyses involving wool, cotton, and synthetic fiber are necessary both at the mill level and at the retail level.

Wool Market Prices and Pricing

Wool and lamb meat are the key products of sheep production and by and large are complementary products. According to Babula (1996), increased domestic lamb quantities historically have led to increased wool quantity and lower domestic wool prices as more lambs are shorn. Likewise, increased domestic lamb prices historically have led to decreased quantities of and higher prices for U.S.-produced wool as fewer ewe lambs are slaughtered and shorn.

Price-Determining Factors

Wool market pricing is based on the physical characteristics of wool as determined by the objective measurements discussed earlier, including fiber diameter, variability of fiber diameter, clean wool yield, staple length, staple strength, quantity and type of vegetable matter, and number of black fibers present. The price for wool received by growers is also affected by the extent of grower preparation of wool for market and the method of sale. Wool market price quotes are on a clean wool basis, with fiber diameter, staple length, and staple strength being the primary factors affecting market price of clean wool. Other factors such as vegetable matter content and other contaminants can result in price discounts.

World wool market prices are primarily established by Australian market prices due to their dominance in world wool markets. Australia produces approximately 50 percent of the world's wool sold at auction. Although Australian Wool Exchange (AWEX) weekly wool market price quotes are based on AUS cents/kg for clean wool across micron grades, money transfers by wool buyers are usually based on U.S. dollars. The exchange rate of the U.S. and Australian dollars, therefore, can have a major impact on world market prices and sales. Wool prices in most countries are based on Australian export prices as adjusted for wool type, quality standards for preparation and marketing, adequate buyer network and wool supplies for a competitive market, and distance (freight costs) to processors and end users (Wilcox, 2007).

U.S. wools have historically sold at substantial discounts to Australian wools of comparable micron and quality for various reasons, including superior Australian marketing standards, uniformity of larger volumes of wool within grades and types, and the fact that Australia is closer to the major Asian processors of wool than the United States. The most recent complete annual data on comparative wool prices are for 2005 and indicate that U.S. wool prices range from a little over 60 percent to nearly 75 percent of imported Australian wool prices depending on the micron grade (Table 5-4).

In addition to those mentioned above, an additional factor contributing to the differences in U.S. and Australian wool prices is the difference in the marketing windows in the two countries. The largest volume wool sales in Australia are normally in the first 3 months of the year, declining in volume from mid-April through the end of formal sales in August. Prices tend to decline as volume declines and as buyers fill their orders. The peak sale months for wool in the United States are from mid-April through mid-June when prices normally decline in Australia and some buyers may have filled their orders. Market trends in 2007 were an exception to the normal market expectations. As the market year progressed in Australia, both number of sheep shorn and fleece weights clearly were lower than in previous years and the demand for quality wools was solid. As a result, market prices increased almost every week from early January through June 1, 2007. Prices in the U.S. wool market in mid-2007, with most wool sold from April through June, were estimated by U.S. markets to be 25 to 35 percent higher than the previous year, suggesting that U.S. market prices were no more than 10

TABLE 5-4 Average Prices (U.S.$/Kg) of U.S. Wool and Imported Australian Wool, Cleaned and Delivered to Charleston, SC, by Micron Grade, 2005

	Micron Grade				
Country	19	21	23	25	Total Clip Ave.
Australia	7.01	5.87	5.60	4.98	5.45
United States	5.03	4.32	3.88	3.11	3.75
U.S. as a % of Australian price	71.6	73.7	69.2	62.3	68.4

Source: Based on data from McDonnell (personal communication).

to 15 percent lower than Australian prices at that time. World wool market analysts predict that wool prices will likely remain at their current near record highs for the next 2 to 3 years due to reduced raw wool supplies, with the only concern being that continued price increases may encourage substitution of other lower-cost fibers for wool in apparel garments (WOOLNEWS.net, 2007).

The Effect of Market Preparation Method on Wool Prices

As mentioned earlier, U.S. wool producers have lagged those in other major producing countries in both preparing their wools for sale and in classing their wool in accordance with international standards. Given that Australian wool prices are usually higher than U.S. wool prices (for similar types) as shown by Hager (2003) and given that some Australian wool is quite similar to some U.S. wool, an important question for U.S. wool producers is whether these price differences are due to differences in the extent of preparation of wool for marketing. Kott (1997) contended that maximizing the returns to wool production requires producers not only to grow the wool but also to harvest and package the wool properly and then market it properly. If that is the case, then the question is how much of a premium, if any, is received for skirted and classed wools, as typically sold by Australian producers, as compared to OB wools as typically sold by U.S. producers.

Lupton et al. (1989) conducted some of the earliest research on the effects of skirting and classing on wool prices and concluded that skirting could be profitable when applied to fine-wool fleeces when prices are at high levels. They also concluded that the financial incentive to skirt wool fleeces is reduced as wool prices decrease, as skirting costs increase, and when wool is most coarse. Lupton et al. (1996) compared clean prices of skirted and classed wool to OB wool over a 4-year period ending in 1996 using Texas Agricultural Experiment Station sheep flocks in San Angelo, Texas. They found that skirted and classed wool prices were higher by 6.6 percent to 26.9 percent per year over OB wool (equivalent to $0.19 to $0.42 per kilogram). The potential to add value to wool by skirting and classing is attributed to the fact that less sorting of skirted and classed wool is required when the wool clip reaches the textile mills. The resulting labor cost savings then could be passed back to producers in the form of higher prices. Pfeiffer and Lupton (1999), however, found that skirted and classed wool may not produce more net income to producers than selling wool in OB form. Another factor to be considered is the way wool is presented to buyers. One experiment conducted by Lupton et al. (1993) presented buyers with subjective measurements for all wool lots. Objective measurements were available on only half the lots. The wool lots that were accompanied by objective measurements consistently received higher prices.

Hedonic Wool Price Analysis

Given the less than conclusive results of past research and the importance of the issue to the profitability of wool production, a hedonic price model was developed (Hager, 2003; Anderson et al., 2007) to test the hypothesis that skirting and classing wool generally produce higher prices compared to OB wool. The model measures the premiums/discounts among different levels of preparation and wool types, controlling for seasonality, year, region, average fiber diameter (AFD), and grease weight (GW, lot size). The data for the analysis came from a comprehensive survey sent to wool warehouses and pool sales across the United States.[2] The data included 8,589 observations on skirted and classed as well as OB wool sales over a 10-year period starting in January 1993 and ending in January 2002. Clean wool prices were gathered, noting region, season (month of year), year, wool preparation, wool type, AFD, and GW.

The United States was divided into three regions for the analysis: (1) Eastern, (2) Central, and (3) Western. The Eastern region included all states east of the Mississippi River. The Central region was separated from the Western region by a line that ran west of the Dakotas, Nebraska, Kansas, and New Mexico. The regions were chosen on the basis of demographic and market attributes. The Eastern market usually consists of smaller volumes of wool. Eastern producers have few market outlets except in niche areas. The wool produced in this region typically is variable in quality and style. In the Central region, generally more uniform wool in terms of quality, style, and quantity is produced. In this region, most producers raise sheep on privately owned land. Marketing outlets are well established in the Central region, and producers, warehouse operators, and buyers have well-established relationships.

Nearly 80 percent of the data observations were associated with the Central region, about 18 percent with the Western region, and roughly 3 percent with the Eastern region (Table 5-5). The data were separated into the three primary levels of preparation: (1) OB, (2) BOU, and (3) TSC. About 23 percent of the wool sold was identified as OB, 56 percent as BOU, and 21 percent as TSC. The data included sales of 17 different types of wool. The highest percentage of the wool sold (slightly more than 60 percent) was identified as in the Wool Breed, Main Line category. Nearly 9 percent of the observations were associated with the Tender category (fiber content not strong and easily broken) or the Short Line category (staple length shorter than three inches). Wool Breed Bellies and OB Wool Breeds each constituted about 7 percent of the wool sales. Bellies are inferior wool sheared from the belly of the sheep.

---

[2]The data for those groups surveyed without the resources to respond to the survey were collected from them by the research team.

TABLE 5-5 Number of Wool Lots Processed by Warehouses and Pools Surveyed by Region, Year, Level of Preparation, and Wool Types, January 1992 to January 2002

	Number of Lots Processed	%
Region		
Western	1,555	18.1
Central	6,762	78.3
Eastern	281	3.3
Year		
1990	1	0.0
1991	7	0.1
1992	5	0.1
1993	434	5.1
1994	426	5.0
1995	436	5.1
1996	408	4.8
1997	537	6.3
1998	657	7.7
1999	821	9.6
2000	1,630	19.0
2001	1,447	16.9
2002	1,768	20.6
Level of Preparation		
Original Bag (OB)	1,941	22.6
Bellies Out Untied (BOU)	4,812	56.0
Table Skirted Classed (TSC)	1,836	21.4
Wool Type		
Wool breed (wool type from BOU and TSC)		
Main line	5,305	61.8
Tender or short line	736	8.6
Bellies	606	37.1
Pieces	146	1.7
Stains	78	0.9
Locks	446	5.2
Clothing	77	0.9
Main line lamb	220	2.6
Meat breed		
Main line	133	1.6
Bellies	1	0.0
Wool types from OB		
Wool breeds	577	6.7
Meat breeds (white face)	100	1.2
Meat breeds (black face)	91	1.1
Hair or cross bred	42	0.5
Wool breed lamb	18	0.2
Meat breed lamb	3	0.0
Black	10	0.1

Source: Anderson et al. (2007). Copyright 2007 by TAMRC (Texas Agribusiness Market Research Center), Texas A&M University. Used with permission.

Some of the original 8,589 observations were eliminated from further consideration in the analysis, including those for operations outside the Western, Central, and Eastern regions. Also, those corresponding to the years 1990, 1991, and 1992 were eliminated since data for those years were not available consistently across all respondents. Missing observations pertaining to U.S. clean price, average fiber diameter, and grease weight (lot weight) were discarded as well. Thus, the number of useable observations for the analysis was 8,533.

The model postulates that premiums and discounts associated with the wool price received by growers are determined by the level of preparation method and the wool type, controlling for region, year, season, AFD, and GW (lot size).[3] Past research considered prices only to be a function of wool preparation. The hedonic wool price model accounts for the three most prevalent levels of preparation (OB, BOU, and TSC) and various wool types, including wool from wool breeds (main line, tender or short line, bellies, pieces, stains, locks, clothing, and main line lamb); wool from meat breeds (main line and bellies); and wool from OB (white-face and black-face meat breeds, hair or cross bred, wool breed lamb, meat breed lamb, and black). The model explains about 83 percent of the variation in U.S. wool prices. The detailed estimation results, including the estimated coefficients and their associated p-values, are provided in the Appendix (Table 5A-1). The following is a summary of the key results.

Seasonal Effects

The months of April, May, July, and August were not significantly different from the base month of September. The month corresponding to the highest U.S. clean prices was June, roughly 8 percent higher and statistically different than those of September (Figure 5-4). Wool prices received by producers also were higher in May and July relative to September but not significantly so. In accordance with prior expectations, wool prices received by producers from January to March as well as from October to December were significantly lower than those in September. The range of differences was from 5.9 percent lower in March to 17.4 percent lower in January. Unequivocally, seasonality in U.S. clean prices for wool is evident.

Yearly Effects

Consistent with prior expectations, U.S. clean wool prices were highest in 1995 and 1997 (Figure 5-5). Controlling for other factors, prices in 1995

[3]Details on the model specification and estimation results are provided in the appendix to this chapter.

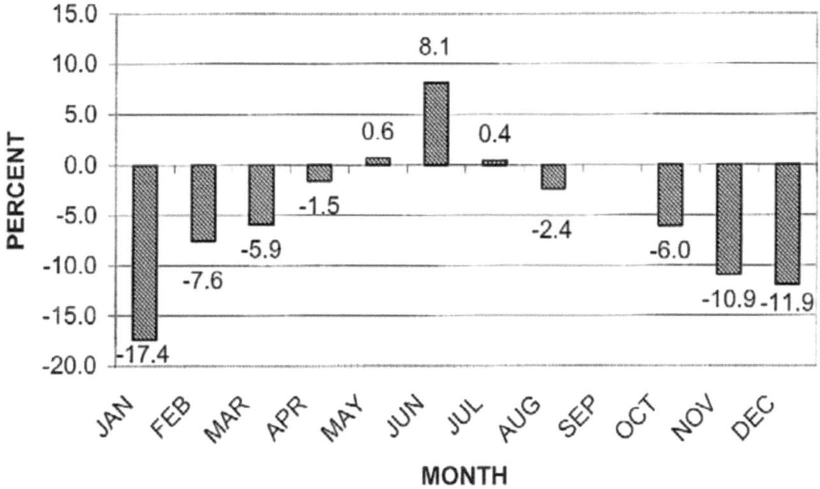

FIGURE 5-4 Percentage difference in U.S. clean wool price by month relative to the base month of September.
Source: Anderson et al. (2007). Copyright 2007 by TAMRC (Texas Agribusiness Market Research Center), Texas A&M University. Used with permission.

were significantly higher by 17.7 percent relative to the base year of 1997. Prices in all remaining years from 1993 to 2002 were significantly lower relative to the base year of 1997. Annual price differences ranged from 11.8 percent lower in 1996 to 52.2 percent lower in 2000.

Regional Effects

As expected, U.S. clean wool prices received by producers were discounted by 7.9 percent and 9.8 percent, respectively, in the Eastern and Western regions relative to those in the Central region (Figure 5-6). Clearly, regional price differences were evident.

Effects of Level of Preparation

In line with most prior research studies, prices of TSC wool were significantly higher than OB wool by slightly more than 8 percent (Figure 5-7). Although prices of BOU wool were higher by about 2 percent relative to OB wool, this difference was not statistically different from zero. Importantly, as the level of preparation of wool increases, U.S. clean wool prices increase.

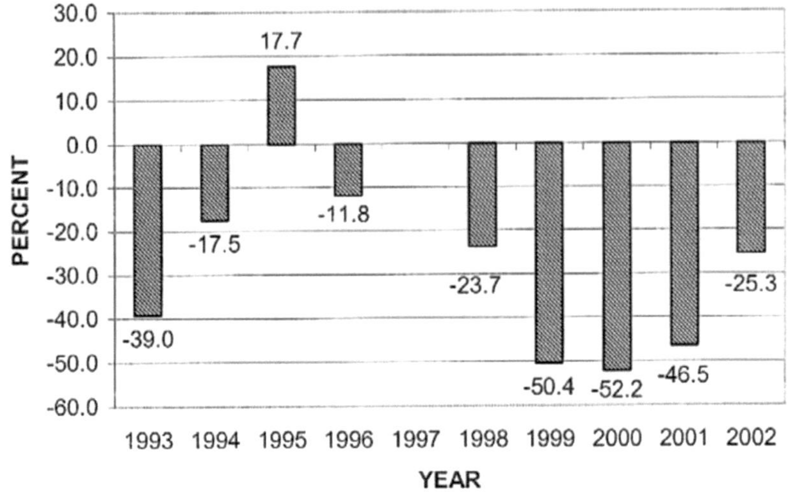

FIGURE 5-5 Percentage difference in U.S. clean wool price by year relative to the base year of 1997.
Source: Anderson et al. (2007). Copyright 2007 by TAMRC (Texas Agribusiness Market Research Center), Texas A&M University. Used with permission.

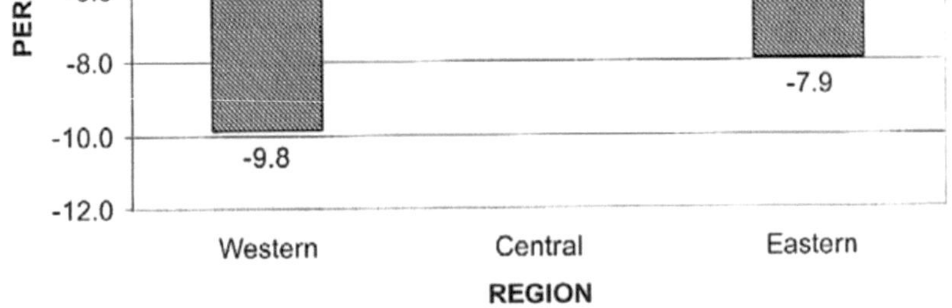

FIGURE 5-6 Percentage difference in U.S. clean wool prices by U.S. region relative to the base region of the Central United States.
Source: Anderson et al. (2007). Copyright 2007 by TAMRC (Texas Agribusiness Market Research Center), Texas A&M University. Used with permission.

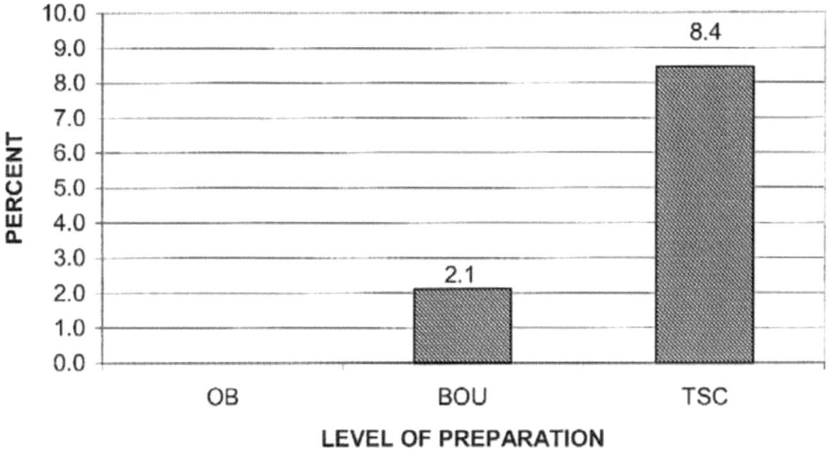

FIGURE 5-7 Percentage difference in U.S. clean wool price by level of preparation relative to the base preparation of original bag (OB).
Source: Anderson et al. (2007). Copyright 2007 by TAMRC (Texas Agribusiness Market Research Center), Texas A&M University. Used with permission.

Effects of Wool Type

As expected, U.S. clean prices of TSC and BOU Main Line wool were higher by 23.5 percent over the base category of OB wool breed (Figure 5-8). U.S. clean prices of TSC and BOU clothing and main line lamb also were higher by 22.0 percent and 15.4 percent, respectively, over OB wool breed. Wool prices of TSC and BOU bellies, pieces, stains, and locks, all lower-quality types, were discounted from slightly more than 25 percent for bellies to slightly more than 6 percent for locks relative to prices of OB wool breeds, the reference category.

Significant differences in wool types from OB were evident as well. Relative to prices associated with the base wool type (wool breeds from OB), prices of other wool types from OB were significantly lower, ranging from roughly 16 percent lower for wool breed lamb to nearly 70 percent lower for black wool (Figure 5-9). Prices of OB wool breed and those from meat breeds, either main line or bellies, were not statistically different. Clearly, U.S. clean prices differ significantly among wool types with relatively large premiums and discounts among wool types relative to OB wool breed types.

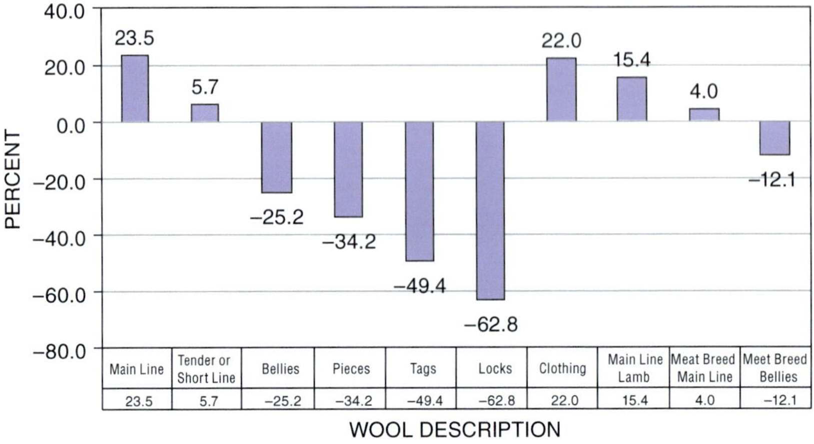

FIGURE 5-8 Percentage difference in U.S. clean wool price by wool types relative to the base OB-wool breed.
Source: Anderson et al. (2007). Copyright 2007 by TAMRC (Texas Agribusiness Market Research Center), Texas A&M University. Used with permission.

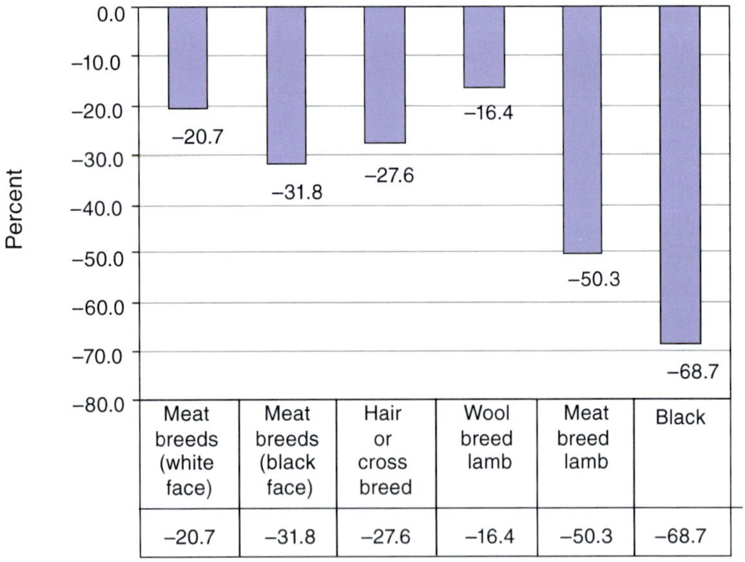

FIGURE 5-9 Percentage difference in U.S. clean wool price by wool breed types relative to the base OB-wool breed.
Source: Anderson et al. (2007). Copyright 2007 by TAMRC (Texas Agribusiness Market Research Center), Texas A&M University. Used with permission.

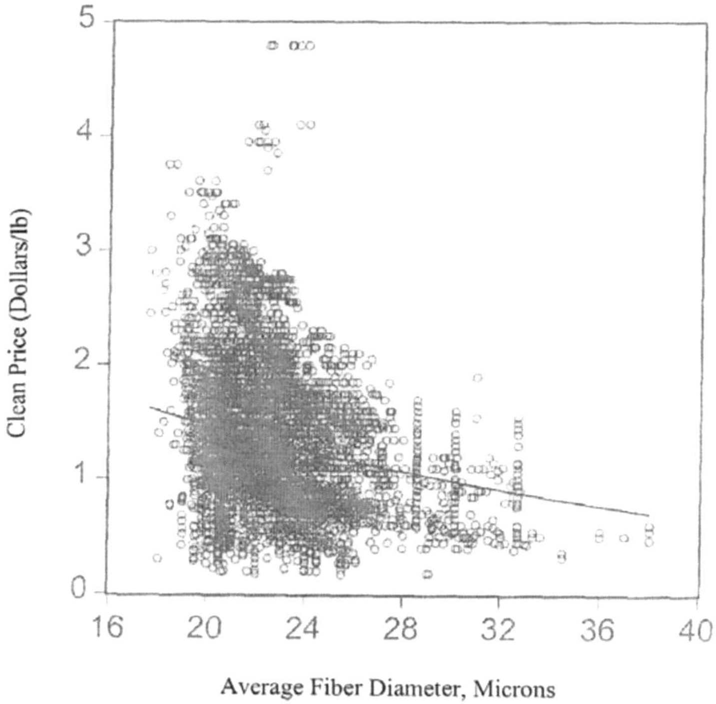

FIGURE 5-10 Relationship between U.S. clean wool price and average fiber diameter based on the sample of 8,533 observations (1 lb = 0.4536 kg).
Source: Anderson et al. (2007). Copyright 2007 by TAMRC (Texas Agribusiness Market Research Center), Texas A&M University. Used with permission.

Effects of Average Fiber Diameter

As hypothesized, U.S. clean prices and AFD were negatively related (Figure 5-10). The elasticity of clean price to AFD was estimated to be –1.416, meaning that, controlling for all other influences on clean prices, a 10 percent change in AFD (e.g., a change from the sample mean of 22 microns to either 20 microns or 24 microns) leads to nearly a 14.2 percent change in price in the opposite direction (e.g., a change from the sample mean of \$2.98/kg to either \$2.56/kg or \$3.40/kg). Thus, U.S. clean wool prices are highly sensitive to changes in AFD.

Lot Size Effects

Again, as hypothesized, clean wool price and lot size, as measured by GW, were positively related (Figure 5-11). The elasticity of clean price to

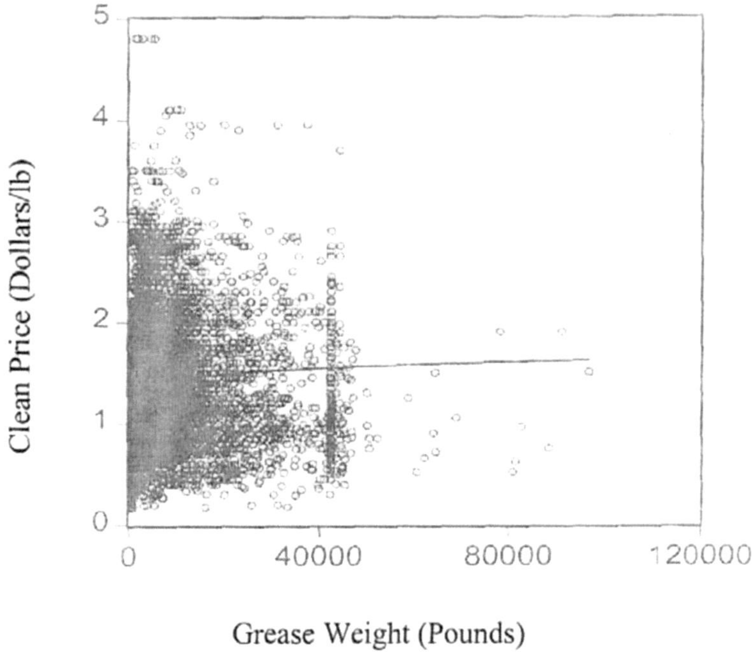

FIGURE 5-11 Relationship between lot size as measured by grease weight and U.S. clean wool price based on the sample of 8,533 observations (1 lb = 0.4536 kg). Source: Anderson et al. (2007). Copyright 2007 by TAMRC (Texas Agribusiness Market Research Center), Texas A&M University. Used with permission.

GW was estimated to be 0.0162. Hence, a 10 percent change in lot size (e.g., a change from the sample mean of 3,851 kg to either 3,465 kg or 4,237 kg) leads to a 0.16 percent change in clean wool price. Although, this elasticity is statistically significant, practically speaking, U.S. clean wool prices were not heavily influenced by lot size.

These results provide a basis for determining premiums or discounts relative to the current practice of marketing OB wool. Clearly, producers who do minimal preparation of their wool prior to sale are losing substantial premiums paid for wool that has been more prepared for sale. Certainly, the type of wool sold also makes a difference in the price received for wool by producers. The TSC and BOU clothing and main line lamb receive substantial premiums over OB wool breed.

Unfortunately, the U.S. wool industry has no consistent means of recording lot descriptions and sales to provide a consistent dataset for calculating premiums and discounts available for wool of differing characteristics. Some warehouses and pools only keep subjective descriptions of wool characteristics while others maintain records on objective measurements of wool. Additionally, historical records are not always kept, or, if kept, they are not maintained efficiently with a consistent database system across warehouses and pools. Moreover, disclosure concerns keep some warehouses and pools from being willing to share information pertaining to sales. Additional information on factors for which records are generally not maintained, such as vegetable matter content, staple length, staple strength, and fiber color, are likely also to be important in determining the premiums paid and discounts deducted from U.S. wool prices.

WOOL POLICIES AND REGULATIONS

The first effort by the government to support the domestic wool industry was included in the provisions of the Buy American Act of 1933. That act required manufacturers of worsted goods for the U.S. Army to use domestic wools if available in grades needed and not unreasonably higher in price than foreign wools (Hyson, 1947). In 1940, as World War II loomed closer, the army began placing large orders for military fabrics containing wool. When domestic supplies proved insufficient to meet military needs in late 1940, the government began allowing the use of some imported wool.

The principal government policy in support of the U.S. wool industry, however, came as the result of intense lobbying by wool producers, marketers, and manufacturers in the 1950s, which spurred the U.S. Congress to commission various studies of the U.S. wool industry. A primary conclusion of the studies was that imports of low-cost wool yarns, fabrics, and apparel goods were causing significant damage to domestic wool processing and manufacturing industries (National Wool Growers Association, 1965). The studies precipitated the passing of the National Wool Act in 1954 (P.L. 102-130), which was signed into law by President Eisenhower. The rationale for the support of wool production rested on the premise that wool was "an essential and strategic commodity" that was not produced in sufficient quantities to meet domestic needs.

The National Wool Act increased tariffs on all wool and woolen products from \$0.116/kg to \$0.19/kg clean wool and established an incentive payment program for growers to be paid out of wool tariff revenues (USDA, 1999). The incentive portion included the establishment of a target price for raw wool and a payment to producers on the basis of the percentage difference between the national average market price and the target price. As a result, producers of better wool clips that earned higher market prices than inferior clips received larger incentive payments.

In 1955, Congress authorized a "self-help" program to permit growers to check-off funds from their incentive payments to be used to improve the production of lamb and wool and promote improved marketing of lamb and wool products to consumers. The authorization required that a sheep producer referendum be conducted that had to be approved by at least two-thirds of the sheep producers who owned at least two-thirds of the sheep in the United States. The referendum passed in 1956, resulting in the development of the American Sheep Producers Council with funding and management oversight through the USDA (National Wool Growers Association, 1965).

Despite the incentive payments to growers, U.S. sheep numbers continued to decline from the highs achieved in the 1940s, as noted in Chapter 1. In the early 1990s, President Clinton requested that Congress repeal the National Wool Act. In response, Congress approved a phaseout of the program over a 2-year period beginning in 1992, ending 42 years of federal support to the U.S. wool industry (USDA, 1999). Wool production declined markedly after incentive payments were terminated in 1994. The wool share of total revenues from sheep production fell to only about 10 percent after the termination of the Wool Act compared to 20–25 percent in years when the Wool Act was in force. Over the life of the Wool Act, the incentive payments to growers were less than imported wool tariff revenues. Because the tariffs on imported wool were not repealed when the Wool Act was terminated in 1994, the wool tariff revenues were diverted to the U.S. Treasury general fund.

To investigate the market implications of the Wool Act, Whipple and Menkhaus (1990) developed an econometric simulation model that accounted for the jointness of lamb meat and wool production. They found that between 1960 and 1985, the wool incentive program had a number of positive effects on wool and lamb markets over what would have been the case in the absence of the program, including (1) 26 percent more sheep in the national breeding flock, (2) 29 percent more production of wool, (3) 17 percent more lamb production, (4) 30 percent more producer revenue from lamb and wool, (5) 23 percent lower lamb imports, (6) 6 percent lower wool imports, (7) 14 percent more lamb consumption, and (8) slightly more (about 1 percent) wool consumption.

On the negative side for producers but on the positive side for consumers, Whipple and Menkhaus (1990) found that that as a result of the program, the retail price of lamb was 7 percent lower and the wholesale price of wool 3 percent lower than would have been the case without the program. The lower prices led to several other effects on the demand side of the wool market, including (1) 8 percent lower consumer lamb expenditures, (2) 9 percent lower consumer wool expenditures, (3) nearly 10 percent lower revenue accruing to lamb exporters, and (4) 5 percent lower revenue accruing to wool exporters. Importantly, government costs associated with the Wool

Act were in excess of $100 million. The Whipple and Menkhaus (1990) analysis concluded that the Wool Act had positive net benefits for lamb and wool consumers as well as sheep and lamb producers. Interestingly, they also concluded that the gains to lamb and wool consumers exceeded the gains to sheep and lamb producers as a result of the program. Lamb and wool exporters, on the other hand, suffered losses. Because of the rather sizeable increase in government costs associated with the incentive program, they estimated that the net national welfare effect of the Wool Act was a loss of between $25.7 million and $43.4 million.

In 1974, the United States entered into the MFA with other developed textile-importing countries and developing textile-exporting countries. Beginning in the 1930s, the United States and other developed textile-importing countries established a growing number of quotas and other restrictions on textile imports from Japan, Hong Kong, Pakistan, India, and other developing countries. The MFA pulled together all these restrictions into one multilateral agreement and established import quotas for specific countries and products when imports threatened to disrupt domestic fiber markets (MacDonald and Vollrath, 2005). Under the MFA, quotas were set to increase annually at a target of 6 percent, although lower rates were often negotiated with major textile exporters. As discussed earlier, the MFA was replaced by the WTO ATC in 1995, which established a schedule for eliminating the MFA quotas and lowered textile and clothing tariffs.

According to MacDonald and Vollrath (2005), the MFA increased textile and clothing production, reduced prices for clothing, and lowered textile consumption in the United States, the European Union (EU), and other developed textile-importing countries. They indicate that the MFA quotas added 5 to 10 percent to clothing prices paid by U.S. consumers. The phased elimination of the MFA quotas under the ATC has had the opposite effect, allowing greater imports of wool and other fiber textile imports and leading to lower prices and production along the fiber and textile supply chains in the United States and other textile-importing countries. China and other textile-exporting countries have benefited from the elimination of the MFA with increasing textile production and prices.

Established by the U.S. Congress in 2000, the American Wool Trust has provided $9 million in funding through 2006 for activities aimed at increasing the competitiveness of American wool. The trust is administered by the ASI Wool Council, which includes representatives from the production, research, marketing, processing, manufacturing, and merchandising sectors of the wool industry. The trust supports producer communications, raw wool quality improvement, introduction of new raw wool measurement technologies to improve quality assurance for U.S. wools, product development and new uses for wool, research to develop improved wool processing technologies, improved market and price information, and improving international marketing opportunities for U.S. wool. It also works closely on needs and new products for the U.S. military, which is the largest domestic user of American wool.

The Farm Security and Rural Investment Act of 2002 reinstated wool price supports through marketing assistance loans (MALs) and loan deficiency payments (LDPs) for the 2002 to 2007 crop years (USDA, 2004). Commodity loan programs in general allow producers to receive a loan from the government at a commodity-specific loan rate per unit of production by pledging production as loan collateral. Wool and mohair nonrecourse MALs are 9-month loans that provide production financing to wool producers and facilitate the orderly marketing of wool throughout the year. The loan rates for 2002–2007 are $2.20/kg for graded wool and $0.88/kg for nongraded wool.

Instead of selling the wool and mohair immediately after shearing, a nonrecourse loan allows a producer to store the production, pledging the wool as collateral. The producer can then redeem the loan and sell the commodity when market conditions are more favorable in order to secure a higher market price. If the producer is unable to repay the loan, he or she can deliver the quantity of wool or mohair pledged as collateral to the USDA Commodity Credit Corporation (CCC) as full payment for the loan at maturity regardless of the market value of the wool at the time. In this way, marketing loans are repaid at less than principal plus accrued interest and other charges, with repayment of some portion of the relevant interest and principal being waived (USDA, 2004). Producers may also purchase commodity certificates and exchange a commodity certificate with outstanding loan collateral in repayment of marketing assistance loans. Commodity certificates are negotiable certificates that the CCC can exchange for a commodity owned or controlled by the CCC.

Instead of obtaining a loan on graded or ungraded wool, producers may request LDPs that are payable at the loan rate that would have been received for the lot of wool, less the announced repayment amount for wool of that quality (USDA, 2004). Loan deficiency payment rates for nongraded wool have ranged from $0.29/kg to $0.41/kg in 2007, with only wool finer than 20 microns eligible for the higher-graded wool rate. Unshorn pelts also are also eligible to receive a loan deficiency payment of $0.88/kg.

The United States is not the only wool-producing country to provide price support for producers. Price support programs in Australia, New Zealand, and South Africa set wool prices well above market levels, leading to growing world stockpiles in the 1980s and early 1990s. During the 1990s, however, these stockpiles were gradually placed on the world market. Along with the collapse of the former Soviet Union in the early 1990s, which contributed to a notable decline in wool demand, and the repeal of the U.S. National Wool Act, the release particularly of Australian wool from

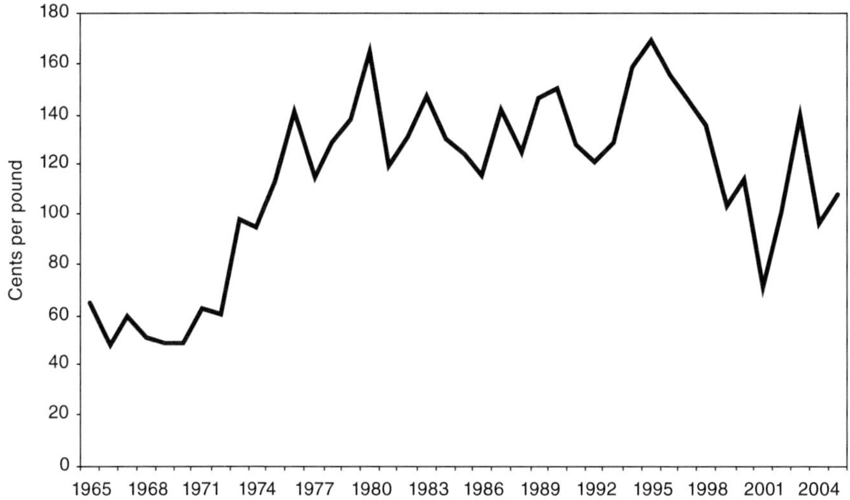

FIGURE 5-12 U.S. farm price of wool, 1965–2005.
Note: 1 lb = 0.4536 kg.
Source: Based on data in USDA (2006).

government stockpiles by the Australian Wool Council (AWC) depressed world wool prices from about 1995 through 2000. With the liquidation of the Australian stockpiles in August 2001, wool prices began to rebound (Figure 5-12).

MAJOR ACCOMPLISHMENTS, OPPORTUNITIES, AND CHALLENGES OF THE U.S. WOOL INDUSTRY

With the major transitions that have occurred in the sheep industry over time, the reductions in domestic wool manufacturing, changing consumer tastes and preferences, and the shift of textile manufacturing to China, India, and the developing economies in Eastern Europe, the U.S. wool industry has been under siege and in an almost constant state of adjustment. The industry has made some progress, however, in responding to these pressures and faces a number of opportunities. The industry still faces many challenges as noted by several industry representatives in response to a survey requesting their input on the key opportunities and challenges facing the U.S. wool industry.

Major Accomplishments of the U.S. Wool Industry

- *Increases in wool preparation to international standards.* The Anderson et al. (2007) hedonic wool price analysis reported that from 1992 to 2002, an average of 26 percent of the wool was marketed as OB, 56 percent as BOU, and 21 percent as TSC. The 2007 survey of major wool warehouses as presented in Table 5-1 reported that most warehouses received < 15 percent of their wool from growers as OB, 40 to 75 percent as BOU, and 30 to 50 percent as TSC. Several of the warehouses also reported that a significant portion of the wool in the BOU category had some of the tags, skirts, and vegetable matter removed even though it was not TSC. These data clearly indicate grower response to market premiums for improved preparation and from producer education programs sponsored by the American Wool Trust.
- *New research and product developments.* Industry support for collaborative research with funding from the American Wool Trust, U.S. military research laboratories, other federal research grants, and private industry have resulted in fabrics and garments that are machine washable, more breathable, and less prickly when worn against the skin; shrinkproof; and flame resistant. Research has also demonstrated the value of wool blends with cotton and certain synthetic fabrics (ASI, 2007c).

Major Opportunities and Challenges Facing the U.S. Wool Industry

To get a sense of the opportunities and challenges facing the wool industry from an industry perspective, a questionnaire was sent to several key leaders in the wool industry, including market analysts, wool warehouse and wool pool managers, major wool buyers for domestic and international markets, key wool research scientists and extension specialists, and key producers. While the sample was by no means representative of all participants in the wool industry at all levels in terms of demographics or other characteristics, those who responded to the survey are among the most knowledgeable in the wool industry about how the industry functions and where it is headed.[4] According to the survey respondents, the following are among the most important opportunities currently facing the U.S. wool industry:

[4]The following 10 people responded to a questionnaire on challenges and opportunities for the future of the wool industry: (1) Ronald L. Cole, USDA AMS Livestock & Grain Market News, Greely, CO; (2) Mike Corn, manager and co-owner, Roswell Wool, Roswell, NM; (3) Glen Fisher, wool warehouseman (Ret.) and industry leader, Sonora, TX; (4) Dr. Rodney Kott, extension sheep and wool specialist, Montana State University, Bozeman, MT; (5) Dr. Chris Lupton, wool research scientist, San Angelo Research and Extension Center, Texas A&M University; (6) Tom McDonnel, McDonnel & Associates, Douglas, WY; (7) Ron Pope, manager, PMCI Wool Marketing, Mertzon, TX; (8) Larry Prager, Manager, Center of the Nation Wool, Belle Fouche, SD; (9) Dr. Robert Stobart, sheep and wool scientist, University of Wyoming,

- *International demand growth.* Growth in the wool textile industries in China and other developing countries is also helping spur growth in income and purchasing power in these countries and a growing internal demand for wool apparel goods. International market development activities could help widen the market for U.S. wool textiles in these growing markets. As incomes increase in these countries, the demand for apparel and other wool textiles could well outstrip their ability to both export and fill their growing internal demand. Market development activities could effectively introduce a broader array of consumers in those markets to U.S. wool textiles through the efforts of the American Wool Trust with funding through the USDA Foreign Agriculture Service, along with the efforts of other groups to attract additional foreign buyers and provide technical assistance and training to U.S. wool producers, marketing organizations, and textile manufacturers.
- *Grower price premiums from improved wool clip preparation.* Although the percentage of U.S. wool that is prepared for market according to international market standards has increased, a tremendous opportunity exists for producers to significantly enhance their returns to wool production by dedicating more time and effort to further preparing their wool. Research presented here suggests that producers can receive an 8 percent or higher price premium by table skirting and classing their wool, rather than marketing it as OB wool. Some reputation clips that have been using TSC for ≥ 5 years are receiving 25 percent or higher premiums for their wool.
- *New research developments.* A number of new research developments provide some promise of greater competitiveness and profitability in the wool industry, including the Kroy superwash process and enzyme research, which has produced machine-washable wool fabrics that are shrinkproof and more user friendly and can encapsulate wool with resins to develop breathable, waterproof, stain-repellant, machine-washable garments for the U.S. military. Other promising research results include refined wool-carded battings for use as sound absorption in automobiles and new wool products for the U.S. military in collaboration with U.S. military research laboratories.
- *Retail wool advertising.* Important opportunities may exist for additional advertising of wool at the retail level that emphasizes that wool is a natural product and renewable resource from sheep that can be used to enhance the environment through targeted or prescribed grazing practices to control invasive nonnative plant species. Other strengths of wool fabrics, such as their flame resistance and their use in blends to enhance synthetic fiber products, could be effective as well.
- *Niche market growth.* Although not potentially as great as emerging

---

Laramie, WY; and (10) Don Van Nostran, general manager, Mid States Wool Growers Cooperative, Canal Winchester, OH.

ethnic lamb markets, specialty wools for hand spinners, yarn for weavers and knitters, and other wools such as naturally colored wools have a small but growing market. Emphasis in this market is on natural products with increasing interest in organically grown wools. Wool may be sold as individual fleeces, yarn, fabric, or finished products. Mini-mills, ranging in one-time capacity of 20 to 100 tonnes clean wool, with products from yarn to fabric or finished products such as sweaters, blankets, and other apparel goods, represent a growing new market for wool (see Chapter 7).

The industry also faces many challenges for the future, among which are the following suggested by the survey respondents:

- *Re-invigorating the industry.* The decline in sheep numbers over the last several decades has been accompanied by reductions in wool marketing entities and related infrastructure. The result has been fragmented selling systems, fewer domestic wool buyers, distance-to-market challenges for producers, reduced numbers of buyers, and greater concentration of the wool processing industry. If the industry is to avoid further downsizing and to achieve some growth, either new market opportunities will need to be found or effective means of enticing current consumers to purchase more lamb and wool (or some of both) will be necessary.
- *Easing the shortage of qualified sheep shearers and wool classers.* A worldwide shortage of qualified sheep shearers is becoming a limiting factor for many producers wishing to implement or expand their sheep enterprises. Proper shearing and wool classing by trained professionals is critical to achieving maximum value from a wool fleece. The sharp decline in U.S. sheep numbers, however, has also drastically reduced the number of shearers and discouraged many from learning the trade. The industry has compensated over the years by bringing in qualified shearers from Australia and elsewhere in their off-season to harvest the wool from U.S. sheep. However, this practice has become less common in recent years because of a growing shortage of shearers in those countries and the difficulty of obtaining the necessary work visas in a timely manner for foreign sheep shearers in the post-9/11 era. New requirements and delays in obtaining work visas for sheep shearers and H2A visas for sheep herders are a major challenge for producers and shearing crew operators. Australia and New Zealand have developed coordinated training and certification programs for shearers to improve or maintain the quality and value of the wool fiber harvested from their sheep. For example, AWI is spending about $13 million over 3 years to improve the training of shearers and develop better shearing technology in a bid to stop a massive shearer shortage in that country. While various shearer training programs are also available in this country, a major challenge is enticing people to become wool harvesting professionals through

enhanced promotion and training of sheep shearing and wool classing as occupations with financial support to trainees and producer hosts of training programs.

• *Encouraging growers to prepare their wool to international market standards.* The income earned from wool by sheep producers currently ranges from < 5 percent to 30 percent of their gross income from sheep production. Although many progressive producers are committed to producing a quality wool clip, most generally view wool as a byproduct and are more committed to increasing lamb production since this is their major income source. Increasingly in the future, competing in international markets and obtaining competitive prices for wool will require U.S. growers to produce, harvest, prepare, and market their wool to international standards. Encouraging growers in this direction is one of the primary educational challenges for the industry leadership and organizations. Effective educational programs would emphasize the price determinant factors most critical to wool clip value and the ASI Code of Practices for Wool Clip Preparation.

• *Reducing the contamination from hair sheep breeds.* As some growers turn from wool breeds to hair breeds, a continuing challenge for the wool clip is contamination with hair, kemp, and colored fibers in the process of grading up from traditional breeds to hair sheep breeds through mating hair breed males to traditional wool breed females, which normally requires three to four generations (e.g., ½, ¾, etc.).

• *Increasing support for wool research and development.* As national sheep inventories have declined, research support for the industry has declined and focused increasingly on lamb rather than wool production and marketing. Additional research support is critically needed in a number of areas in the wool industry, including (1) genetic improvement of the range maternal breeds for increased production of both wool and lamb; (2) improvements in marketing standards to enhance the competitiveness of U.S. wools with those of Australia and other major wool producing countries; (3) new wool product development to meet changing consumer needs and growing challenges from competing fibers; (4) collaboration with military research and product development programs; (5) the use of bioclip chemical defleecing as an alternative to shearing in the United States; and (6) research focusing on own-price, cross-price, and income elasticities for wool both domestically and internationally to support pricing, promotion, policy, and other strategic decision making in the industry. Demand analyses involving wool, cotton, and synthetic fibers are necessary at both the mill and retail levels.

Survey respondents also suggested that several other issues will continue to challenge the future growth and development of the U.S. wool industry, including changing practices and regulations relative to public land sheep grazing permits, the impact of increasing private land market prices on sheep production, and the decline in support for sheep research, education, and extension programs at the land-grant universities and federal agencies.

REFERENCES

Anderson, D. P., O. Capps, Jr., E. E. Davis, and S. D. Hager. 2007. Wool Price Differences by Preparation in the United States. Consumer and Product Research Report No. CP-02-07, Texas Agribusiness Market Research Center, Texas A&M University, College Station. June.

ASI (American Sheep Industry Association). 2002. SID Sheep Production Handbook: Wool Chapter. Englewood, CO: American Sheep Industry Association.

ASI. 2007a. Wool Marketing. Online at: http://www.sheepusa.org/index.phtml?page=site/text&nav_id=b5cd92c158e527a90be72c1ce8be84a2. Accessed August 23, 2007.

ASI. 2007b. China dominates U.S. wool export market. December 14 edition, weekly newsletter. Englewood, CO. Online at: http://www.sheepindustrynews.org.

ASI. 2007c. American Wool Trust Annual Report to Congress. Englewood, CO: American Sheep Industry Association.

ATMI (American Textile Manufacturers Institute). 2003. The China threat to world textile and apparel trade. Washington, DC. Online at: http://www.ncto.org/textilecrisis/china.pdf. Accessed August 23, 2007.

Babula, R. A. 1996. An empirical examination of U.S. lamb-related import and domestic market relationships near the farmgate. J. Int. Food Agr. Marketing 8(2):65–82.

Capps, Jr., O. and G. W. Williams. 2006. The Economic Effectiveness of the Cotton Checkoff Program. Commodity Market Research Report No. CM-02-06, Texas Agribusiness Market Research Center, Texas A&M University, College Station. November.

Capps, Jr., O., D. A. Bessler, G. C. Davis, J. P. Nichols, C. G. Anderson, and E. G. Smith. 1997. Economic Evaluation of the Cotton Checkoff Program. Department Technical Report 97-2, Department of Agricultural Economics, Texas A&M University, College Station.

Clements, K. W., and Y. Lan. 2001. World fibers demand. J. Agric. Appl. Econ. 33:1–23.

Dickerson, K. 1999. Textiles and Apparel in the Global Economy, Third Edition. Englewood Cliffs, NJ: Prentice-Hall.

Donald, J. R., F. Lowenstein, and M. S. Simon. 1963. The Demand for Textiles in the United States. Technical Bulletin No. 1301, Economic Research Service, U.S. Department of Agriculture, Washington, DC.

Dudley, G. E. 1974. U.S. Textiles Fiber Demand: Price Elasticities in Major End-user Markets. Technical Bulletin No. 1500, Economic Research Service, U.S. Department of Agriculture, Washington, DC.

Hager, S. D. 2003. Determining price differences among different classes of wool from the United States and Australia. MS Thesis. Department of Agricultural Economics, Texas A&M University, College Station. December.

Hyson, C. D. 1947. Maladjustments in the wool industry and need for new policy. J. Farm Econ. 29:425–456.

Kott, R. 1997. Wool: What is it worth? Sheep Ind. J. 1:15.

Lowenstein, F. 1952. Factors affecting the domestic mill consumption of cotton. Agric. Econ. Rev. 4:44–51.

Lupton, C. J., F. A. Pfeiffer, and N. E. Blakeman. 1989. Optimizing the value of grease wool through preparation and marketing. SID Res. J. 5:1–20.

Lupton, C. J., F. A. Pfeiffer, N. E. Blakeman, D. N. Ueckert, and J. E. Huston. 1992. Effects of skirting on yield, fineness, and value of wool from fine-wool range ewes. J. Anim. Sci. 70:2657–3664.

Lupton, C. J., F. A. Pfeiffer, and N. E. Blakeman. 1993. Economic impact of pre-sale fiber measurements on prices paid for wool. Sheep Goat Res. J. 9:35–37.

Lupton, C. J., F. A. Pfeiffer, and S. Byrns. 1996. Adding value to wool clips by fleece skirting and classing. Research Reports: Sheep and Goat, Wool and Mohair 57:51–52.

MacDonald, S., and T. Vollrath. 2005. The Forces Shaping World Cotton Consumption after the Multifiber Arrangement. Report No. CWS-05c-01. Electronic Outlook Report. Economic Research Service. U.S. Department of Agriculture. Washington, DC. April. Online at: http://www.ers.usda.gov/publications/cws/apr05/cws05c01/cws05c01.pdf. Accessed August 22, 2007.

Murray, B. C., R. H. Beach, W. J. White, C. Viator, N. Piggott, and M. Wohlgenant. 2001. An Economic Analysis of the Cotton Research and Promotion Program. Project No. 8024. Research Triangle Institute Intenational, Research Triangle Park, NC.

National Wool Growers Association. 1965. Men, Sheep and 100 Years. The National Wool Grower: Special Publication. Salt Lake City, UT.

NCTO (National Council of Textile Organizations). 2007. Trade and jobs. Online at: http://www.ncto.org/tradejobs/. Accessed August 23, 2007.

Pan, S., and S. Mohanty. 2005. Technical documentation of the world fiber model. Cotton Economics Research Institute, Department of Agricultural and Applied Economics. Texas Tech University. Lubbock.

Pfeiffer, F. A., and C. J. Lupton. 1999. Results of skirting and classing on the value of wool from fine-wool range ewes. J. Anim. Sci. 77(Suppl. 1):245 (Abstr.).

Shui, S., J. C. Beghin, and M. Wohlgenant. 1993. The impact of technical change, scale effects, and forward ordering on U.S. fiber demands. Am. J. Agric. Econ. 75:632–641.

Stennis, E. A., M. Pinar, and A. J. Allen. 1983. The futures market and price discovery in the textile industry. Am. J. Agric. Econ. 65:308–310.

USDA (U.S. Department of Agriculture). 1999. Economic Impact of the Elimination of the National Wool Act. Economic Research Service. Report prepared for the U.S. House of Representatives Committee on Appropriations, as requested in House report 105–588, page 16 (1998). Online at: http://www.ers.usda.gov/Briefing/Sheep/WoolActStudy.pdf. Accessed November 10, 2007.

USDA. 2004. Wool and Mohair Marketing Assistance Loan and Loan Deficiency Payment Program. Program Fact Sheets. Farm Service Agency. Washington, DC., December. Online at: http://www.fsa.usda.gov/FSA/webapp?area=home&subject=prsu&topic=col-nl-wm. Accessed May 5, 2008.

USDA. 2006. Cotton and Wool Situation and Outlook Yearbook. November. Economic Research Service. Washington, DC.

USDA. 2007. Quick Stats. National Agricultural Statistics Service. Online at: http://www.nass.usda.gov/Data_and_Statistics/Quick_Stats/index.asp. Accessed August 8, 2007.

USDC (U.S. Department of Commerce). 2007. Adjustment of import limits for certain cotton, wool, man-made fiber, silk blend and other vegetable fiber textiles and textile products produced or manufactured in the People's Republic of China. Committee for the Implementation of Textile Agreements, Office of Textiles and Apparel, Washington, DC, August. Online at: http://www.otexa.ita.doc.gov/fr2006/chin13(07-07).htm. Accessed May 5, 2008.

Whipple, G. D., and D. J. Menkhaus. 1990. Welfare implications of the Wool Act. Western J. Agric. Econ. 15:33–44.

Wilcox, C. 2007. Woolmark, Inc. report on the economic status of wool production, marketing, manufacturing and consumer demand. Presented at the ASI annual meeting in San Antonio, TX, January.

WOOLNEWS.net. 2007. Monthly Newsletter. January through June. Online at: http://www.woolnews.net.

APPENDIX

The Hedonic Wool Price Model

The statistical model employed in the analysis is given by the following equation:

$$\begin{aligned}
&\log \text{U.S. Clean Price}_{it} = \alpha_0 + \alpha_1\text{January} + \alpha_2\text{February} + \alpha_3\text{March} \\
&+ \alpha_4\text{April} + \alpha_5\text{May} + \alpha_6\text{June} + \alpha_7\text{July} + \alpha_8\text{August} + \alpha_9\text{October} \\
&+ \alpha_{10}\text{November} + \alpha_{11}\text{December} + \alpha_{12}\text{YR1993} + \alpha_{13}\text{YR1994} \\
&+ \alpha_{14}\text{YR1995} + \alpha_{15}\text{YR1996} + \alpha_{16}\text{YR1998} + \alpha_{17}\text{YR1999} + \alpha_{18}\text{YR2000} \\
&+ \alpha_{19}\text{YR2001} + \alpha_{20}\text{YR2002} + \alpha_{21}\text{WESTERN} + \alpha_{22}\text{EASTERN} \\
&+ \alpha_{23}\log \text{AFD}_{it} + \alpha_{24}\log \text{GW}_{it} + \alpha_{25}\text{OU} + \alpha_{26}\text{TSC} + \alpha_{27}\text{WT MAINLINE} \\
&+ \alpha_{28}\text{WTTENDERORSHORT LINE} + \alpha_{29}\text{WTBELLIES} + \alpha_{30}\text{WTPIECES} \\
&+ \alpha_{31}\text{WTSTAINS} + \alpha_{32}\text{WTLOCKS} + \alpha_{33}\text{WTCLOTHING} \\
&+ \alpha_{34}\text{WTMAINLINELAMB} + \alpha_{35}\text{WTMBMAINLINE} \\
&+ \alpha_{36}\text{WTMBBELLIES} + \alpha_{37}\text{WTOBMBWHITEFACE} \\
&+ \alpha_{38}\text{WTOBMBBLACK FACE} + \alpha_{39}\text{WTOBHAIRORCROSS BRED} \\
&+ \alpha_{40}\text{WTOBWOOLBREEDLAMB} + \alpha_{41}\text{WTOBMEATBREEDLAMB} \\
&+ \alpha_{42}\ \text{WTOBBLACK} + \varepsilon_i
\end{aligned}$$

where the variables in the model include the following:

January, February, etc.	Seasonal dummy or indicator variables
YRyyyy	Dummy variables corresponding to year yyyy
WESTERN, EASTERN	Regional indicator variables
BOU, TSC	Level of preparation indicator variables
AFD	Average fiber diameter (microns)
GW	Grease weight of the lot (pounds)

Wool-type indicator variables (wool breed (types from BOU and TSC), namely Main Line; Tender or Short Line; Bellies; Pieces; Stains; Locks; Clothing; and Main Line Lamb; Meat Breed, namely Main Line and Bellies; and Wool types from OB, namely Meat Breeds (White Face); Meat Breeds (Black Face); Hair or Cross Bred; Wool Bred Lamb; Meat Breed Lamb; and Black).

The base year and month for the analysis were chosen to be 1997 and September, respectively. The Central region was chosen to be the base region. The reference categories for level of preparation and wool type were Original Bag and Original Bag Wool Breeds. Original Bag corresponds to the lowest level of preparation and the Original Bag Wool Breed corresponds to the highest quality wool for the OB level of preparation.

U.S. clean prices are hypothesized to be the highest in the third quarter of the year when wool supply is less abundant. The majority of the world wool production is clipped and sold during the first and fourth quarters of the year. A large proportion of U.S. wool is clipped in April and May. U.S. wool prices are expected to be higher in 1995 and 1997 relative to other years. Prices in the Eastern and Western regions of the United States are expected to be lower compared to prices in the Central region. Marketing outlets for wool in the Central region have been well established relative to those of other regions. Also, more uniform wool in terms of quality, style, and quantity generally is produced in the Central region relative to other regions.

Importantly, BOU and TSC prepared wools are expected, a priori, to command a premium to OB wool. As well, BOU and TSC Main Line wool and BOU and TSC Tender or Short Line wool are expected to command a premium over wool types from OB. Further, average fiber diameter (AFD) is hypothesized to be inversely related to U.S. clean price. Finally, the lot size as measured by grease weight is expected to be positively related to U.S. clean price. The closer a lot is to a truckload, the less money buyers spend on transportation per kilogram.

The model was used to examine price differences for U.S. wools by preparation and by type using data collected from warehouses and pool sales across the United States over the period 1993 to 2002. The goal was to determine premiums/discounts in wool prices by preparation and type, controlling for season, year, region, average fiber diameter, and lot size. The hedonic price model explains about 83 percent of the variation in U.S. wool prices (Table 5A-1). The estimated coefficients and their associated p-values are provided in Table 5A-1). The level of significance chosen for this analysis to conduct statistical tests is 0.01, given the rather sizeable sample of 8,533 observations. Given that the dependent variable is the logarithm of U.S. clean price, the interpretation of the estimated coefficients for each of the qualitative variables (season, year, region, level of preparation, and wool type) is in terms of percentage changes. To calculate the premium/discount or the percentage difference relative to the base or reference category for each of the qualitative variables, the transformation exp(> i-1)*100% can be used, where > i is the estimated coefficient associated with the ith indicator variable.

Also, note that the U.S. clean price, AFD, and grease weight (GW) of the lot size are expressed in terms of logarithms. Consequently, the estimated coefficient of AFD and of GW in the hedonic price model represent elasticities, the percentage change in U.S. clean price due to a unit percentage change in AFD and GW, respectively.

The model explained about 83 percent of the variation in U.S. wool prices over the study period. Seasonality in U.S. clean wool prices was evident. Wool prices received by producers from January to March as well

TABLE 5A-1 Estimated Coefficients and P-Values in the Hedonic Price Model

		Estimated Coefficients	Premium/Discount Relative to Base	P-value
Month	January	–0.1913	–17.4	<0.001
	February	–0.0789	–7.6	<0.001
	March	–0.0608	–5.9	<0.001
	April	–0.0156	–1.5	0.212
	May	0.0065	0.6	0.532
	June	0.0779	8.1	<0.001
	July	0.0039	0.5	0.756
	August	–0.0240	–2.4	0.081
	September	Base	Base	Base
	October	–0.0624	–6.0	<0.001
	November	–0.1154	–10.9	<0.001
	December	–0.1267	–11.9	<0.001
Year	1993	–0.4947	–39.0	<0.001
	1994	–0.1922	–17.5	<0.001
	1995	0.1629	17.7	<0.001
	1996	–0.1260	–11.8	<0.001
	1997	Base	Base	Base
	1998	–0.2702	–23.7	<0.001
	1999	–0.7013	–50.4	<0.001
	2000	–0.7379	–52.2	<0.001
	2001	–0.6249	–46.5	<0.001
	2002	–0.2921	–25.3	<0.001
Level of Preparation	Original Bag	Base	Base	Base
	Bellies Out Untied	0.0209	2.1	0.270
	Table Skirted Classed	0.0811	8.4	<0.001
Region	Central	Base	Base	Base
	Western	–0.1036	–9.8	<0.001
	Eastern	–0.0823	–7.9	<0.001
Wool Breed (Wool Types from BOU and TSC)	Main Line	0.2114	23.5	<0.001
	Tender/Short Line	0.0551	5.7	0.013
	Bellies	–0.2903	–25.2	<0.001
	Pieces	–0.4179	–34.2	<0.001
	Stains	–0.6808	–49.4	<0.001
	Locks	–0.9894	–62.8	<0.001
	Clothing	0.1986	22.0	<0.001
	Main Line Lamb	0.1432	15.4	<0.001
Meat Breed	Main Line	0.0394	4.0	0.174
	Bellies	–0.1288	–12.1	0.545
Wool Types from OB	Wool Breed	Base	Base	Base
	Meat Breed (White Face)	–0.2325	–20.7	<0.001
	Meat Breeds (Black Face)	–0.3826	–31.8	<0.001
	Hair or Cross Bred	–0.3226	–27.6	<0.001
	Wool Breed Lamb	–0.1790	–16.4	<0.001
	Meat Breed Lamb	–0.6988	–50.3	<0.001
	Black	–1.1606	–68.7	<0.001
Log of Average Fiber Diameter		–1.4160		<0.001
Log of Grease Weight		0.0162		<0.001
Constant		4.8090		<0.001
$R^2 = 0.8303$				

as from October to December were significantly lower from 5.9 percent to 17.4 percent than prices in September. Wool prices in June were roughly 8 percent higher than those of September. As expected, U.S. clean wool prices were highest in 1995 and 1997. Prices in remaining years from 1993 to 2002 were significantly lower, from 11.8 percent to 52.2 percent relative to the base year of 1997. Further, U.S. clean wool prices were discounted by 7.9 percent and 9.8 percent, respectively, in the Eastern and Western regions of the United States relative to the Central region.

In line with prior research, prices of TSC wool were significantly higher than OB wool by slightly more than 8 percent. Significant differences among wool types also were evident. In particular, U.S. clean prices of TSC and BOU Main Line Wool were higher by 23.5 percent over the OB wool breed. Significant differences were noted as well among wool types from OB. Among wool types, the premiums/discounts relative to OB wool breed type were quite large in magnitude.

U.S. clean wool prices were found to be highly sensitive to changes in AFD. The elasticity of the clean wool price with respect to AFD was estimated to be about –1.42. Lot size, as measured by GW, also positively affected U.S. clean prices. The elasticity of clean price with respect to lot size was estimated at 0.16.

6

The U.S. Dairy Sheep Industry[1]

Dairy sheep production is a new agricultural venture in the early stages of becoming an economically important agricultural industry in the United States. The first U.S. dairy sheep flocks and the first commercial dairy sheep farms were established in the mid-1980s with nondairy breeds of sheep because true dairy sheep were not present in North America until the early 1990s. European dairy sheep genetics of the East Friesian (EF) and Lacaune (LA) breeds were first imported into North America by Canada in 1992 and 1996, respectively (Thomas et al., 2001) and subsequently into the United States from Canada. Initial research in the United States showed that EF-crossbred ewes produced almost twice as much milk as domestic nondairy ewes (Thomas et al., 1999, 2000). The majority of dairy sheep farms in North America now milk crossbred ewes containing ≥ 50 percent EF and/or LA breeding. The proportion of dairy sheep genetics in flocks is increasing.

The largest concentrations of dairy sheep farms in the United States are found in two regions: (1) the Upper Midwest, specifically northwestern Wisconsin and east-central Minnesota and (2) New York and New England, specifically Vermont, New Hampshire, and Maine. There is excellent interaction between U.S. and Canadian dairy sheep producers. Most Canadian dairy sheep farms are located in southern Ontario and southern Quebec near the U.S. border (Figure 6-1).

---

[1]Much of the information in this chapter is based on the report by Thomas (2004), which summarizes the results of a survey of dairy sheep producers in the United States and Canada conducted in 2003–2004.

FIGURE 6-1 Location of dairy sheep farms in the United States and Canada in 2003.
Source: D. L. Thomas, personal communication (2004). Copyright 2004 by David L. Thomas. Used with permission.

DAIRY SHEEP MILK PRODUCTION

Few official records are kept at the state or national level on the population of dairy sheep or their production. The Dairy Sheep Association of North America (DSANA) and the University of Wisconsin–Madison surveyed dairy sheep producers in the United States and Canada in 2003–2004 (Thomas, 2004; Thomas and Haenlein, 2005). The Wisconsin Agricultural Statistics Service surveyed Wisconsin producers in 2005 (Wisconsin Agricultural Statistics Service, 2006). Thomas (2004) estimated that 44 farms milked ewes in the United States in 2003. The greatest numbers of producers were in Wisconsin (14) and Vermont (10).

Table 6-1 presents the estimated number of dairy sheep producers in the eastern and western regions in the United States. According to Thomas (2004), the average U.S. flock size in 2003 was 145 milking ewes with little difference between the East and West (Table 6-1). However, there were large differences between regions in the variation in flock size. Flocks in the East

TABLE 6-1 Estimated Number of U.S. Dairy Sheep Producers and Number of Milking Ewes per Flock, 2003

U.S. Region	State	Estimated Number of Producers	Milking Ewes Per Flock (number)		
			Ave.	Median	Range
East	Maine	2	—	—	—
	New Jersey	1	—	—	—
	New York	5	—	—	—
	Pennsylvania	1	—	—	—
	South Carolina	1	—	—	—
	Virginia	1	—	—	—
	Vermont	10	—	—	—
	Total East	**21**	**138.1**	54	5–850
					—
West	California	1	—	—	—
	Colorado	1	—	—	—
	Iowa	1	—	—	—
	Minnesota	3	—	—	—
	Missouri	2	—	—	—
	Nebraska	1	—	—	—
	Wisconsin	14	—	—	—
	Total West	23	149.4	**145**	80–305
Total United States		**44**	**145.4**	100	5–850

Source: Adapted from Thomas (2004).

region were more variable in size than flocks in the West region as indicated by the range in flock size. The median values presented in Table 6-1 are the flock sizes that are in the middle of the range. In the East region, half of the flocks were smaller than 54 ewes; in the West region, half of the flocks were smaller than 145 ewes. Overall, half of the flocks in the United States milked fewer than 100 ewes in 2003.

Milk production per ewe in the flocks of producers in the two regions surveyed by Thomas (2004) is presented in Table 6-2. He reported large differences between the two regions. Respondents in the East region reported milk production per ewe at 212 kg whereas those in the West region reported per-ewe production at 140 kg. Production per ewe was much more variable among flocks in the East region than among flocks in the West region. The median milk production among all flocks was 146 kg milk/ewe, with an average production of 174 kg/ewe. In the 2005 Wisconsin survey, the 11 licensed dairy sheep producers in Wisconsin reported average milk production per ewe of 168 kg, higher than the 140 kg/ewe production reported for the West region in the Thomas (2004) survey.

Table 6-3 presents the Thomas (2004) estimate of the total number of

TABLE 6-2 Milk Production per Ewe in Flocks of Survey Respondents, 2003

U.S. Region	Flocks (number)	Total Milk (kg)	Total Ewes (number)	Milk Per Ewe (kg)		
				Average	Median	Range
East	9	282,674	1,332	212.2	142	5–600
West	8	167,216	1,195	139.9	94	54–222
Total U.S.	17	449,890	2,527	178.0	146	5–600

Source: Adapted from Thomas (2004).

TABLE 6-3 Estimated Number of Ewes Milked and Milk Production in the United States, 2003

U.S. Region	Flocks (number)	Total Milking Ewes (number)	Total Milk Production (kg)
East	21	3,088	593,248
West	23	3,390	555,434
Total U.S.	44	6,478	1,148,682

Source: Adapted from Thomas (2004).

ewes milked and the total amount of sheep milk produced in the United States in 2003. Each of the 27 flocks that did not respond to the survey were estimated to have the same average number of ewes (145.4) and average milk production per ewe (178.0 kg) as the respondent flocks for a total estimated milk production of each nonrespondent flock of 25,881 kg. Based on this assumption, 6,478 ewes in the United States produced an estimated 1,148,682 kg of milk in 2003.

DAIRY SHEEP MANAGEMENT SYSTEMS

Types of sheep management and milking systems in use are quite variable among dairy sheep operations. Presented in Figure 6-2 is the annual cycle for a dairy sheep operation lambing in the late winter or early spring. Some dairy sheep flocks receive only pasture during the grazing season. Others are grazed on pasture but supplemented with concentrates. Still others are fed harvested roughage and concentrates in confinement. Starting to milk ewes after weaning the lambs at 30 to 60 days postpartum (DY30 system) was quite common in the early 1990s. An increasing number of

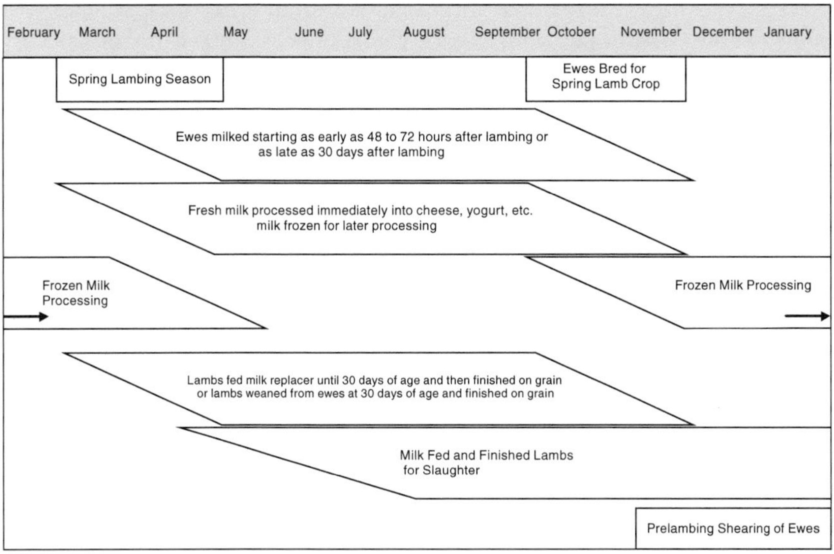

FIGURE 6-2 Annual dairy sheep and lamb cycle.

producers, however, milk the ewes from shortly after parturition. Ewes may be milked only once per day during early lactation while the lambs are nursing and then switched to twice-per-day milking after the lambs are weaned at 30 days of age (MIX system). Alternatively, the ewes may be milked twice-per-day from shortly after parturition with the lambs reared on milk replacer (DY1 system). Research in the United States has shown that the MIX system results in the greatest net returns if milk is sold on weight or volume (McKusick et al., 2001). However, milk from MIX ewes during the 30 days when they are nursing their lambs has a significantly lower fat percentage than milk from DY1 or DY30 ewes and would receive a price discount if sold on the basis of fat content.

Almost all farms use machine milking. Milking systems vary from elevated platforms, with cascading yoke stanchions and milking into buckets, to double-24 milking parlors with a pit for the milkers and several milking units attached to a pipeline, to a carousel milking parlor.

DAIRY SHEEP MILK QUALITY

The U.S. Grade "A" Pasteurized Milk Ordinance (PMO) requires sheep milk at the farm to have a bacterial count of not more than 100,000/ml of milk and a somatic cell count of not more than 750,000/ml of milk

(USDHHS, 2002). Unlike goat milk that has a naturally high somatic cell count (Thomas and Haenlein, 2005), sheep milk produced under sanitary conditions does not have difficulty meeting these minimum standards. Measures of milk composition and quality from 355,000 liters of sheep milk marketed cooperatively in the United States in 2002 and 2003 were 453,000 somatic cells/ml, 41,000 bacteria/ml, 6.2 percent fat, 4.9 percent protein, and 17.1 percent total solids (Thomas, 2004).

DAIRY SHEEP MILK MARKETING, TRANSPORTATION, AND PRICING

The biggest concern for someone considering entry into dairy sheep production is the marketing end of the business. The problem generally is not a lack of demand for sheep milk but rather the production of small amounts of sheep milk on individual farms that are great distances from a milk processor willing to process sheep milk. One solution to this problem is for dairy sheep producers to pool their milk and ship larger quantities of milk periodically to processors. The Wisconsin Sheep Dairy Cooperative (WSDC) is the only U.S. dairy sheep milk marketing cooperative. The cooperative collects milk from its members and markets it to processors. The cooperative is increasing the amount of its milk that is custom-made into cheese with the cheese marketed by the cooperative as well. This adds value to the producers' milk and results in greater net returns.

Cooperatives require a tremendous commitment of volunteer time on the part of members, especially in the initial years of the organization. The effort to establish the WSDC occurred at the same time that producers were establishing their own dairy sheep farms and is now paying dividends. For example, the WSDC has seen over a 20-fold increase in the amount of milk sold in 2007 compared to 1996 when it was established (Figure 6-3). Demand for its milk continues to grow. Without the WSDC, the dairy sheep industry in the Upper Midwest of the United States would not be as viable as it is today.

Much of the milk sold from farms is first frozen in plastic bags in large commercial freezers on the farm. Bags of frozen milk are accumulated on the farm and shipped in large quantities in refrigerated trucks to processors. Research has shown that this milk can be frozen at –27°C for at least 12 months with no detrimental effects on processing characteristics (Wendorff, 2001). The ability to freeze milk and make a quality product from the thawed milk has allowed small producers who are great distances from processors to enter the industry. A small producer can accumulate the milk produced from his flock in a freezer during the milking season and send the milk individually or along with that of other producers to a processor in a large shipment only once or a few times per year. Since sheep are seasonal

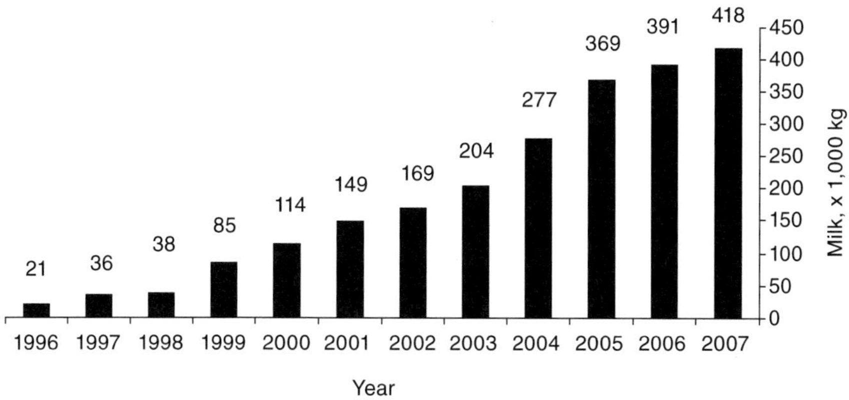

FIGURE 6-3 Milk sales by the Wisconsin Sheep Dairy Cooperative (WSDC), 1996–2007.
Sources: Sales through 2003 from Thomas (2004). Sales for 2004 through 2007 from WSDC Board of Directors (personal communication). Copyright 2004 and 2007 by David L. Thomas. Used with permission.

breeders and most sheep milk is produced during the spring and summer, a frozen stockpile of milk also allows processors access to milk year-round.

Frozen milk, however, is not without its problems. The costs of installation and maintenance of a large commercial freezer and of freezer bags are large expenses for dairy sheep producers. Processors experience increased costs in storage of frozen milk and in the amount of time required to thaw milk prior to processing compared to fluid milk (Clark, 2004; Cook, 2004). Some frozen-thawed milk separates and has a large amount of sediment that does not allow its use in all sheep milk products. Failure to keep the milk cold enough while frozen on the farm or partial thawing during transportation to the processor may be the cause (Clark, 2004). In areas where there is a large concentration of dairy sheep farms in close proximity to a major processor, the accumulation of large quantities of fresh milk during peak lactation periods is possible for shipping directly to processors in bulk tank trucks. This is greatly preferred by processors (Cook, 2004). As the industry grows and more milk becomes available, fluid shipments of milk will increase. For example, the WSDC shipped 63 percent of its milk to processors in fluid rather than frozen form in 2007 (WSDC, personal communication, 2007). However, there will still be a need for some frozen milk to sustain processors during times of the year when sheep milk is not produced and processors are large distances from producers.

According to the Thomas (2004) survey, the price for dairy sheep milk sold to processors ranged from $1.20 to $1.65 per kg in 2003. Producers in Wisconsin received an average price of $1.22 per kg for sheep milk in 2005 (Wisconsin Agricultural Statistics Service, 2006). The vast majority of the milk is sold by weight with no premiums or discounts for milk composition or quality. Value-based pricing of sheep milk could become a reality in the future as the supply of milk increases. Of course, milk cannot be sold if it does not meet the minimum quality standards for bacterial and somatic cell counts established by the federal government.

DAIRY SHEEP MILK PROCESSING

The few sheep milk processors in North America appear to be pleased to have sheep milk available. Most have increased the amount of sheep milk they process each year as the supply slowly increases. The largest commercial U.S. sheep milk processor produced 115,000 kg of cheese and 4,500 kg of yogurt in 2003 (Thomas, 2004). A sheep milk cheese was the champion specialty cheese from more than 725 cow, goat, and sheep milk cheeses entered in the American Cheese Society's 2004 annual competition. The maker of the champion cheese runs a commercial processing plant and purchases dairy sheep milk from the WSDC. Several other U.S. sheep milk cheeses have received major honors at national and international cheese competitions in the last few years.

Farms that are not members of a marketing cooperative have a number of marketing options, including selling their milk directly to a commercial processor, processing their milk into cheese or other products on their farms, having their milk custom processed into cheese by a commercial processor, or selling their milk to another dairy sheep producer who processes their own milk plus purchased milk.

The lack of local commercial processing factories for sheep milk in most areas has led many sheep producers to make cheese on their farms in small batches and market it directly to individuals, food stores, and restaurants. Of the 24 respondents to the Thomas (2004) survey who had milked sheep between 2000 and 2003, 11 (46 percent) processed milk on their farms into value-added products (Table 6-4). The percentage of producers in the East region that processed milk on their farms (71 percent) was much higher than the percentage that processed milk on their farms in the West region (10 percent). This difference in the percentage of farms processing their own milk in the two regions may be because the WSDC was formed in 1995, giving producers in the West region a marketing outlet for fluid milk. In other areas, there were no options except to process the milk produced. Also, the northeast United States may have had more of a tradition of producing farmstead cheeses than the West region. Wisconsin requires all cheese makers to be licensed. Passing an examination and a one-year apprenticeship

TABLE 6-4 Proportion of Survey Respondents That Process Milk on Farm, Purchase Milk from Other Producers for Processing, or Wish to Start Processing Milk, 2003

	Number (%)			
	Process Milk on the Farm		Do Not Now Process Milk on the Farm but Wish to in the Future	
U.S. Region	Yes	No	Yes	No
East	10 (71)	4 (29)	3 (75)	1 (25)
West	1 (10)	9 (90)	3 (33)	6 (67)
Total U.S.	11 (46)	13 (54)	6 (46)	7 (54)

Source: Adapted from Thomas (2004).

are required to obtain a license to make cheese in Wisconsin. These requirements do not encourage producers to become farmstead cheese makers.

Of the 13 producers in the Thomas (2004) survey that did not process milk on their farms, almost half expressed a desire to process milk in the future. However, there were large differences among regions in the desire to enter into processing of milk. About 75 percent of the nonprocessing farms in the East region wished to process in the future, whereas only 33 percent of the nonprocessing farms in the West region wished to process in the future (Table 6-4). This result again seems to indicate the satisfaction that West region producers have with their milk marketing alternatives and, perhaps, the lack of good outlets for fluid sheep milk in the East region at the present time.

DOMESTIC AND IMPORTED DAIRY SHEEP MILK SUPPLY

The United States is the largest importer of sheep milk cheese in the world. Approximately half the world exports of sheep milk cheese in 2005 came to the United States (FAO, 2007). Sheep milk cheese imports to the United States have increased from 14,476 metric tons (tonnes) in 1985 to 33,359.5 tonnes in 2005, a 30 percent increase in 20 years (FAO, 2007). The five European countries of Italy, France, Bulgaria, Greece, and Spain accounted for 93.4 percent of the sheep cheese exports to the United States in 2005 with Italy alone accounting for 56.7 percent of those exports (Table 6-5).

The total U.S. sheep milk production was estimated at 1,149 tonnes in 2003. If total production increased at the same rate as the volume of milk marketed by the WSDC from 2003 to 2005 (+ 81 percent), total U.S. sheep

TABLE 6-5 U.S. Sheep Milk Cheese Imports by Exporting Country, Quantity, and Value, 2005

	Volume		Value	
	Tonnes	%	$1,000	Per kg
Italy	18,910.0	56.7	$101,227	$5.35
France	4,660.0	14.0	$27,855	$5.98
Bulgaria	3,553.0	10.7	$11,348	$3.19
Greece	2,245.0	6.7	$15,085	$6.72
Spain	1,803.0	5.4	$18,552	$10.29
All other countries	2,188.5	6.6	$11,156	$5.10
Total	33,359.5	100.0	$185,223	$5.55

Source: FAO (2007).

milk production in 2005 may have been 2,080 tonnes. This amount of sheep milk could produce approximately 416 tonnes of sheep milk cheese (5 kg of sheep milk per kilogram of sheep milk cheese). Therefore, domestic production of sheep milk cheese in 2005 was probably less than 1.3 percent of the total supply available to U.S. consumers, suggesting considerable potential for a large increase in domestic production to displace some of the imported product.

Imported dairy sheep products present some pricing challenges for domestic products. The average landed value of imported sheep milk cheese in the United States was $5.55 per kg in 2005 (Table 6-5). At current values of U.S. sheep milk ($1.20/kg to $1.65/kg) and 5.0 kg of sheep milk required to produce 1 kg of sheep milk cheese, the milk alone in 1 kg of domestic sheep milk cheese costs between $6.00/kg and $8.25/kg. Obviously, domestic sheep milk cheeses cannot compete with imported sheep milk cheeses on price in the current economic environment. Continued growth in production of domestic sheep milk cheese will depend on the ability of U.S. cheese makers to continue to produce high-quality specialty sheep milk cheeses that command a higher price than imported products.

MAJOR ACCOMPLISHMENTS, OPPORTUNITIES, AND CHALLENGES OF THE U.S. DAIRY SHEEP INDUSTRY

Major Accomplishments of the U.S. Dairy Sheep Industry

- *Developing industry.* The dairy sheep industry has developed from nothing in the early 1980s to a small but growing industry with potential for continued growth. As with all new industries, the U.S. dairy sheep industry owes its existence to the hard work of a few pioneer producers and processors.

- *Domestic sheep milk cheeses now available.* Prior to the early 1980s, there were no domestic sheep milk cheeses available to U.S. consumers. Now, cheese cases in many specialty cheese stores and even stores of national grocery chains contain domestic sheep milk cheeses. This is due to the production of high-quality milk by producers, the manufacture of high-quality cheeses by processors, and the promotion of these cheeses by both national (e.g., The American Cheese Society, http://www.cheesesociety.org/) and state (e.g., Vermont Cheese Council, http://www.vtcheese.com) organizations.
- *Formation of milk marketing cooperative.* The Wisconsin Sheep Dairy Cooperative (http://www.sheepmilk.biz/), a sheep milk marketing cooperative for producers in the Upper Midwest and the only such marketing cooperative in the United States has served as a catalyst for the continued growth of the industry in that region.
- *Dairy sheep producer organization established.* The Dairy Sheep Association of North America (DSANA) was established in 2002 to foster the North American dairy sheep industry (http://www.dsana.org/index.php). The association publishes a quarterly newsletter and has taken over the organization of the annual Great Lakes Dairy Sheep Symposium started by the University of Wisconsin–Madison and the Wisconsin Sheep Breeders Cooperative in 1995. The symposium now rotates each year among sites in Wisconsin, the northeastern United States, and southern Ontario and Quebec, Canada. Speakers include North American scientists, producers, and processors, as well as international experts. The proceedings of these symposia contain the most up-to-date information available to the North American dairy sheep industry. Past proceedings can be viewed at http://www.uwex.edu/ces/animalscience/sheep/.

Major Opportunities of the U.S. Dairy Sheep Industry

- *Large domestic market for sheep milk cheeses.* The greatest opportunity for the dairy sheep industry is the increasing consumption of sheep milk cheese by U.S. consumers. The consumption of sheep milk cheeses in the United States, as measured by imports, has increased by 30 percent in the 20-year period from 1985 to 2005. In 2005, it was estimated that < 1.3 percent of total sheep milk cheese availability was domestically produced.
- *Demand for locally raised food.* The growing movement among consumers to eat locally raised and produced products should also be good news for the domestic sheep milk industry.
- *Viable option for small farms.* Dairy sheep production is an agricultural enterprise that requires a relatively small amount of land, and sheep are small enough to be handled safely by most members of the family. The

enterprise is a viable alternative for small-scale farmers. Returns to labor and management per year, including capital costs, are estimated to be between $51/ewe (high debt load) and $110/ewe (low debt load) for a 300-ewe flock producing and marketing milk, lambs, and cull sheep (Berger, 2002).

Major Challenges of the U.S. Dairy Sheep Industry

- *No genetic improvement program.* There are no regional or national genetic improvement programs for dairy sheep traits in the United States. Annual increases in production that have been noted in most flocks over time have been a result of improved management and an increase in the percentage of dairy breeding in the ewe flock. However, as the industry matures, there will be a desperate need for proven sires with high estimates of genetic value for economically important traits. The industry should establish a national genetic improvement program for dairy sheep. The two logical organizations to carry out these genetic evaluations are either the Animal Improvement Laboratory of USDA, which conducts the genetic evaluations for dairy cattle and dairy goats, or the National Sheep Improvement Program, which conducts the genetic evaluations for meat and wool sheep.
- *Limited research and extension support.* There is limited research and extension support for dairy sheep production, sheep milk processing, and sheep milk product marketing. The only dairy sheep research flock in North America was established in 1995 at the University of Wisconsin–Madison at the Spooner Agricultural Research Station. This flock is composed of approximately 300 milking ewes of various percentages and combinations of East Friesian and Lacaune breeding. The University of Vermont has an Extension Small Ruminant Dairy Specialist to work with producers of dairy goats and sheep in Vermont. In addition, the University of Guelph, Cornell University, the University of Vermont, and the University of Wisconsin–Madison have research and/or extension programs in sheep milk processing and/or dairy sheep production. Potential producers or processors in states other than those listed above are very limited in the information and assistance they can obtain to enter the industry.
- *Few marketing options for sheep milk.* There is a lack of marketing options for most U.S. sheep dairy producers. The Wisconsin Sheep Dairy Cooperative serves the milk marketing needs of the majority of the producers in Wisconsin and Minnesota. Potential producers in other areas of the United States are at a great disadvantage because there is no ready market for the milk they would like to produce. Their only option is to become a sheep milk producer and a farmstead cheesemaker and marketer, but these skills may not be present among the personnel (generally the family members) to be a farmer, processor, and marketer. While there appears to be a

demand for more sheep milk, the infrastructure is not in place throughout the United States to get milk from the producers to the processors.

- *Lower-priced imported sheep milk cheeses.* The lower price of imported sheep milk cheeses relative to domestic sheep milk cheeses will remain a continual challenge. The domestic industry must compete with imported product by continuing to develop cheeses that are unique compared to imported cheeses, by producing higher-quality cheeses or the perception of higher-quality cheeses, and by capitalizing on the movement among consumers to eat more locally produced foods.

REFERENCES

Berger, Y. M. 2002. Economics of winter milking for medium to large dairy sheep operations. Pp. 118–126 in Proceedings of the 8th Great Lakes Dairy Sheep Symposium. University of Wisconsin–Madison, Dept. of Anim. Sci. Online at: http://www.ansci.wisc.edu/Extension-New%20copy/sheep/Publications_and_Proceedings/res.html. Accessed April 17, 2008.

Clark, J. T. 2004. Processing of frozen milk—Current procedures and difficulties encountered. Pp. 153–156 in Proceedings of the 10th Great Lakes Dairy Sheep Symposium. University of Wisconsin–Madison, Dept. of Anim. Sci. Online at: http://www.ansci.wisc.edu/Extension-New%20copy/sheep/Publications_and_Proceedings/res.html. Accessed April 17, 2008.

Cook, S. 2004. Marketing of sheep milk—Problems faced by processors. Pp. 157–158 in Proceedings of the 10th Great Lakes Dairy Sheep Symposium. University of Wisconsin-Madison, Dept. of Anim. Sci. Online at: http://www.ansci.wisc.edu/Extension-New%20copy/sheep/Publications_and_Proceedings/res.html. Accessed April 17, 2008.

FAO (Food and Agriculture Organization of the United Nations). 2007. FAOSTAT data base. Online at: http://faostat.fao.org/. Accessed September 30, 2007.

McKusick, B. C., D. L. Thomas, and Y. M. Berger. 2001. Effects of weaning systems on commercial milk production and lamb growth of East Friesian dairy sheep. J. Dairy Sci. 84:1660–1668.

Thomas, D. L. 2004. Overview of the dairy sheep sector in Canada and the United States. Pp. 166–180 in Proceedings of the 10th Great Lakes Dairy Sheep Symposium. University of Wisconsin–Madison, Department of Animal Science. Online at: http://www.ansci.wisc.edu/Extension-New%20copy/sheep/Publications_and_Proceedings/res.html. Accessed April 17, 2008.

Thomas, D. L., and G. F. W. Haenlein. 2005. Panorama of the goat and sheep dairy sectors in North America. Pp. 23–28 in Proceedings of the International Symposium on the Future of the Sheep and Goat Dairy Sectors, Zaragoza, Spain. Special issue 200501, Part 1. International Dairy Federation. Online at: http://www.fil-idf.org/content. Accessed April 17, 2008.

Thomas, D. L., Y. M. Berger, and B. C. McKusick. 1999. Milk and lamb production of East Friesian-cross ewes in the north central United States. Pp. 474–477 in Milking and Milk Production of Dairy Sheep and Goats. Proceedings of the 6th International Symposium on the Milking of Small Ruminants. F. Barillet and N. P. Zervas, eds. EAAP Publication No. 95. Wageningen, the Netherlands: Wageningen Pers.

Thomas, D. L., Y. M. Berger, and B.C. McKusick. 2000. East Friesian germplasm: Effects on milk production, lamb growth, and lamb survival. Proc. Am. Soc. Anim. Sci., 1999. Online at: http://www.asas.org/jas/symposium/proceedings/0908.pdf.

Thomas, D. L., Y. M. Berger, B. C. McKusick, R. G. Gottfredson, and R. Zelinsky. 2001. Comparison of East Friesian and Lacaune breeding for dairy sheep production systems. Pp. 44–51 in Proceedings of the 7th Great Lakes Dairy Sheep Symposium. University of Wisconsin–Madison, Dept. of Anim. Sci. Online at: http://www.ansci.wisc.edu/Extension-New%20copy/sheep/Publications_and_Proceedings/res.html. Accessed April 17, 2008.

USDHHS (U.S. Dept. of Health and Human Services). 2002. Grade "A" Pasteurized Milk Ordinance, 2001 Revision. Online at: http://www.cfsan.fda.gov/~ear/pmo01toc.html. Accessed April 17, 2008.

Wendorff, W. L. 2001. Freezing qualities of raw ovine milk for further processing. J. Dairy Sci. (E. Suppl.):E74–E78.

Wisconsin Agricultural Statistics Service. 2006. Wisconsin Dairy Sheep Industry Overview. Online at: http://www.nass.usda.gov/Statistics_by_State/Wisconsin/Publications/Dairy/dairysheep.pdf. Accessed April 17, 2008.

7

Alternative and Emerging Markets

The U.S. sheep industry is in the process of transformation forced by economic necessity following years of decline, growing competition from the sheep and textile industries of other countries, and a number of other factors as documented in the preceding chapters. Although the full extent of the changes taking place in the industry is difficult to determine, sheep and lamb inventories indicate growth in eastern and mid-Atlantic states and in the Midwest, where alternative and emerging markets are particularly important (see Chapters 1 and 2). These alternative or "niche" markets represent a growth area for the industry. New and expanded industries include sheep for dairy, purebred flocks used in shows, club lambs, specialty wools, and hair sheep. The growth of these markets is supported through active producer organizations, local marketing systems, statewide fairs, and new technologies, such as the Internet, that have reduced the costs of exchanging information in the market and facilitate identification of buyers and sellers of the products. In addition, the consumer base now includes new ethnic demand for lamb and other meat products, and a growing consumer market for locally produced and organic products.

Although not well documented, alternative and emerging markets are increasing within the sheep and wool industry, as evidenced by an increasing share of lightweight (23–40 kg) feeder lambs and hair sheep now being purchased for slaughter to meet demand from ethnic groups for lighter-weight and younger lamb carcasses. Producers in areas remote from traditional markets but with access to inspected slaughter plants are venturing into direct marketing. Customers may be local friends and neighbors or local buyers that have interests in "local" products (including restaurant and local retail outlets). Organic or specialty products may also be marketed via the Internet or other direct mail-order approaches that meet demand in a broader market. Individual producers have cultivated retail and restaurant outlets for their lamb. Although the alternative and emerging markets are small relative to the traditional industry, they are sectors that are experiencing growth and are creating new markets for lamb.

ETHNIC MARKETS

U.S. lamb consumption has remained steady at about 0.5 kilograms per capita since the mid-1990s, falling from about 3.0 kilograms per capita since the mid-1940s to its present level. However, the average per capita consumption conceals considerable variation by region of the country. For much of the central United States, lamb is rarely considered in the red meat purchase decisions for home consumption and infrequently found in restaurants. The largest U.S. markets for lamb appear to be the East and West coasts, both as household consumption and as food away from home (see Chapter 4). Even in those regions, however, the consumption of lamb is skewed toward religious and ethnic groups in society and away from that part of the population with origins in northern Europe.

For the U.S. population of North European descent, lamb has declined in popularity since World War II (see Chapters 1 and 4). Regaining interest in lamb as an integral part of the diet among this population has proven difficult. Certainly, lamb is slowly entering the restaurant trade, particularly higher-end restaurants. In consumer surveys, restaurants are the primary exposure to lamb for much of this population (ALB, 2007). Nonetheless, the ethnically and culturally diverse character of the East and West coasts appears to be ahead of the American heartland in breaking down this long standing barrier to lamb consumption.

For some religious groups, lamb is a major consumption item on a regular basis and for several significant holidays or holy days. In the Jewish faith, lamb (kosher) is a regular item in the diet. Similarly, lamb is a regular dietary item for Greek and Eastern Orthodox groups and the preferred meat for the celebration of Easter, with the Orthodox Easter occurring at the same time as the Jewish Passover. In the Muslim faith, lamb (halal) is the preferred meat during Ramadan and the Eidu al-fitr holiday at the end of Ramadan. As well, lamb is a preferred meat for the Eidu al-adha holiday. Several ethnic groups within the United States regularly consume lamb, particularly those of Middle Eastern, North African, Caribbean, southern European, and South Asian origins. Within these groups, lamb consumption on a regular basis seems to persist from generation to generation, following the first-generation immigration to the United States (see, e.g., Larson and Thompson, Undated; USDA, 2006a). Hair sheep lambs are well suited to the ethnic markets because of their smaller carcass size, presence of a tail, and lower likelihood of feedlot finishing. The changing ethnic backgrounds of major U.S. cities away from the New York and California coasts are expanding the populations from these ethnic groups and their associated ethnic markets into areas such as Chicago, Detroit, and Houston.

Seasonal Consumption and Religious/Ethnic Effects

For many U.S. ethnic groups, lamb is consumed on a regular basis throughout the year. In other words, for these groups there is no particular seasonal pattern to their lamb consumption. Although lamb is consumed throughout the year by various religious groups (which often overlap the ethnic groups noted above), lamb is the preferred meat for specific religious holidays. Muslim, Greek, and Eastern Orthodox populations are the only significant ethnic groups that, because of religion, consume lamb during specific periods of the year.

In the Muslim calendar (Alnaseej, 2007), two major periods are associated with lamb consumption. The first is the month of Ramadan (the ninth month of the Hijri calendar). Then, Eidu al-fitr is celebrated on the first day of the month immediately following Ramadan. The second period is the Eidu al-adha, which occurs on the 10th day of Dhul Hijjah, the last month of the Islamic calendar, celebrating the end of the annual Hajj. Eidu al-adha is the most important Moslem holy period for the consumption of lamb. Because the Hijri calendar does not contain the same number of days as the Gregorian calendar, the holy periods for Muslims change each year, advancing 9 to 12 days annually, taking 31 to 33 years to return to the same month in the Gregorian calendar. For Muslims, the consumption is mostly lamb, although some older ovine animals also enter this market. Because of the Eastern and Greek Orthodox population, as well as a large and growing U.S. Muslim population, the question is whether these holy periods affect the seasonal consumption or disappearance of lamb in the United States.

The U.S. Census is prevented by law from collecting information regarding religious affiliation. As a result, little information on religious affiliation is available on a consistent annual basis. One source is a study of religious affiliation in the adult population comparing 1990 and 2001 (USCB, 2007). By telephone survey, adults were asked to indicate their religion, without prompt or verification, or membership in a religious group. The results for the Jewish population indicated 3.1 million adults for 1990 and 2.8 million adults for 2001 (1.8 percent and 1.4 percent, respectively, of the adult population). For Muslims, there were 527,000 adults in 1990 and 1.1 million adults in 2001 (0.3 percent and 0.5 percent of the adult population, respectively). Although Eastern Orthodox groups made up only 0.3 percent of the U.S. population in 1990, their growth rate was about 2.7 percent annually. Whereas the total Jewish population is declining both as a percentage of the U.S. population and in absolute numbers, the Muslim population is growing in numbers as well as a percentage of the total U.S. population.

While kosher foods have been widely available in North America for several years, the availability of halal foods has increased substantially over the past two or three decades. As a result, it is much easier today for Muslims to find or access the products they want than in earlier years. For this reason, a trend toward greater use of lamb in Muslim holy periods today than earlier should be apparent. As well, generally in North America, Muslims have become more conscious of their own religious rites and stricter in eating habits over the past several years. For both these reasons, as well as the growth in the Muslim population, a trend toward greater consumption of lamb during Muslim holy periods over time should be evident in the published data.

Because there is relatively little information on the importance of the ethnic lamb market, an analysis of the effects of ethnic markets on aggregate (national level) U.S. lamb demand was conducted as a part of this study. The methodology used for the analysis, including the econometric models and data, are presented in the appendix to this chapter. For the analysis, monthly data over the 1970–2006 period were assembled to explore the impact of Orthodox and Muslim holy periods on lamb slaughter. The data represented head (thousands) of lambs and yearlings slaughtered each month, including both federally inspected slaughter and non-federally inspected slaughter. One feature of the slaughter data is the strong, downward trend over the years. U.S. domestic lamb and yearling slaughter stood at nearly 10 million head in 1970 and at 2.6 million head in 2006. Because of the religious nature of lamb demand, particularly in certain periods of the year, price is likely to be far less of a factor affecting demand during those periods than would be the case generally for lamb. Also, the substitution of lamb for other meats is likely to have little effect on demand during the religious periods.

The data were used to evaluate long-term average seasonal trends (by month) that are presumed to occur because of the religious holy days. The Orthodox Easter holiday and Muslim holy days of Ramadan and Eidu al-fitr occur at varying times (though specific dates) during the year. A question of primary interest was whether the impact of Muslim and Christian/Orthodox religious events has grown stronger over time. The analysis of the monthly data supports the premise that religious holy days, both Christian and Orthodox Easter and Muslim religious events, affect the national lamb market. These holidays lead to additional slaughter of lambs and yearlings. The same was the case for the Eidu al-adha holiday. The effect of both the Muslim religious events and Christian/Orthodox Easter periods is considerably larger after the 1990 period than before. Purchases of lamb were found to be sensitive to price, even during the holiday periods for the ethnic periods examined.

The primary conclusions of the analysis of the slaughter data, therefore, are that Muslim holiday periods and Christian and Orthodox Easter affect slaughter levels of lamb and yearlings and that the impact of these holidays appears to be increasing with time. The increases estimated are on the order of 7 to 8 percent of the monthly lamb slaughter and 1.6 to 1.7 percent of annual disappearance. To the extent that some lamb is purchased directly from farms and not recorded in official tallies of slaughter, these results could underestimate the impact of these religious periods.

Other ethnic markets of growing importance in some regions of the country include Hispanic, Italian, and Greek communities. A survey conducted in the Reno, Nevada area, found 25 percent of Asian and Hispanic households purchased lamb during a one-month period, compared to 1 percent of white, Caucasian households (Lotterman, 1993). Another recent study of consumer preferences in the southern U.S. market for goat meat found that Hispanic consumers and those buying lamb are more likely to purchase goat meat (Knight et al., 2006).

Supply and Marketing Channels

While virtually all major food retail chains offer lamb cuts on a continuous basis at the meat counter, purchases of lamb in many ethnic and religious market segments are largely from individually owned, local butcher shops, operated in most cases by families from the ethnic or religious population these stores serve. Family, friendship, and community relationships are a significant part of the rationale for purchases from these stores. These operations typically purchase whole lamb carcasses for breaking into retail portions for sale to meet the expectations of the local market they serve. The lamb carcasses often come from small, local, single-site packing plants, specifically serving these market outlets. With the growth of the ethnic and specialty markets, verification of marketing claims and traceability systems in marketing channels are becoming increasingly important in the market for lamb and lamb products (e.g., organic production, halal slaughter, or "naturally raised" claims).

In general, the ranges in live weight of lambs preferred in the ethnic markets may vary both by regional and ethnic markets. The differences in preferred weights influence the retail price. Kosher markets in the Northeast favor 45–57 kg lambs, while the Muslim market favors 27–40 kg lambs (O'dell et al., 2003). A study conducted on the market for West Virginia lamb found the price spread in lighter-weight lambs (32–39 kg feeder lambs) compared to heavier-weight slaughter lambs (45–57 kg) to have increased over the 1996 to 2001 period, from 1 percent higher to 16 percent higher at the end of the period (O'dell et al., 2003).

A detailed study of pricing and market attributes in eastern states indicates the importance of period of time and type of lamb in markets with growing ethnic populations (Singh-Knights et al., 2005). The study used sales transaction data from auction markets in Virginia, Pennsylvania, and West Virginia for the period 1994–2003 in combination with hedonic pricing estimation methods. Their results showed a significant break in pricing relationships across the period. The study also concluded that lighter-weight slaughter lambs and all weight classes of feeder lambs were sold at significant price discounts during the earlier part of the study period (1994–1997). For the later period of the analysis (1998–2003), the study found that premiums were paid for the lighter-weight lambs and lower premiums for heavier-weight lambs. The period of year was also important, with premiums paid for sales during the first two quarters of the year.

Lamb and other sheep products, such as sheep milk cheese and yogurt, can also be found in farmers' markets within large metropolitan areas. Interestingly, the large packing plants for lamb appear to serve the large chain retail outlets almost exclusively with little or no penetration into the single family meat outlets. As well, the sheep cheese and yogurt industry products are rarely supplied through the large retail chains, with most sales through on-site stores, farmers' markets, and small individually owned outlets.

Another source of lamb and mutton for ethnic and religious markets is direct purchases of live animals for immediate slaughter on the farm where they are purchased in many instances. Unlike meat derived from livestock and sold to consumers, which must be inspected and passed at a state or federally inspected facility, animals for direct purchase and on-farm slaughter are sold as live animals to consumers and, therefore, do not require state or federal inspection. Animals are selected live and killed on-site by the purchasers to meet specific religious or ethnic requirements. This practice appears common in the areas surrounding large urban centers. Long-term relationships in these markets are often established between buyer and seller, with the initial contact for the seller identified through family and friends within the ethnic or religious grouping.

Even though lamb is available from foreign as well as domestic sources, there is no tariff line specifically identifying kosher or halal lamb and lamb product imports. As a result, there is no information on the proportion of imported lamb that meets halal or kosher requirements, even though lamb is imported and labeled for these markets. For many in the Jewish and Muslim communities, freshly killed lamb under kosher and halal conditions, respectively, is the product of choice for the religious periods. As a result, the expectation is that the demand for lamb during these religious periods is expressed predominantly for domestically sourced product. However, it should be noted that slaughter plants in Australia and New Zealand that cater to the export trade meet halal requirements. Effective market trade in halal products, as is the case for other process claims, requires certification systems or verification of product claims. As an example for halal products, Islamic Food and Nutrition Council of America (IFANCA) is a globally recognized halal certification program that currently certifies a wide range of food products including meat products (IFANCA, 2008). Certification programs such as this are important to both domestic and internationally traded products and are of growing importance in some ethnic and niche markets.

As evidence of the growing importance of the ethnic and religious markets and the associated practice of on-farm slaughter, a number of sites and sources provide instructions for humane (halal) on-farm slaughter methods (e.g., Grandin, 2007; Schoenian et al., 2007). In this, practice, the animals are harvested in the name of Allah (God) without being stunned.

ADDITIONAL ALTERNATIVE MARKETS

A number of other alternative markets for the sheep and lamb industry have emerged, especially in the last 10 years. These product markets offer the opportunity for higher prices and profit to producers through specialized value-added products, such as organic production, specialty cheeses, fine wool, and specialty products valued by consumers as locally produced, organic, gourmet, or having other unique qualities.

The availability of web-based resources has expanded information on marketing sheep and goat products for both producers and consumers. One major effort, for example, is SheepGoatMarketing.info, a joint project of the University of Maryland and Cornell University (Schoenian et al., 2007). The web project originated from the Northeast Sheep and Goat Marketing Project, a USDA-funded effort to improve the marketing infrastructure for sheep and goat producers in the 12 northeastern states. Although the website identifies sellers for wool products (fleece, roving, yarn, and pelts), cheese, milk, and animals for vegetation control, the major market target of the website is ethnic and religious markets for sheep and goat meat. A review of the processors listed as handling lamb and goats on the website shows more than 150 livestock processors who buy live animals (sheep, lambs, and goats) and process them for resale to wholesale and retail businesses. Many advertise that they buy in small quantity and do custom slaughter. The firms listed are primarily in the Northeast and mid-Atlantic states.

Organic Lamb Market

The rate of growth for certified organic sheep and lambs has outpaced that of other livestock, growing nearly six-fold during the period of 1997–2002 according to USDA (2007a). Nevertheless, the total number of certified organic livestock is still relatively small, although different sources report different numbers. The USDA (2007a) reported that there were 4,471 certified sheep and lambs in 2005, just over 2 percent of all total certified organic livestock. Geisler (2007) reported that there were 5,347 organically certified sheep and lambs in 2005, up from 4,561 in 2003. Although the industry grew rapidly in the period 1997–2002, USDA data indicate little growth in the number of animals being raised in certified organic production systems in recent years (Figure 7-1). Today, most of the organic sheep and lamb livestock are located in the West and Midwest (USDA, 2007a).

As required by USDA standards for organic food implemented in 2002 and administered through the USDA Agricultural Marketing Service (AMS), lamb product sold, labeled, or represented as organic must be under continuous organic management from the last third of gestation forward (USDA, 2007b). Except for approved feed supplements and additives, the total feed ration for the animals must be composed of agricultural products, including pasture and forage that are organically produced. The organic production methods require minimal use of off-farm inputs. Lambs are raised without

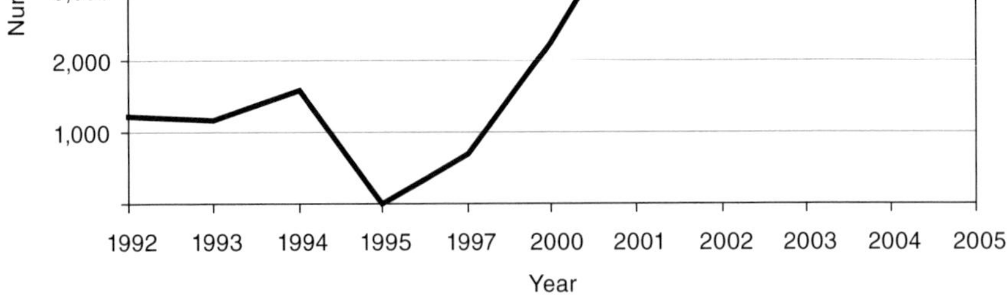

FIGURE 7-1 Inventory of certified organic sheep and lambs in the United States, 1992–2005.
Source: USDA (2007a).

the use of antibiotics or drugs or hormones to promote growth and in production systems that maintain conditions that provide for freedom of movement, reduce stress, and promote animal welfare. These conditions are well suited to lamb production.

Lambs certified as organic must be harvested and processed without artificial ingredients or preservatives, and by processors who maintain systems that prevent contact of product with any substance that is not approved. In practice, this requires that the processors are able to keep harvest and processing of organic animals separate from production lines for conventional product, and limits processing to plants certified as organic. In some areas of the country, including California and other western states, there is sufficient volume of organic lamb so that companies with product under contract may own their own harvest and processing facilities, or operate them under contract. However, in many areas of the country, the volume is not sufficient and organic lambs are custom processed.

Effective in 2005, producers and marketers that operate under a National Organic Program-approved organic system plan and that produce and market only commodities eligible for a "100 percent organic" label are exempt from the assessment for market research and promotion (USDA, 2005).

Although the rules and regulation of organic product claims fall clearly under USDA organic standards, other production or processing claims are less well understood or verified in the market. Marketing claims for "locally grown" or "naturally raised" may well compete directly with organic market claims for some consumers. However, such claims are less well documented and would also require some voluntary certification or audit programs to verify the accuracy and meaning of such claims. The USDA AMS is currently considering a voluntary standard for naturally raised livestock and meat (USDA, 2007c).

Direct Marketing

Many individual lamb producers are taking proactive approaches to sell their own products directly to customers and foodservice operators and bypass traditional marketing channels. The number of individual sheep and lamb producer websites on the Internet has expanded sharply and is evidence of the growing sales through suppliers identified through Internet listings. Lamb sold directly from producer to retail consumers is especially important in some markets, including some large urban markets.

Although the specific number of animals sold through direct, private negotiations is not available, evidence based on the growing number of lamb and sheep inventories in some areas of the country where there are few large processors that handle lamb attest to the growing (though poorly documented) importance of the lamb trade through direct, private party sales. In direct marketing, lambs may be sold as live animals directly to consumers, to custom processors who may custom process the animals for the owners, or to retail grocery or foodservice outlets through state or federally inspected processing facilities.

Lamb sold directly to the freezer market is another growing direct marketing method. In this market, lambs are sold live to customers and arrangements made with a custom slaughter facility for processing and packaging.

Exhibition Animals and Specialty Lamb and Wool Production

Sheep exhibitions and livestock shows were originally designed for producers to exhibit prize animals. Today, the exhibitions and shows continue, and have developed into a segment of the industry that caters to the public partly as a form of entertainment. Many segments of the population see more sheep through exhibitions and livestock shows than they do on the open range or on the farm flocks in rural America. Some producers who participate in these events fit the classic definition of purebred breeders for the development of superior seedstock in the sheep industry. Others are hobbyists who produce and exhibit sheep as a form of recreation. Prices for animals in this arena are high relative to the national norm of commercial lamb and sheep prices and may be inflated by the belief that superior breeding stock should command a premium.

An area that has progressively increased in size and scope and is also in the public eye as a form of entertainment is the Junior Club lamb (market lambs) shows and exhibitions. As with other niche areas, exact numbers are difficult to obtain. W. S. Ramsey (personal communication, 2007) estimated that 4-H and Future Farmers of America (FFA) sheep projects in Texas involve over 40,000 sheep and lambs. Youth who belong to organizations catering to those interested in animal agriculture, such as 4-H and FFA, are allowed to exhibit livestock at county fairs and at regional, state, and national exhibitions. Like the purebred breeders mentioned earlier, sales of lambs and sheep to these markets offer opportunities for producers to cater to a specialty clientele that intend to exhibit their animals in a public arena. Purebred producers, club lamb producers, and breeders who cater to the specialty wool market all have an established niche in the sheep, lamb, and wool industry. Although the size of this market is not a main driver in the sheep, lamb, and wool industry, the availability of specialized marketing opportunities for these animals adds value (significant in some cases) to producers and is available to small- and large-scale producers.

Specialty Wool Markets

Small wool mills, often called mini-mills, produce wool yarn for specialty markets and, in some cases, produce wool fabrics and finished products. A mini-mill may process from 20 to 100 tonnes of scoured wool. One mill purchased approximately 2.3 tonnes of finer raw wool from a western producer at a 40 percent premium over current market price and spun the wool into high-quality fine wool yarn that sold for over $45/kg. Some producers sell their raw wool, often naturally colored wools, to hand spinners that may result in sales of as much as $50–$100 of wool per ewe per year. The number of small retail outlets and hand spinners listed on the Internet is rapidly increasing. Internet-based resources, targeted to cottage industries, offer opportunities to small, niche producers. This includes naturally colored wool and spinning fibers. Although the specialty wool market continues to be an important one for select buyers and sellers, the size of the market is difficult to measure because of the use of Internet and private party sales.

MAJOR ACCOMPLISHMENTS, OPPORTUNITIES, AND CHALLENGES OF ALTERNATIVE AND EMERGING MARKETS

The production of alternative lamb and sheep products represents growing value to the U.S. sheep and lamb industry. The exact value of the various segments is difficult to measure, but its worth is larger than that captured by sales at the auction markets. For example, an addition to the value of lamb sales in live animal markets would be the value of animals sold through direct, private negotiations arranged through Internet, local marketing areas, and neighbor transactions.

Major Accomplishments in Alternative and Emerging Markets

The ability to raise sheep on relatively small farms and the availability of marketing opportunities through the Internet support the growth of these types of informal markets. At the same time and to a limited extent, producer groups and processors handling animals on a larger scale are looking at opportunities to dedicate plant capacity to processing animals for ethnic (e.g., halal) and specialized (e.g., organic) markets. The growth and success of alternative and emerging markets, including specialized trade for ethnic markets, organic markets, dairy, wool, and club lambs, indicate how resourceful the industry has become in adjusting to changes taking place in the traditional markets. However, without better data on the numbers of animals in various segments of the industry, it is difficult to monitor and assess the full scope of industry value.

More specifically, the major accomplishments in alternative and emerging markets include the following:

- *Development and growth of ethnic markets.* The nontraditional markets serving several ethnic groups seem to be growing rapidly, sourcing live lambs across the United States, for direct sale to custom abattoirs or to consumers. Initially found primarily in East Coast and California markets, these markets are becoming established in other metropolitan areas as ethnic groups become larger in these centers and create the sustained demand for such products as halal and kosher meats.
- *Development of specialty and high-end markets.* Some high-end restaurants and specialty meat retailers have begun sourcing whole carcasses, sometimes specifically sourcing quality live animals for custom slaughter, for animals meeting such conditions as sustainable production, "locally grown," organic, and related characteristics relevant to their own customers. They price cuts to meet the local demand and in a way that sells the entire carcass. One restaurant chef the committee interviewed was sourcing live animals, custom slaughtering and moving the entire carcass through the menu of the restaurant, priced in a way to ensure all cuts were used in proportion to the whole carcass.
- *An emerging market for sheep milk products.* A dynamic and growing market for sheep milk products is emerging, particularly in New England and the north central United States. This industry is in its infancy and remains a very small element in the overall sheep and lamb industry. However, with the growth in high-end, exotic retail products, it has already begun to serve as an alternative to imported sheep milk products.
- *A growing market for specialized wool.* The specialized wool markets have been emerging through clubs, fairs, exhibitions, and the Internet for the production of specialty wools, including naturally colored wools. The demand for these wools comes primarily from the cottage industry level, offering quality, hand-made wool products appealing to higher-income families.
- *Introduction and development of new breeds for specialty markets.* New breeds are being introduced into the United States with specific characteristics for the farm operation or to meet specialty product markets. Hair sheep remove the need to shear annually; naturally colored wool sheep offer opportunities in the hand-weaving industry; selected breeds offer higher sustained milk production for the sheep milk industry. Hair sheep, with smaller carcass size, are also well suited to ethnic and other local markets for fresh, whole lamb products.
- *Continued support of club lambs.* Club lambs continue to be popular through 4-H and similar organizations, enabling youth to experience rais-

ing and caring for animals, and competitive presentation at local fairs and exhibitions.

Major Opportunities and Challenges Facing Alternative and Emerging Markets

The traditional ranched and farmed sheep and lamb production systems have been contracting for a long period, whereas the alternative value chains and product pathways represent the growth components of the industry in very recent years. While the traditional value chains for lamb may continue to contract, there are real opportunities for considerable growth in the nontraditional, emerging, and alternative markets for both lamb meat and specialty wools as well as sheep milk products.

The major opportunities for alternative and emerging markets include the following:

- *Sheep milk production and marketing.* Continued growth and development of the sheep milk products market offers both high-quality product and high return in specialized markets. Over time, this emerging market can displace imports of similar products coming principally from Europe. To support this growth, improved breeds for milk production, improved technology for handling sheep milk throughout the year, and expansion of sheep milk dairies and transportation systems can contribute to the rate of growth in this market.
- *Growing cottage industry in hand spinning and weaving.* The cottage industry in hand spinning and weaving offers a very small but growing market for further development. It will require improved and expanded herds of specialty wool sheep, and further development of market infrastructure beyond the local cottage industry character today.
- *Developing access to emerging and alternative markets for the traditional sectors of the industry.* The traditional ranched and farmed sheep industry has had very limited entry into the ethnic and religious markets, markets that seem to be expanding rapidly in the major U.S. metropolitan areas. The major packers have concentrated their efforts on meeting the competition with other meats in the large-footprint food chains. Gaining access to and serving the emerging and alternative markets for ethnic and religious groups, as well as organic meats, could represent an expansion of demand for lamb products for these traditional lamb value chains. Also, lamb is well suited to growing market channels for locally produced meats.

Expanding the alternative and emerging markets offers considerable challenge to the industry and to government. The limited information on these markets, the value chains, and product pathways make it quite difficult to determine market opportunity, size, and characteristics required for considerable growth. Regulatory systems, either by government or as industry-established standards, could prove very useful in market development.

The major challenges facing alternative and emerging markets include the following:

- *Measuring the size of the market.* Anecdotal evidence, marketing information (listed on various Internet-based websites such as SheepGoatMarketing.info), and presentations to the committee as part of this study all suggest that a small but significant amount of lamb marketed in the United States, and especially during religious holidays, is sourced directly from farms or small abattoirs. Lamb marketed in this way is not necessarily inspected through federally or state-inspected facilities, nor recorded in the official USDA slaughter data. As a result, using the USDA-published data on number of head slaughtered may not accurately portray the full extent of meat consumed or the demand occasioned by these religious periods. One implication is that estimates of changes in seasonal demand attributed to religious holidays may be understated. To the extent this segment of the industry continues to grow, the official data will continue to register an industry in decline rather than an industry in transition.
- *Potential barriers to growth of the certified organic market.* Development of the organic market for lamb is spread throughout the United States. Because sheep production often occurs in forage and field locations, the production of animals is well suited to transition to organic production, although some parasite control may become a larger issue as more lambs are finished on grass and producers seek to fill the growing demand for natural, grass-fed, and organic food. The leveling-off of growth in certified organic sheep numbers suggests that there may be barriers to growth of this market. One possibility is that the lack of processing facilities certified to process organic sheep has hampered the growth of organic production. Without use of certified processing methods, the final product cannot be sold as certified organic and fully capture the premium associated with organic production methods.

Another possibility is that increasing consumer interest in locally produced meats or meats produced by "family-owned" farms may have allowed producers to take advantage of higher premiums for local production or "sustainably produced" attributes without needing to fully transition their production practices to organic methods. Consumers have expressed strong interest in "natural" products, and it is easier to meet the natural requirements than to be certified organic. Lamb is well suited for natural production because lambs usually do not receive growth promotants or antibiotics. One major distributor of quality, "high-end" meats (Niman Ranch) for specialty stores and for freezer-ready and restaurant sales, for example, promotes the lamb served as raised on "environmentally sustainable" ranches with no antibiotics or added growth hormones. However, the distributor makes no specific claims of "certified organic" production methods. Measures of the high-end market producing "environmentally sustainable," "naturally raised" lamb product may capture much of the potential organic market. Documenting the size of this segment of the industry is difficult. Some, though not all, transactions take place in direct marketing, and the total number of producers and distributors is small. Voluntary industry standards will help producers distinguish their products in the marketplace and capture added value from some of the production (and processing) practices preferred by some consumers (USDA, 2007c).

- *Marketing challenges for specialty products.* A major challenge for the development of alternative markets is managing growth and size. Providing product to restaurants, national distributors, or even successful local distribution channels requires the ability to meet the demand for product that may, at times, exceed existing capacity. Newly emerging areas of the industry—including producers catering to local or ethnic markets, organic producers, and dairy producers—involve different participants than the traditional industry players. Dairy sheep cheese may require input from imported product or frozen stocks; delivery to restaurant buyers may require the capacity to deliver minimum level of supply; supplies of halal-processed meat have seasonal demands that require delivery of lamb to market. New information technology has been very useful to the development of alternative and emerging markets for the lamb and sheep industry. The ability to identify buyers and sellers through Internet-based resources allows access to local and national markets for specialty products. This new technology widens market access. Successful efforts are often started through public support (federal grants or state extension). Without some university or public support, it may be difficult to sustain the information base and infrastructure. Good data on the size of the market are critical to better understand the emerging industries related to the production of sheep and lamb and to promote its development.
- *Identifying the effects of ethnic change on future demand.* Although based on a simple model of the U.S. demand for lamb in selected periods of the year, the analysis of ethnic demand for lamb reported in this chapter suggests that the U.S. lamb market is clearly undergoing considerably more change in recent years than the persistent decline in sheep and lamb inventories would suggest. Based on the consistency of the findings, the results point to an identified need for substantially more research using more powerful methodologies to explore the reshaping of lamb demand in the United

States. However, in addition to use of more sophisticated model structure, a key component to better understanding the changes in lamb demand rests on having better data. Although some part of the ethnic market may shift to more tractable market channels (processing and retail) as the sector matures, the omission of many direct sales and transactions that may characterize the lamb market proportionately more than other meats indicates the need to look closely at methods for collecting better data on the segment of the industry that serves growing ethnic markets.

- *An integrated approach to addressing research needs.* Research involving all industry segments from production through live animal marketing, processing, distribution, and merchandising to meet consumer needs for these niche markets for lamb should be given priority. An integrated approach to including nontraditional segments of the industry, and value-added production and processing opportunities should be included in research on production and marketing of lamb products.

REFERENCES

Alnaseej, W. 2007. Islamic finder. Online at: http://www.islamicfinder.org/. Accessed April 18, 2008.

ALB (American Lamb Board). 2007. American lamb attitude and usage study 2006: Foodservice evaluation. Jeffrey B. Gross Marketing Research. http://www.americanlambboard.org/. Accessed April 18, 2008.

Geisler, M. 2007. Organic lamb profile. Agricultural Marketing Resource Center. Iowa State University. Ames, February. Online at: http://www.agmrc.org/agmrc/commodity/livestock/lamb/organiclambprofile.htm. Accessed April 18, 2008.

Grandin, T. 2007. Recommended ritual slaughter practices. Department of Animal Science, Colorado State University. Online at: http://www.grandin.com/ritual/rec.ritual.slaughter.html. Accessed June 15, 2007.

Knight, E., L. House, M. C. Nelson, and R. Degner. 2006. An evaluation of consumer preferences regarding goat meat in the South. J. Food Distrib. Res. 37:88–96.

IFANCA (Islamic Food and Nutrition Council of America). 2008. Halal food certification. Online at: http://www.ifanca.org/index.php. Accessed March 13, 2008.

Larson, A., and E. Thompson. Undated. Direct marketing of lamb in niche and ethnic markets. University of Illinois Extension publication, Initiative for the Development of Entrepreneurs in Agriculture, College of Agriculture, Consumer and Environmental Sciences. Online at: http://web.extension.uiuc.edu/iidea/PDF/lambfactsheet.pdf. Accessed April 18, 2008.

LMIC (Livestock Marketing Information Center). 2007. Weekly, monthly live lamb prices: File 3MKTLAMB.xls. Online at (subscription required): http://www.lmic.info/tac/spreadsheets/spreadsheets.html. Accessed May 1, 2007.

Lotterman, E. 1993. Out like a lamb: The decline of the sheep industry offers case study for other sectors. Fedgazette. Federal Reserve Bank of Minneapolis, July. Online at: http://www.minneapolisfed.org/pubs/fedgaz/93-07/sheep.cfm. Accessed April 18, 2008.

O'dell, D., D. J. Marsh, D. Singh, G. F. Plaugher, P. E. Lewis, E. K. Inskeep, and D. Smith. 2003. Final report: West Virginia lamb marketing information project. Marketing and Development Division, West Virginia Department of Agriculture. Charleston. March.

Radwin, M. J. 2007. hebcal.com: Jewish calendar tools. Hebrew date converter. Online at: http://www.hebcal.com/converter/. Accessed April 18, 2008.

Schoenian, S. M., L. Thonney, and T. L. Stanton. 2007. SheepGoatMarketing.info. Joint project of the University of Maryland and Cornell University. Online at: http://sheepgoatmarketing.info/index.cfm. Accessed June 17, 2007.

Singh-Knights, D., D. K. Smith, and M. Knights. 2005. A hedonic analysis of sheep and goat prices in a changing environment: The role of consumers and implications for management. Selected paper, American Agricultural Economics Association Annual Meeting, Providence, RI. July.

USCB (U.S. Census Bureau). 2007. The 2007 statistical abstract, Washington, DC. Online at: http://www.census.gov/compendia/statab/tables/07s0073.xls.

USDA (U.S. Department of Agriculture). 2005. 2002 Farm Bill Provisions: Organic exemption. Agricultural Marketing Service. Online at: http://www.ams.usda.gov/2002farmbill/organicexempt/.

USDA. 2006a. Sheep and Wool: Background. Economic Research Service. Washington, DC. November. Online at: http://www.ers.usda.gov/Briefing/Sheep/ background.htm. Accessed April 18, 2008.

USDA. 2006b. Red Meat Yearbook. Economic Research Service. Washington, DC. Online at: http://usda.mannlib.cornell.edu/usda/ers/94006/livestockprices.xls. Accessed April 18, 2008.

USDA. 2007a. Data Sets: Organic Production. Economic Research Service, Washington, DC. July. Online at: http://www.ers.usda.gov/Data/Organic/. Accessed April 18, 2008.

USDA. 2007b. The National Organic Program. Agricultural Marketing Service, Washington, DC. August. Online at: http://www.ams.usda.gov/nop/indexNet.htm. Accessed April 18, 2008.

USDA. 2007c. United States Standards for Livestock and Meat Marketing Claims, Naturally Raised Claim for Livestock and the Meat and Meat Products Derived From Such Livestock. Federal Register, Vol. 72, No. 228, Wednesday, November 28, 2007. Docket No. AMS-LS-07-0131; LS-07-16. Online at: http://www.ams.usda.gov/lsg/stand/NRnotice1107.pdf. Accessed April 18, 2008.

APPENDIX

This appendix describes an initial analysis conducted to examine the effects of ethnic markets on U.S. lamb demand, specifically, the impact of Christian, Orthodox, and Muslim holy periods on lamb slaughter. The slaughter data for commercial lamb and yearlings by month were assembled for the period 1970–2004. The data represented head (thousands) of lambs and yearlings slaughtered each month, including both federally inspected slaughter and non-federally inspected slaughter.

Data Issues

An immediate difficulty is that U.S. domestic lamb and yearling slaughter stood at nearly 10 million head in 1970 with a strong downward trend to 2.6 million head in 2004. To remove the time trend effect in lamb and yearling slaughter, a monthly index of slaughter was calculated for each month for each year so that the sum of the 12 monthly indexes in each year summed to 1200. These indexes were then used to calculate the slaughter level in each month as a prediction, based on the average seasonality by month in the data for the entire period. The difference between the actual and predicted slaughter levels was then used as the dependent variable in the regressions. Effectively, the dependent variable displays any variation from the long-term average seasonal trends by month in the data, presumed to occur because of the religious holy days.

Because the data on slaughter are available for the Gregorian calendar months, the dates for Orthodox and the Muslim holy days needed to be converted to Gregorian dates. The Hijri calendar commonly used within North America was used for the Muslim date conversions. Similarly, the dates for Orthodox Easter were calculated by converting the Hebrew calendar dates to Gregorian dates.[1] For the Orthodox and Christian Easter, a dummy variable was constructed with a value of 1 for the month in which the 9 days prior to the two Easter dates occurred, and 0 otherwise. When the two 9-day periods stretched over 2 months, the dummy variable was assigned the proportion of days falling in each of the 2 months. In all cases, the dummy variable summed to 1 for the 2 months. Several lead times other than 9 days were examined, ranging from 4 to 15 days. Based on R^2 and F-values for the regressions, 9 days was selected, giving nearly identical results with an 8-day lead time. Little difference in the coefficients was found over the various lead times although the R^2, F-values, and t-tests were highest using a 9-day lead.

[1]Orthodox Easter follows the Hebrew calendar, falling on the same day as Jewish Passover. For date conversion from Hijri to Gregorian calendars, see Alnaseej (2007). For Gregorian and Hebrew date conversions, see Radwin (2007).

For Ramadan, a dummy variable was constructed with a value of zero for months with no days during Ramadan and Eidu al-fitr and a sum of 1 for months in which Ramadan occurred. Since Ramadan usually falls across two Gregorian months, the proportion of days of Ramadan plus Eidu al-fitr in the Gregorian month was used for the dummy variable. That is, if 20 days of Ramadan occurred in May, then the dummy for May would be 20 divided by 29, where 29 is the number of days from the beginning of Ramadan to the end of Eidu al-fitr. Similarly, June would have the remainder of the Ramadan period with a dummy variable value of 9 divided by 29. The month in which Eidu al-adha fell was assigned a value of 1 and zero otherwise.

As well, the datasets were divided into two parts, before 1991 and after 1990 for the Easter, Eidu al-adha, and Ramadan-Eidu al-adha dummy variables. Dummy variables were constructed for each of the Easter, Eidu al-adha, and Ramadan-Eidu al-adha variables, the first for all months before January 1991 and the second for all months from January 1991 to the end of the dataset, December 2004. The purpose was to determine if the impact of these events was growing stronger over time. The issue is whether there is a growing trend to consume lamb during these holy periods.

One of the difficulties in examining the data is that while Passover remains within the period of March–April each year, Ramadan advances several days per year over the entire year. Whenever Ramadan and Passover coincide in the same months, separating the effects of each is problematic. As a result, two additional Ramadan dummy variables were constructed. The first of these (Ramadan outside Passover) was for the months in which Ramadan and Passover did not overlap and the second (Ramadan Passover overlap) was for those months when the two periods overlapped.

Because of the religious nature of the demand for lamb during these religious holidays, price was expected to be far less of a factor affecting demand during those periods than is the case generally for lamb. As well, substitution of lamb for other meats is not considered. To the extent that income or price play a role in the additional demand during these religious periods, they may be expressed in the quality and source of the cuts purchased, limiting the impact of price during these periods. Nonetheless, the USDA price series for slaughter lambs was included in all of the equations (USDA, 2006b).

Model

Four equations were formulated and estimated using OLSL:

$$Z_{ij} = \alpha + \gamma P_{ij} + \beta_1 D_{1ij} + \beta_2 D_{2ij} + \beta_3 D_{3ij} + \beta_4 D_{4ij} + \beta_5 D_{5ij} + \beta_6 D_{6ij} + \varepsilon_{ij} \quad (1)$$

$$Z_{ij} = \alpha + \gamma P_{ij} + \beta_1 D_{1ij} + \beta_2 D_{2ij} +$$

$$\beta_3 D_{3ij} + \beta_4 D_{4ij} + \beta_5 D_{5ij} + \beta_7 D_{7ij} + \beta_8 D_{8ij} + \varepsilon_{ij} \qquad (2)$$

$$Z_{ij} = \alpha + \gamma P_{ij} + \beta_1 D_{1ij} + \beta_2 D_{2ij} + \beta_3 D_{3ij} + \beta_4 D_{4ij} + \beta_5 D_{5ij} + \beta_9 D_{9ij} + \beta_{10} D_{10ij} + \varepsilon_{ij} \qquad (3)$$

$$Z_{ij} = \alpha + \gamma P_{ij} + \beta_1 D_{1ij} + \beta_2 D_{2ij} + \beta_3 D_{3ij} + \beta_4 D_{4ij} + \beta_5 D_{5ij} + \beta_9 D_{9ij} + \varepsilon_{ij} \qquad (4)$$

where:

Y_{ij} = number of head (thousands) slaughtered in month i and year j, for i = 1,2,3,...,12 and j = 1970, 1971, 1972,...,2004.

Y_j = $\sum_{i=1}^{12} Y_{ij}$ = number of head (thousands) slaughtered in year j

Y_i = $\sum_{j} Y_{ij}$ = number of head (thousands) slaughtered in month i for all years.

$(Y_{ij}/Y_j)*1200$ = I_{ij} for all i and j = monthly index of slaughter head numbers such that: $\sum_{i=1}^{12} I_{ij}$ = 1200 for all j.

$I_{ij} * Y_j/1200$ = calculated number of head (thousands) of lambs and yearlings slaughtered, based on long term seasonal/monthly slaughter patterns.

Z_{ij} = $Y_{ij} - I_{ij} * Y_j/1200$ which equals the difference between actual and calculated slaughter by month.

P_{ij} = USDA monthly price series for slaughter lambs.

D_{1ij} = dummy variable, 1 for any month in which the nine days prior to Orthodox and Christian Easter occurs for the years 1970 to 1990, 0 otherwise; where the nine days falls across two months, the proportion of days in each month is used.

D_{2ij} = dummy variable, 1 for any month in which the nine days prior to Orthodox and Christian Easter occurs for the years 1991 to 2004, 0 otherwise; when the nine days falls across two months, the proportion of days in each month is used.

D_{3ij} = dummy variable, 1 for any month in which Eidu

al-adha occurs for the years 1970 to 1990, 0 otherwise.

D_{4ij} = dummy variable, 1 for any month in which Eidu al-adha occurs for the years 1991 to 2004, 0 otherwise.

D_{5ij} = dummy variable, the proportion of days of the month within Ramadan plus Eidu al-fitr, 0 otherwise.

D_{6ij} = dummy variable, the proportion of days of the month within Ramadan plus Eidu al-fitr for the years 1970 to 2004, 0 otherwise.

D_{7ij} = dummy variable, the proportion of days of the month within Ramadan plus Eidu al-fitr for the years after 1990, 0 otherwise.

D_{8ij} = dummy variable, the proportion of days of the month within Ramadan plus Eidu al-fitr when Ramadan plus Eidu al-fitr does not occur during the same month as Orthodox and Christian Easter, 0 otherwise.

D_{9ij} = dummy variable, the proportion of days of the month within Ramadan plus Eidu al-fitr, 0 when Ramadan plus Eidu al-fitr occurs during the same month as Orthodox and Christian Easter, 0 otherwise.

Three price series were used with each of the equations with the exception of equation (2) in the case of the Sioux Falls price series. Since the price series began in 1990, equations (1) and (2) were virtually identical. In no case was the coefficient β_7 significant in equation (3). As a result, equation (4) was included without the D_{7ij} dummy variable. The results of selected regressions are shown in Table 7A-1.

Results

The price coefficients across all equations are highly significant in all cases and all within a very narrow range of –0.192 to –0.219. This indicates that purchases of lamb are quite sensitive to price for the holy events. This result was somewhat unexpected as noted earlier. The consequence is that considerably more research needs to be done to explore the substitutability of lamb with other meats during these periods as one means of determining price strategies for distributors. Similarly, the complementarity of lamb with other foods also needs further exploration.

For the Easter variables (D_{1ij} and D_{2ij}), the results show that the effect of Easter is insignificant in the earlier period (1970 to 1990) but highly significant and much larger in the later period (1991 to 2004) across all four equations. As well, the coefficients for the later period are considerably

TABLE 7A-1 Regression Results of Religious Holidays on Lamb Disappearance

	Equation (1)			Equation (2)		
Variable	Coefficient	Standard Error	t-Value	Coefficient	Standard Error	t-Value
Intercept	9.88	5.58	1.77	10.80	5.68	1.90
USDA Price	–0.20	0.08	–2.43	–0.22	0.09	–2.56
Easter 1970–1990	1.94	7.63	0.25	2.30	7.64	0.30
Easter 1991–2004	42.84	9.19	4.66	42.23	9.22	4.58
Eidu al-adha 1970–1990	3.34	6.67	0.50	3.20	6.67	0.48
Eidu al-adha 1991–2004	18.67	8.38	2.23	19.00	8.39	2.26
Ramadan	8.67	6.56	1.32			
Ramadan 1970–1990				4.10	8.28	0.50
Ramadan 1991–2004				15.58	10.08	1.55
Ramadan Outside Easter						
Ramadan Easter Overlap						
Multiple R	0.28			0.28		
R^2	0.08			0.08		
Adjusted R^2	0.06			0.06		
Standard Error	29.92			29.93		
F	5.89			5.16		
Observations	420			420		

larger and statistically different from the coefficients for the earlier period. This pattern of insignificance for the period 1970 to 1990 and greater significance for the period 1991 to 2004 was also found for the coefficients for Ramadan and Eidu al-adha (D_{3ij} and D_{4ij}). In the case of the Ramadan coefficients, these were only statistically significant for periods when Easter and Ramadan did not overlap in time.

For equation (3), the coefficient for "Ramadan Easter overlap" is unexpectedly negative although insignificant. As a result, this variable was dropped from the equation and reestimated as equation (4). The coefficient for Ramadan outside the Easter period is nearly identical to the coefficient in equation (3) and highly significant in both cases.

For the Easter period after 1990, the results suggest that an additional 42,000 to 46,000 head of lambs are consumed because of Easter. This represents about 8 percent of the average disappearance in March and April and about 1.7 percent of annual disappearance based on harvest levels in 2004. Similarly, there appears to be an additional demand for 18,000 to 19,000 lambs in the months when Eidu al-adha occurs after 1990, representing about 7 percent of average monthly disappearance. In the case of Ramadan for the period 1991–2004, there is an additional demand of 16,700 to

Equation (3)			Equation (4)		
Coefficient	Standard Error	t-Value	Coefficient	Standard Error	t-Value
8.88	5.56	1.60	9.06	5.57	1.63
−0.19	0.08	−2.30	−0.20	0.08	−2.39
4.69	7.66	0.61	3.39	7.62	0.44
46.34	9.24	5.01	44.38	9.17	4.84
3.71	6.63	0.56	3.96	6.64	0.60
18.03	8.33	2.16	18.87	8.33	2.27
16.74	7.28	2.30	17.23	7.29	2.36
−20.35	13.40	−1.52			
0.30			0.30		
0.09			0.09		
0.08			0.07		
29.74			29.79		
5.99			6.58		
420			420		

17,250 lambs, about 7 percent of average monthly disappearance, whenever Ramadan and Easter do not occur in the same months.

Across all equations, the R^2 is low as expected because of the construction of the independent variable. Nonetheless, the F-value is consistently significant in all cases.

In general, there appears to be a discernable impact of the Muslim holiday periods and the Christian and Orthodox Easter holiday periods on the disappearance levels of lamb and yearlings. As well, the impact of these holidays appears to be increasing with time. To the extent that some lamb is purchased directly from farms and not recorded in official tallies of slaughter, these results underestimate the impact of these religious periods. As well, because of the construction of the independent variable, there will also be some (minor) underestimation of the religious event impacts.

Other datasets were examined as well using the above methodology, although the results are not shown here. Prices for slaughter lambs from the San Angelo, Texas, and Sioux Falls, South Dakota markets were substituted for the USDA price in the equations. These price datasets were constructed from price series obtained through the courtesy of the Livestock Marketing Information Center (LMIC, 2007). In these cases, coefficients and their significance levels were quite similar to those reported above.

Appendix A

Committee Statement of Task

A committee of experts will prepare a report based on a study of the economic development and current status of the U.S. sheep industry. The study will examine the structure of the industry, including the nature of the businesses in the industry, the production and marketing systems used in each sector, and the challenges of disease, predation, sheep biology, and genetic resources on production efficiency and competitiveness. The study will examine each product and service sector of the industry in detail, including meat, wool, live animals, pelts, milk, other byproducts, and supporting businesses, and describe the influence of regulation, product pricing, market demand patterns and projections, and international trade issues affecting commercial activities. The study will also explore the role of land stewardship in sheep production. The committee's report will summarize the findings of the study and identify major accomplishments of the past and challenges to the industry in the future. However, the committee will not make recommendations related to policy issues, such as regulatory matters, industry support, or international trade.

Appendix B

Abbreviations and Acronyms

ABARE	Australian Bureau of Agriculture and Resource Economics
ABS	Australian Bureau of Statistics
ADAA	Animal Drug Availability Act
AFD	Average fiber diameter
AFO	Animal feeding operation
AI	Artificial insemination
AIN	Animal identification number
ALB	American Lamb Board
ALBC	American Livestock Breeds Conservancy
AMA	Alternative marketing arrangement
AMS	Agricultural Marketing Service (USDA)
APHIS	Animal and Plant Health Inspection Service (USDA)
ARC	Agricultural Research Council (UK)
ARS	Agricultural Research Service (USDA)
ASGC	American Sheep and Goat Center
ASI	American Sheep Industry Association
ASPC	American Sheep Producers Council
ATC	Agreement on Textiles and Clothing
ATMI	American Textile Manufacturers Institute
AUD	Australian dollar
AVMA	American Veterinary Medical Association
AWC	Australian Wool Council
AWEX	Australian Wool Exchange
AWI	Australian Wool Innovation

BBC	British Broadcasting Corporation
BCR	Benefit:cost ratio
BCTRC	Boneless, closely trimmed, retail cuts
BI	Bioelectrical impedance
BLM	Bureau of Land Management
BOU	Bellies out untied
BSE	Bovine spongiform encephalopathy
CAFO	Concentrated animal feeding operation
CCC	Commodity Credit Corporation
CERI	Cotton Economics Research Institute
C-FARE	Council on Food, Agricultural, and Resource Economics
CFR	Code of Federal Regulations
CJD	Creutzfeldt-Jakob disease
CL	Caseous lymphadenitis
CNS	Central nervous system
COOL	Country-of-origin labeling
CP	Crude protein
CR4	Concentration ratio (four-firm)
CRIS	Current Research Information System
CSREES	Cooperative State Research, Education, and Extension Service (USDA)
CWD	Chronic wasting disease
CWFP	Clean wool fibers present
cwt	Hundredweight (100 pounds)
DCP	Demand constant price
DNA	Deoxyribonucleic acid
DSANA	Dairy Sheep Association of North America
EA	Environmental assessments
EF	East Friesian
EIA	Environmental impact analysis
ELRRPP	Ewe Lamb Replacement and Retention Payment Program
EPA	Environmental Protection Agency (U.S.)
EPD	Expected progeny difference
EPT	Elasticities of price transmission
ERS	Economic Research Service (USDA)
ESA	Endangered Species Act
EU	European Union
FAO	Food and Agriculture Organization of the United Nations
FDA	Food and Drug Administration (U.S.)

FFA	Future Farmers of America
FFI	Fatal familial insomnia
FMD	Foot and mouth disease
FOB	Free on board
FSA	Farm Service Agency
FSE	Feline spongiform encephalopathy
FSIS	Food Safety and Inspection Service (USDA)
FSVM	Food supply veterinary medicine
FTE	Full-time equivalent
FWS	Fish and Wildlife Service (U.S. Department of the Interior)
FY	Fiscal year
GATT	General Agreement on Tariffs and Trade
GIN	Group identification number
GSS	Gerstmann-Sträussler-Scheinker disease
GW	Grease weight
HAACP	Pathogen reduction/hazard analysis and critical control point
HRI	Hotel, restaurant, and institution
ID	Identification
IFANCA	Islamic Food and Nutrition Council of America
LA	Lacaune
LDP	Loan deficiency payment
LMAs	Long-term marketing arrangements
LMR	Livestock mandatory reporting
LRP	Livestock risk protection
LVS	Lamb vision system
MAL	Marketing assistance loan
MAP	*Mycobacterium avium*, subspecies *paratuberculosis*
MBM	Meat and bone meal
MFA	Multifiber arrangement
MPR	Mandatory price reporting
MUMS	Minor Use and Minor Species (Act)
NAHMS	National Animal Health Monitoring System
NAHSA	North American Hair Sheep Association
NAIS	National Animal Identification System
NAS	National Academy of Sciences
NASS	National Agricultural Statistics Service (USDA)
NCERA	North Central Education and Research Activity

NCTO	National Council of Textile Organizations
NEPA	National Environmental Policy Act
NFCS	Nationwide Food Consumption Survey
NLFA	National Lamb Feeders Association
NIAA	National Institute for Animal Agriculture
NIDT	National Identification Development Team
NLPA	National Livestock Producers Association
NMCS	National Meat Case Study
NPB	National Pork Board
NPPC	National Pork Producers Council
NRC	National Research Council
NSEP	National Scrapie Eradication Program
NSIIC	National Sheep Industry Improvement Center (USDA)
NSIP	National Sheep Improvement Program
NZD	New Zealand dollar
OB	Original bag
OECD	Organisation for Economic Co-operation and Development
OGP	Optical grading probe
OIE	World Organization for Animal Health
OJD	Ovine Johne's disease
OLSL	Ordinary least squares level
OPP	Ovine progressive pneumonia
P.L.	Public law
PDL	Polynomial distributed lag
PMCI	Producers Marketing Cooperative, Inc.
PMO	Pasteurized Milk Ordinance
RBA	Reserve Bank of Australia
RE	Ram epididymitis
RMA	Risk Management Agency
RTI	Research Triangle Institute International
RTU	Real-time ultrasound
SEAC	Spongiform Encephalopathy Advisory Committee (United Kingdom)
SFCP	Scrapie Flock Certification Program
SGRJ	Sheep & Goat Research Journal
SID	Sheep Industry Development
SKU	Stock keeping units
SNZ	Statistics New Zealand
SOSS	Scrapie ovine slaughter surveillance

SPA	Standardized performance analysis
SSQA	Sheep Safety and Quality Assurance
TAHC	Terrestrial Animal Health Code
TAMRC	Texas Agribusiness Market Research Center
TB	Tuberculosis
TES	Threatened and endangered species
TME	Transmissible mink encephalopathy
TOBEC	Total body electrical conductivity
TRQ	Tariff rate quota
TSC	Table skirted and classed
TSE	Transmissible spongiform encephalopathy
UK	United Kingdom
USAHA	U.S. Animal Health Association
USCB	U.S. Census Bureau
USC	U.S. Code
USD	U.S. dollar
USDA	U.S. Department of Agriculture
USDC	U.S. Department of Commerce
USDHHS	U.S. Department of Health and Human Services
USDI	U.S. Department of the Interior
USFS	U.S. Forest Service
USITC	U.S. International Trade Commission
vCJD	Variant Creutzfeldt-Jakob disease
VIA	Video image analysis
VSFCP	Voluntary Scrapie Flock Certification Program
WERA	Western Education and Research Activity
WHO	World Health Organization
WLB	Wyoming Livestock Board
WRA	Western Range Association
WSDC	Wisconsin Sheep Dairy Cooperative
WTO	World Trade Organization

Appendix C

Committee Member Biographies

Gary W. Williams *(chair)* is professor and coordinator of the Texas Agribusiness Market Research Center in the Department of Agricultural Economics at Texas A&M University. Williams received a B.S. (1974) in economics from Brigham Young University and an M.S. (1977) and Ph.D. (1981) in agricultural economics from Purdue University. He has been a member of the American Agricultural Economics Association since 1976, and served on the editorial council of the *Journal of Agricultural and Resource Economics* from 2001 to 2003. From 1992 to 1994, he served on the National Research Council's Committee on Livestock Disease Eradication: Bovine Tuberculosis. Williams' research has focused on lamb production and consumption and the effects of policy on domestic and international trade. He is an expert in agricultural economics and the sheep industry and lamb market.

DeeVon Bailey is professor and extension marketing specialist in the Department of Economics at Utah State University. Bailey received a B.A. (1980) in economics, an M.S. (1981) in agricultural economics, both from Utah State University, and a Ph.D. (1983) in agricultural economics from Texas A&M University. He has received awards from the American Agricultural Economics Association (1997) and the Western Agricultural Economics Association (1997, 2005, and 2006) for outstanding extension projects. Bailey has also received the top research award offered by Utah State University (2006) and USU's top extension award (2003). Bailey is an expert in consumer preferences and meat traceability programs in red meat markets. He is familiar with the sheep industry in the Northwest and Intermountain areas of the United States.

Oral Capps, Jr. is a professor in the Department of Agricultural Economics at Texas A&M University. Capps received a B.S. (1975) in mathematics, an M.S. (1977) in agricultural economics, an M.S. (1979) in statistics, and a Ph.D. (1979) in agricultural economics, all from Virginia Polytechnic Institute and State University. He served as president of the Southern Agricultural Economics Association from 1992 to 1993 and has received many awards for both his research and teaching, including the American Agricultural Economics Association Distinguished Teaching Award in 1999, the American Council on Consumer Interests' Applied Consumer Economics Award (co-recipient) in 1999, and the *Agricultural and Resource Economics Review* Outstanding Journal Article Award (co-recipient) in 2000. Capps' areas of expertise include the economics of health and nutrition, agribusiness, consumer demand analysis, agricultural marketing, evaluation of commodity checkoff programs, and applied econometrics.

Linda A. Detwiler is assistant director for the Virginia-Maryland Regional College of Veterinary Medicine at the University of Maryland and is a private animal health consultant. Detwiler received a B.S. (1980) in dairy science from Delaware Valley College of Science and Agriculture and a D.V.M. (1984) from the Ohio State University. She has worked in private food animal practice, but has spent the bulk of her career overseeing public animal health programs for USDA APHIS. She has chaired several advisory groups for international organizations including the World Health Organization (WHO) and the World Organization for Animal Health, as well as national governments. These groups include WHO's Consultation on Public Health and Animal Transmissible Spongiform Encephalopathy Diseases (co chair, 1999) and the Working Group on Bovine Spongiform Encephalopathy in Sheep (2001). Detwiler has experience with sheep diseases with the primary focus on prion diseases such as scrapie and has assisted the sheep industry in its efforts to control scrapie beginning in 1985.

Hudson A. Glimp is Edwin L. Wiegand Professor Emeritus at the University of Nevada–Reno (UNR). Glimp received a B.S. (1960) in animal science and an M.S. (1961) in animal nutrition, both from Texas A&M University, and a Ph.D. (1964) in animal nutrition from Oklahoma State University. Before moving to UNR, Glimp worked in sheep research for the USDA and private firms. From 1987 to 1990, he was the director of the USDA Sheep Experiment Station in Dubois, Idaho. He served as the UNR's sheep extension specialist for 15 years. Glimp is currently the coordinator of a UNR research station, Rafter 7 Ranch, which specializes in Merino sheep breeding programs and has produced the premier wool of the United States for the past several years.

Timothy Hammonds is president and CEO of the Food Marketing Institute (FMI), a nonprofit organization that provides guidance to the food distribution industry through research, education, and programs in industry relations, policy, and consumer information. Hammonds received his B.S. (1966), M.B.A. (1967), and Ph.D. (1970) from Cornell University. He has served on many committees for the National Research Council, including the Committee on Nutrition Components of Food Labeling, Committee on Technological Options to Improve Nutritional Attributes of Animal Products, and Committee on Food Consumption Patterns. He currently serves as chairman of the board of the National Partnership for Food Safety Education. Before moving to FMI in 1975, Hammonds taught agricultural economics at Oregon State University.

Douglas D. Hedley is the executive director of the Canadian Faculties of Agriculture and Veterinary Medicine, an organization of deans and presidents of faculties of agriculture and veterinary medicine in Canada. These faculties represent researchers, educators, and scientists who investigate questions about agriculture, food, health, and environment. Hedley earned a B.S. (1965) from the University of Guelph and an M.S. (1968) and Ph.D. (1970) in agricultural economics from Michigan State University. In addition to extensive work as a scholar and policy advisor in Nigeria and Indonesia, he has held several positions for Agriculture Canada since 1972, among them assistant deputy minister, Programs Branch in Agriculture and Agri-food Canada. From 1997 to 2000, Hedley served as president of the International Association of Agricultural Economists and was the founding editor of *Agricultural Economics*.

Helen H. Jensen is a professor of economics and head of food and nutrition policy research in the Center for Agricultural and Rural Development at Iowa State University. Jensen received a B.S. in economics from Carleton College, an M.S. in agricultural and applied economics from the University of Minnesota, and a Ph.D. in agricultural economics from the University of Wisconsin–Madison. She has served on five committees and panels for the National Research Council, including the Panel on Animal Health and Veterinary Medicine, Committee on Biological Threats to Agricultural Plants and Animals, and the Committee on Assessing the Nation's Framework for Addressing Animal Diseases. Currently she is serving on the board of directors of the American Agricultural Economics Association and the American Council of Consumer Interests, as well as the editorial boards of *Food Economics, Agricultural Economics*, and *Agribusiness: An International Journal*. Jensen is an expert in consumer consumption patterns and food safety.

Paul S. Kuber is an assistant professor in the Department of Animal Sciences at The Ohio State University. Kuber received a B.S. (1991) from California State University–Fresno, an M.S. (1993) from the University of Nebraska, and a Ph.D. (2001) from Washington State University. He has worked in the lamb slaughter and processing industries in both California and Australia. Kuber's areas of expertise are meat science, particularly fresh and processed meat quality; consumer perception and preference; and product development.

David L. Thomas is a professor in the Department of Animal Sciences at the University of Wisconsin–Madison. Thomas received a B.S. (1971) in meat and animal science from the University of Wisconsin–Madison and an M.S. (1975) in animal science and a Ph.D. (1977) in animal breeding, both from Oklahoma State University. In 2003, he received the Award for Recognition of Outstanding Contributions to the Dairy Sheep Industry of North America from the Dairy Sheep Association of North America. He has received three awards from the American Society of Animal Science: the Animal Breeding and Genetics Award (2003), the Animal Management Award (2004), and the Bouffault International Animal Agriculture Award (2005). Thomas' areas of expertise include sheep genetics and dairy sheep production.

Appendix D

Recent Publications of the Board on Agriculture and Natural Resources

POLICY AND RESOURCES

Achievements of the National Plant Genome Initiative and New Horizons in Plant Biology (2008)
Agricultural Biotechnology and the Poor: Proceedings of an International Conference (2000)
Agriculture, Forestry, and Fishing Research at NIOSH (2008)
Agriculture's Role in K-12 Education (1998)
Air Emissions from Animal Feeding Operations: Current Knowledge, Future Needs (2003)
Animal Biotechnology: Science-Based Concerns (2002)
Animal Care and Management at the National Zoo: Final Report (2005)
Animal Care and Management at the National Zoo: Interim Report (2004)
Animal Health at the Crossroads: Preventing, Detecting, and Diagnosing Animal Diseases (2005)
Biological Confinement of Genetically Engineered Organisms (2004)
California Agricultural Research Priorities: Pierce's Disease (2004)
Changes in the Sheep Industry in the United States: Making the Transition from Tradition (2008)
Countering Agricultural Bioterrorism (2003)
Critical Needs for Research in Veterinary Science (2005)
Designing an Agricultural Genome Program (1998)
Diagnosis and Control of Johne's Disease (2003)
Direct and Indirect Human Contributions to Terrestrial Carbon Fluxes (2004)

Ecological Monitoring of Genetically Modified Crops (2001)
Emerging Animal Diseases: Global Markets, Global Safety: Workshop Summary (2002)
Ensuring Safe Food: from Production to Consumption (1998)
Environmental Effects of Transgenic Plants: The Scope and Adequacy of Regulation (2002)
Exploring a Vision: Integrating Knowledge for Food and Health (2004)
Exploring Horizons for Domestic Animal Genomics (2002)
Frontiers in Agricultural Research: Food, Health, Environment, and Communities (2003)
Future Role of Pesticides for U.S. Agriculture (2000)
Genetically Engineered Organisms, Wildlife, and Habitat: A Workshop Summary (2008)
Genetically Modified Pest-Protected Plants: Science and Regulation (2000)
Global Challenges and Directions for Agricultural Biotechnology (2008)
Incorporating Science, Economics, and Sociology in Developing Sanitary and Phytosanitary Standards in International Trade (2000)
National Capacity in Forestry Research (2002)
National Research Initiative: A Vital Competitive Grants Program in Food, Fiber, and Natural-Resources Research (2000)
Predicting Invasions of Nonindigenous Plants and Plant Pests (2002)
Professional Societies and Ecologically Based Pest Management (2000)
Publicly Funded Agricultural Research and the Changing Structure of U.S. Agriculture (2002)
Safety of Genetically Engineered Foods: Approaches to Assessing Unintended Health Effects (2004)
Scientific Advances in Animal Nutrition: Promise for a New Century (2001)
Scientific Criteria to Ensure Safe Food (2003)
Status of Pollinators in North America (2007)
The National Plant Genome Initiative (2002)
The Scientific Basis for Estimating Emissions from Animal Feeding Operations: Interim Report (2002)
The Scientific Basis for Predicting the Invasive Potential of Nonindigenous Plants and Plant Pests in the United States (2002)
The Use of Drugs in Food Animals: Benefits and Risks (2000)

ANIMAL NUTRITION PROGRAM—NUTRIENT REQUIREMENTS OF DOMESTIC ANIMALS SERIES AND RELATED TITLES

Mineral Tolerance of Animals: Second Revised Edition (2005)
Nutrient Requirements of Beef Cattle, Seventh Revised Edition, Update (2000)

Nutrient Requirements of Dairy Cattle, Seventh Revised Edition (2001)
Nutrient Requirements of Dogs and Cats (2006)
Nutrient Requirements of Horses: Sixth Revised Edition (2007)
Nutrient Requirements of Nonhuman Primates, Second Revised Edition (2002)
Nutrient Requirements of Small Ruminants: Sheep, Goats, Cervids, and New World Camelids (2007)
Nutrient Requirements of Swine, Tenth Revised Edition
Scientific Advances in Animal Nutrition: Promise for a New Century (2001)
The First Seventy Years 1928-1998: Committee on Animal Nutrition (1998)
The Scientific Basis for Estimating Emissions from Animal Feeding Operations: Interim Report (2002)

Further information and prices are available from the National Academies Press website at http://www.nap.edu/. To order any of the titles above, go to http://www.nap.edu/order.html or contact the Customer Service Department at (888) 624-8373 or (202) 334-3313. Inquiries and orders may also be sent to the National Academies Press, 500 Fifth Street, NW, Lockbox 285, Washington, DC 20055.